OPTIMIZATION FOR CHEMICAL AND BIOCHEMICAL ENGINEERING

Discover the subject of optimization in a new light with this modern and unique treatment. Includes a thorough exposition of applications and algorithms in sufficient detail for practical use, while providing you with all the necessary background in a self-contained manner. Features a deeper consideration of optimal control, global optimization, optimization under uncertainty, multiobjective optimization, mixed-integer programming and model predictive control. Presents a complete coverage of formulations and instances in modelling where optimization can be applied for quantitative decision-making. As a thorough grounding to the subject, covering everything from basic to advanced concepts and addressing real-life problems faced by modern industry, this is a perfect tool for advanced undergraduate and graduate courses in chemical and biochemical engineering.

Vassilios S. Vassiliadis is a senior lecturer in the Department of Chemical Engineering at the University of Cambridge. He is also the CEO and CTO of the spin-out company, Cambridge Simulation Solutions LTD.

Walter Kähm a former PhD student under Vassilios S. Vassiliadis, is now a process engineer in the chemical sector.

Ehecatl Antonio del Rio Chanona is a lecturer and head of the optimisation and machine learning for the process systems engineering group in the Department of Chemical Engineering and the Centre for Process Systems Engineering (CPSE) at Imperial College London.

Ye Yuan is currently a professor at Huazhong University of Science and Technology.

CAMBRIDGE SERIES IN CHEMICAL ENGINEERING

Series Editor
Arvind Varma, *Purdue University*

Editorial Board
Juan de Pablo, *University of Chicago*
Michael Doherty, *University of California-Santa Barbara*
Ignacio Grossman, *Carnegie Mellon University*
Jim Yang Lee, *National University of Singapore*
Antonios Mikos, *Rice University*

Books in the Series

Baldea and Daoutidis, *Dynamics and Nonlinear Control of Integrated Process Systems*

Chamberlin, *Radioactive Aerosols*

Chau, *Process Control: A First Course with MATLAB*

Cussler, *Diffusion: Mass Transfer in Fluid Systems, Third Edition*

Cussler and Moggridge, *Chemical Product Design, Second Edition*

De Pablo and Schieber, *Molecular Engineering Thermodynamics*

Deen, *Introduction to Chemical Engineering Fluid Mechanics*

Denn, *Chemical Engineering: An Introduction*

Denn, *Polymer Melt Processing: Foundations in Fluid Mechanics and Heat Transfer*

Dorfman and Daoutidis, *Numerical Methods with Chemical Engineering Applications*

Duncan and Reimer, *Chemical Engineering Design and Analysis: An Introduction 2E*

Fan, *Chemical Looping Partial Oxidation Gasification, Reforming, and Chemical Syntheses*

Fan and Zhu, *Principles of Gas-Solid Flows*

Fox, *Computational Models for Turbulent Reacting Flows*

Franses, *Thermodynamics with Chemical Engineering Applications*

Leal, *Advanced Transport Phenomena: Fluid Mechanics and Convective Transport Processes*

Lim and Shin, *Fed-Batch Cultures: Principles and Applications of Semi-Batch Bioreactors*

Litster, *Design and Processing of Particulate Products*

Marchisio and Fox, *Computational Models for Polydisperse Particulate and Multiphase Systems*

Mewis and Wagner, *Colloidal Suspension Rheology*

Morbidelli, Gavriilidis, and Varma, *Catalyst Design: Optimal Distribution of Catalyst in Pellets, Reactors, and Membranes*

Nicoud, *Chromatographic Processes*

Noble and Terry, *Principles of Chemical Separations with Environmental Applications*

Orbey and Sandler, *Modeling Vapor-Liquid Equilibria: Cubic Equations of State and their Mixing Rules*

Pfister, Nicoud, and Morbidelli, *Continuous Biopharmaceutical Processes: Chromatography, Bioconjugation, and Protein Stability*

Petyluk, *Distillation Theory and its Applications to Optimal Design of Separation Units*

Ramkrishna and Song, *Cybernetic Modeling for Bioreaction Engineering*

Rao and Nott, *An Introduction to Granular Flow*

Russell, Robinson, and Wagner, *Mass and Heat Transfer: Analysis of Mass Contactors and Heat Exchangers*

Schobert, *Chemistry of Fossil Fuels and Biofuels*

Shell, *Thermodynamics and Statistical Mechanics*

Sirkar, *Separation of Molecules, Macromolecules and Particles: Principles, Phenomena and Processes*

Slattery, *Advanced Transport Phenomena*

Varma, Morbidelli, and Wu, *Parametric Sensitivity in Chemical Systems*

Wolf, Bielser, and Morbidelli, *Perfusion Cell Culture Processes for Biopharmaceuticals*

Optimization for Chemical and Biochemical Engineering

Theory, Algorithms, Modeling and Applications

Vassilios S. Vassiliadis

University of Cambridge

Walter Kähm

LANXESS Deutschland GmbH

Ehecatl Antonio del Rio Chanona

Imperial College London

Ye Yuan

Huazhong University of Science and Technology

CAMBRIDGE
UNIVERSITY PRESS

CAMBRIDGE
UNIVERSITY PRESS

University Printing House, Cambridge CB2 8BS, United Kingdom

One Liberty Plaza, 20th Floor, New York, NY 10006, USA

477 Williamstown Road, Port Melbourne, VIC 3207, Australia

314–321, 3rd Floor, Plot 3, Splendor Forum, Jasola District Centre, New Delhi – 110025, India

79 Anson Road, #06–04/06, Singapore 079906

Cambridge University Press is part of the University of Cambridge.

It furthers the University's mission by disseminating knowledge in the pursuit of education, learning, and research at the highest international levels of excellence.

www.cambridge.org
Information on this title: www.cambridge.org/9781107106833
DOI: 10.1017/9781316227268

First published 2020

Printed in the United Kingdom by TJ Books Limited, Padstow Cornwall

A catalogue record for this publication is available from the British Library.

Library of Congress Cataloging-in-Publication Data
Names: Vassiliadis, Vassilios S., author. | Rio Chanona, E. Antonio del, author. |
    Yuan, Ye (Chemical engineer), author. | Kähm, Walter, author.
Title: Optimization : theory, algorithms and applications in chemical engineering /
    Vassilios S. Vassiliadis, Walter Kähm, Ehecatl Antonio del Rio Chanona, Ye Yuan.
Description: New York : Cambridge University Press, 2020. | Series: Cambridge series
    in chemical engineering | Includes bibliographical references and index.
Identifiers: LCCN 2020017896 (print) | LCCN 2020017897 (ebook) |
    ISBN 9781107106833 (hardback) | ISBN 9781316227268 (ebook)
Subjects: LCSH: Chemical engineering. | Mathematical optimization. |
    Algorithms. | Nonlinear functional analysis.
Classification: LCC TP155 .V38 2020 (print) | LCC TP155 (ebook) | DDC 660–dc23
LC record available at https://lccn.loc.gov/2020017896
LC ebook record available at https://lccn.loc.gov/2020017897

ISBN 978-1-107-10683-3 Hardback

Contents

Contents

Contents

Contents

Notation

AFM	atomic force microscope
AKPZ	anisotropic KPZ equation
a_0	lattice constant
$c_q(\ell)$	q-th order correlation function
d_E	embedding dimension
d_f	fractal dimension
L	system size
$\equiv$	*defined* to be equal
$\sim$	*asymptotically* equal (in scaling sense)
$\approx$	*approximately* equal (in numerical value)

Preface

This book is the result of a decade of teaching the Masters course "Optimization" at the University of Cambridge, during which period a lot of material has been collected and taught. The philosophy behind the lecture notes and now this book is that teaching and research are strongly connected and should never be separated. As such, upcoming research topics in literature have to be included always, especially in such a fast-moving area of research as the one dealt with in this book.

In addition to the standard lecture notes of the Masters course, a deeper consideration of optimal control, global optimization, optimization under uncertainty, multiobjective optimization, mixed-integer programming and model predictive control are included here, which are the areas of increased interest in recent years. The mathematics of the topics covered in this book can be complex in their own right, but it is attempted to give enough information to be applied to chemical engineering problems.

The use of optimization techniques in chemical engineering has a long history, with a profound impact by Professor Roger Sargent. Since the inception of Process Systems Engineering (PSE) by Professor Roger Sargent, the field has expanded to areas outside classical chemical engineering, e.g. biotechnology. This is possible due to the flexible nature with which the concepts of PSE can be used for problems with real-world application. This book tries to teach these fundamental concepts.

The preparation of this book took longer than initially planned, but nevertheless it is with great pleasure that we can now share what we think is an exciting and extremely interesting area of research, for industry and academia. Each chapter presenting new concepts and ideas is followed by exercises for the reader to test the understanding. None of the topics shown in this book can be covered in extensive detail, as this would exceed the scope. As such, further reading recommendations are given to guide the reader. In addition, the material covered is backed with references from recent literature to put into context how the concepts presented in this book are used in "the real world."

PART I

Overview of Optimization

Applications and Problem Formulations

1

Introduction to Optimization

1.1 STATEMENT OF GENERAL MATHEMATICAL PROGRAMMING (MP) (OPTIMIZATION) PROBLEM

A general optimization problem statement is presented in this section. The general form is that of the selection of an optimal set of system variables satisfying the modeling equations of the system considered.

The objective function is given by

$$\min_{x} \text{ or } \max_{x} \ f(x), \tag{1.1a}$$

where $x \in \mathbb{R}^n$ is a real-valued n-dimensional vector of system variables that have to be chosen optimally, and $f(x) : \mathbb{R}^n \mapsto \mathbb{R}^1$ is a scalar function we wish to either minimize or maximize. For example, we might wish to

- minimize production cost and
- maximize the profit of a process.

subject to the following equality constraint:

$$h(x) = 0, \tag{1.1b}$$

where $h(x) : \mathbb{R}^n \mapsto \mathbb{R}^{m_h}$ is a vector of m_h functions that have to be satisfied at the solution of the optimization problem. For example,

- Material and energy balances
- Equilibrium relations
- Physicochemical property estimators

The inequality constraint is given by

$$g(x) \leq \text{ or } \geq 0, \tag{1.1c}$$

where $g(x) : \mathbb{R}^n \mapsto \mathbb{R}^{m_g}$ is a vector of m_g functions that have to be satisfied at the solution of the optimization problem. For example,

- Quality constraints, concentration limitations of by-products and pollutants
- Operating condition limitations, *e.g.* pressure and temperature bounds
- Availability of raw materials

Often, we may have simple bounds explicitly imposed on the system variables x:

$$x^L \leq x \leq x^U. \tag{1.1d}$$

Variations of the general optimization problem presented in Equations (1.1a)–(1.1d) are presented in the next few subsections.

1.1.1 Unimodal vs Multimodal Objective Functions

We consider the case here of minimizing some objective function $f(x)$ without any constraints within a region of the variables x. Unimodal functions exhibit one extremum (either maximum or minimum), whereas multimodal functions have many extrema.

The case of one-dimensional functions is given in Figures 1.1 and 1.2.

The case of two-dimensional functions is given in Figures 1.3 and 1.4.

We will deal with such cases and learn to distinguish the type of functions when we discuss convexity in Chapter 3.

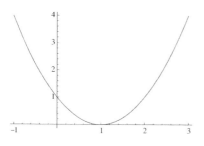

Figure 1.1 Unimodal function in one dimension.

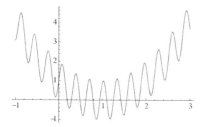

Figure 1.2 Multimodal function in one dimension.

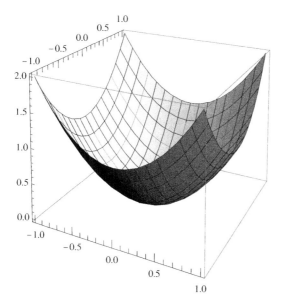

Figure 1.3 Unimodal function in two dimensions.

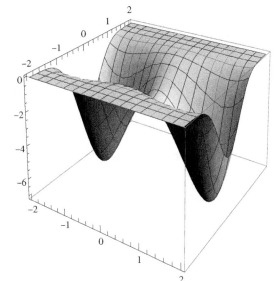

Figure 1.4 Multimodal function in two dimensions.

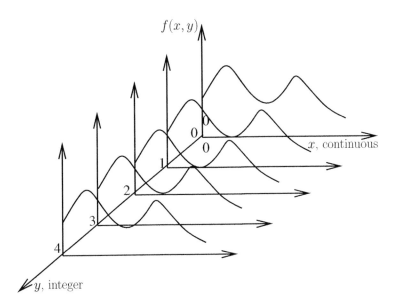

Figure 1.5 Mixed Integer function example.

1.1.2 Variable Types: Continuous, Integer, and Binary

In many real-world applications, we may wish to model systems for which the variables are not all continuous. Integer variables might be used to

- Count indivisible numbers of quantities, such as

  - number of people in a shift
  - the number of trays in a distillation column
  - the number of heat exchangers, *etc.*

- In the simplest integer form the variables might be binary
    - to reflect the presence of a unit or not
    - to reflect logic constraints (0/1 operations)

This type of variables results in *discrete optimization problems,* which need special techniques to be solved. In effect, the problems become discontinuous, as shown for the case of one variable being continuous and one being an integer in Figure 1.5.

1.2 APPLICATIONS IN TECHNOLOGICAL AND SCIENTIFIC PROBLEMS

Optimization is a very widely used tool for many real-world applications and in cutting-edge technologies. Some examples of such applications are given in the next few subsections.

1.2.1 Predicting the 3D Structures of Complex Molecules: Molecular Conformation and the Protein Folding Problem

The protein folding problem is set out here:

- Determine the 3D structure of the protein, given its linear sequence of amino acids
- Variables are the relative locations of the amino acids (angles, distances)
- Objective is the total energy of the conformation

    1. bending energy,
    2. bond stretching energy,
    3. bond torsion energy,
    4. electrostatic energies on amino acids.

- Constraints are the sequence and distance metrics of the amino acids

The characteristic of these problems is that their complexity grows quickly, in particular

- The problem is multimodal, *i.e.* there are multiple local minima of the energy objective function
- The number of local minima grows exponentially with the problem size

The solution of these problems falls within the category of global optimization methods, which are specially designed to guarantee the global optimality.

Global optimization, see Chapter 27, with rigorous guarantees of optimality is difficult – otherwise the formulation is fine and we would be able to predict *ab initio*[1] the structure of any given protein, and to design *de novo*[2] arbitrary proteins with prespecified functionality!

1.2.2 Designing Simpler Molecules: Organic Solvents and Refrigerants

The task here is to design organic molecules such that

- they have thermodynamic properties within specified tolerance margins

---

[1] *I.e.,* using basic micro-scale mechanisms *to predict* the larger-scale properties.
[2] *"from the start; new; not present previously;" i.e.* using fundamental microscale mechanisms to design *a given* larger-scale structure.

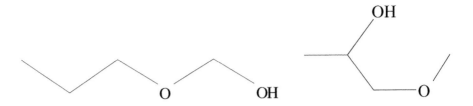

Figure 1.6 Skeletal formulae for propoxymethanol and 1-methoxy-2-propanol.

- they obey certain structural constraints (*e.g.* exclusion of given groups or types of bonds)
- they involve about 16 carbon atoms
- the applications are

  - new solvent design
  - new refrigerant design,
  - new monomers to produce polymers with desired properties

- the design is based on the selection of functional groups to include in the new molecules, such as CH3-, -CH2-, -OH, halogens, aromatic rings, *etc.*

The optimization problem formulations also involve continuous and binary variables. The following is an example of replacing an old solvent with new molecules:

- Ethyl glycol: CH3-CH2-O-CH2-CH2-OH

  - Environmentally problematic
  - Safety and health issues in the working environment of paint and ink industries
  - Toxic substance; serious health hazard

    o Respiratory problems
    o Infiltration into the blood stream can lead to liver problems

- New molecules obtained with optimization method, trying to match a list of thermodynamic properties are shown in Figure 1.6

Although this problem is also of combinatorial nature, the formulation results in an optimization problem with a smaller number of variables and it can be solved very efficiently, which will be considered in Chapter 21.

1.2.3 Refinery Operations: The Case of Pooling-Blending Problems

The blending problem is a typical case of refinery decision-making problem, as set out here:

- A number of *r* streams of distillate products (from crude oil distillation; raw materials to be blended)
- A number of *p* pools where the raw materials will be "pooled"
- A number of *b* blends (the final products)
- Target is to minimize cost of blends produced
- while satisfying blending constraints on concentrations (minimum octane number, maximum concentration of impurities, *e.g.* sulphur)
- and satisfying market demand (minimum production level)

Raw materials Pools Blends

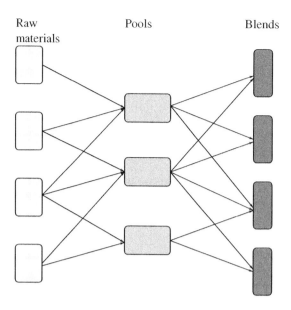

Figure 1.7 Sample pooling-blending flow chart.

A typical flow sheet of a pooling-blending problem is shown in Figure 1.7.
The model equations of this system are,

- Component-wise balances over the pools and blends
- Total flow balances over pools and blends
- Typical network flow problem, to be covered in Chapter 15

A simpler form, the pure blending problem, will be examined in Chapter 13.
The pooling-blending problem

- Is nonlinear, involving *bilinear terms, i.e.* product forms between two unknowns in the component balances,
- hence it belongs to the class of multimodal optimization problems
- and to guarantee its global optimality special solution procedures must be used

1.2.4 Parameter Estimation Problems (Model Fitting to Experimental or Plant Data)

Parameter estimation is a very important area of application of optimization methods. A number of uses are

- Fit a theoretical model to experimental data
- Fit a large nonlinear model to plant data for

  - simulation purposes
  - online optimizing control systems

The typical fitting criterion is the minimization of the sum of squares of the perpendicular distances of experimental data-points from the fitted model line, in one dimension, or the response "hypersurface" of a multidimensional model.

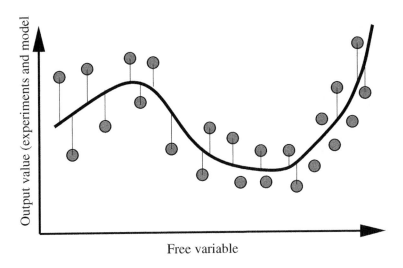

Figure 1.8 One-dimensional model fitting example.

The 1-D case is shown in Figure 1.8.

We shall examine a special case of fitting in detail in Chapter 4, the case of a linear model and least squares error minimization. A more general form of fitting, again with linear models, will be examined in Chapter 14.

1.2.5 Model Predictive Control (MPC) and Optimizing Control

Model Predictive Control (MPC) is the solution of Optimal Control problems in a continuous matter.

The key characteristics of this area of applications are the following,

- Identify the process model (by parameter estimation as in the previous section)
  - a process "drifts" with time (the process changes always), so there is a need to monitor the model's predictive ability and to reestimate its parameters
- Use an optimizer to predict a sequence of future control actions to regulate a process
- Implement the first period control actions predicted by the optimizer

Schematically, MPC is shown in Figure 1.9.

1.2.6 Process Design: The Case of the Synthesis of Binary Distillation Column Sequences

The design of separation sequences, for multicomponent mixtures, is a combinatorial optimization problem. Such problems can be formulated as optimization problems with both continuous and binary variables (logic variables).

Consider the case of a ternary mixture, of three components A, B, and C, listed in order of decreasing volatility. The possible sequences for the separation of the three components are given in Figure 1.10.

For a five-component mixture, Figure 1.11 shows the number of sequences and separations for the mixture A, B, C, D, and E, listed in order of decreasing volatility.

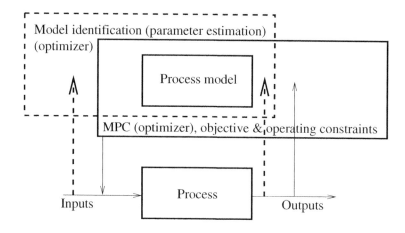

Figure 1.9 Schematic of an MPC algorithm.

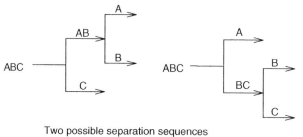

Two possible separation sequences
for a three-component mixture.

Figure 1.10 Two possible separation sequences for a three-component mixture.

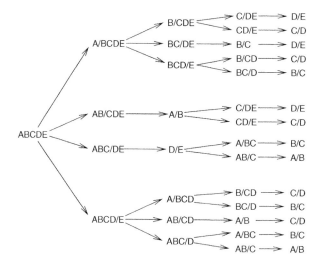

Figure 1.11 Possible separation sequences for a five-component mixture.

Table 1.1 Combinatorial series for number of sequences [1]

# Components N	# Separations $\dfrac{(N-1)N(N+1)}{6}$	# Sequences $\dfrac{(2(N-1))!}{N!\,(N-1)!}$
2	1	1
3	4	2
4	10	5
5	20	14
10	165	4,862
20	1,330	1,767,263,190
30	4,495	1,002,242,216,651,368

For any number of components we get a combinatorial number of sequences, which is illustrated in Table 1.1 [1].

In practice, heuristics (rules of thumb) derived from engineering practice are employed. Nonetheless, it has been proved that these problems can be handled by integer optimization [2].

1.2.7 Scheduling and Maintenance Planning: Heat Exchanger Networks (HEN) Cleaning Scheduling Subject to Fouling

Heat exchanger networks are an indispensable part of any modern processing plant. They serve to regulate temperatures by removing or adding heat to process streams. The major task is to design such networks so that heat is not wasted but recovered from a hot stream that needs cooling into a cold stream that needs heat, so as to minimize the external cooling and heating duties of the plant.

HEN maintenance is thus very important for the efficient operation of the heat recovery network, and this is necessary as in most processes heat exchangers are subject to fouling. Fouling is thus a major problem in such systems and quite often heat exchanger units are taken offline for cleaning.

The task in this area of application is as follows,

1. Given a HEN that is subject to fouling (a potentially highly interconnected and large network)
2. plan ahead the cleaning actions

   - which unit to clean
   - when to clean it

3. while minimizing the total costs, *i.e.* cleaning cost and use of external heating and cooling

An example network [3] is given in Figure 1.12, containing 25 highly interconnected units that comprise the crude oil preheat train in a refinery.

A proposed cleaning schedule is indicated in Figure 1.13, which contains the plan for 36 cleaning periods (months). A cleaning action at the beginning of each period is indicated by a black square.

Finally, the inlet temperature to the the furnace of the system is shown for the above cleaning schedule in Figure 1.14. Note how the temperature drops when there

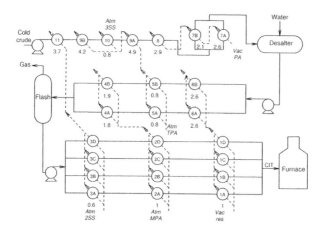

Figure 1.12 Sample process flow diagram of a heat exchanger network.

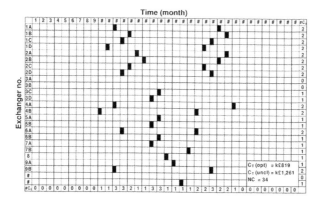

Figure 1.13 Cleaning actions for heat exchanger network.

is a heat exchanger that is taken offline for cleaning, and how the temperature is raised after that action as a clean unit is now restored in the network. This variation of the crude oil inlet temperature into the furnace is directly related to how much we have to pay to bring the oil to the target exit temperature after the furnace.

1.2.8 Routing Problems: The Travelling Salesman Problem (TSP) and the Vehicle Routing Problem (VRP)

This is a class of problems with a combinatorial nature, belonging to the general class of network optimization problems. These can be handled efficiently with integer optimization methods and are highlighted here as important applications of optimization (see Chapter 13 for more detail).

The Travelling Salesman Problem (TSP) is defined as follows

- given a list of cities
- their pairwise distances

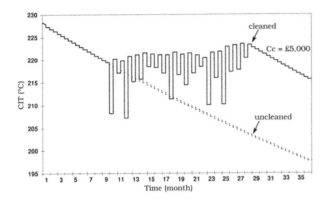

Figure 1.14 Inlet temperature to the furnace with cleaning schedule included.

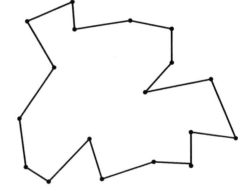

Figure 1.15 Sample Traveling Salesman Problem (TSP) solution.

- the task is to find the shortest possible tour that visits each city exactly once
- and returns to its starting city (closed tour)

A typical solution to this problem is shown in Figure 1.15.

The Vehicle Routing Problem (VRP) is a generalization of the TSP problem, as follows,

- given N customers (demand points)
- given M delivery vehicles
- minimize the total travel cost from the depot to the demand points
- by assigning a sub-tour to each vehicle (maybe not all will be needed)
- with potential time-window constraints for each delivery (earliest arrival time, latest arrival time)
- vehicles may have capacity contraints
- each vehicle returning to the depot once it completes its sub-tour

A typical VRP solution is shown in Figure 1.16.

13

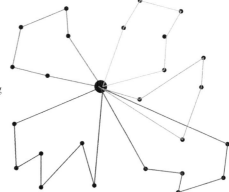

Figure 1.16 Sample Vehicle Routing Problem (VRP) solution.

1.3 ALGORITHMS AND COMPLEXITY

Searching for the global optimum, unless the optimization problem has special properties, is a combinatorially explosive task.

1.3.1 Exponential or Combinatorial Problem Complexity

- consider the case of a one-dimensional problem, where we search by evaluating the objective function at M points
- if the problem is two-dimensional then with the complete enumeration method we would need M^2 points to evaluate the function
- and for N-dimensional problems we will need M^N points, which is clearly growing exponentially
- for $N = 100$ and $M = 10$ we require[3] to evaluate the function at 10^{100} points; the universe is said to contain 10^{80} particles only!!!
- such an approach would run in "exponential time," and would be a very bad algorithm to try

The sampling problem is demonstrated in Figure 1.17.

It is noted that this holds even though the original objective function (or optimization problem) may involve only continuous variables.

In the case of discrete, and in particular binary, variables, then the problem is by nature combinatorial and

- If there are N binary variables, such that $y_i \in \{0, 1\}$, $i = 1, 2, \ldots, N$, then there are 2^N possible solutions to the problem and
- we say that the problem is of *exponential complexity*

Figure 1.18 demonstrates the solution space for small dimensionality problems.

Another way to view the effects of higher dimensional problems is the *loss of information with increase in dimensions.*

- consider a function $f(x)$ in one dimension

---

[3] It is noted that ten points in each variable interval of interest is really coarse: highly nonlinear functions with steep changes will require much finer resolution.

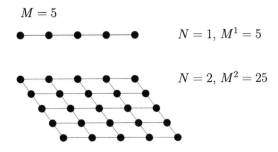

$M = 5$

$N = 1, M^1 = 5$

$N = 2, M^2 = 25$

Figure 1.17 Algorithmic complexity with number of dimensions.

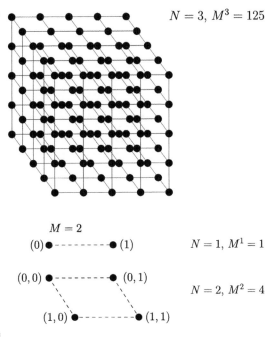

$N = 3, M^3 = 125$

$M = 2$

$(0) \bullet\!-\!-\!-\!-\!-\!-\!-\!\bullet (1)$ $N = 1, M^1 = 1$

$(0,0) \bullet\!-\!-\!-\!-\!-\!-\!\bullet (0,1)$

$N = 2, M^2 = 4$

$(1,0) \bullet\!-\!-\!-\!-\!-\!-\!\bullet (1,1)$

Figure 1.18 Solution space for small dimensionality problems.

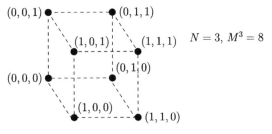

$N = 3, M^3 = 8$

- consider a point $x^{(0)}$ and a distance $\pm\dfrac{\delta}{2}$ around it, contained in an overall interval of interest Δ, such that $\delta < \Delta$
- the value of $f(x^{(0)})$ is within ε in this box, as shown in Figure 1.19 (we choose the value of δ so that we satisfy this error bounding)
- The volume fraction of the search space characterized by the single point function evaluation within error ε is

$$r = \frac{\delta}{\Delta}$$

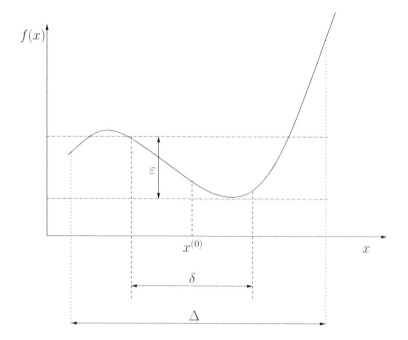

Figure 1.19 Loss of information for one-dimensional problem.

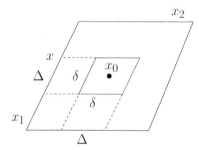

Figure 1.20 Loss of information in two dimensions.

- As the dimensions increase from $1, 2, 3, \ldots, N$
  - maintaining the same perturbation $\frac{\delta}{2}$ around a nominal point $x^{(0)}$ and an overall interval of interest Δ for all new dimensions (as in Figures 1.20 and 1.21), we get:

$$r = \left(\frac{\delta}{\Delta}\right)^N$$

So the overall information conveyed by a single function evaluation, viewed as a fraction of overall search space characterized, decays exponentially with the number of dimensions ($\frac{\delta}{\Delta} < 1$).

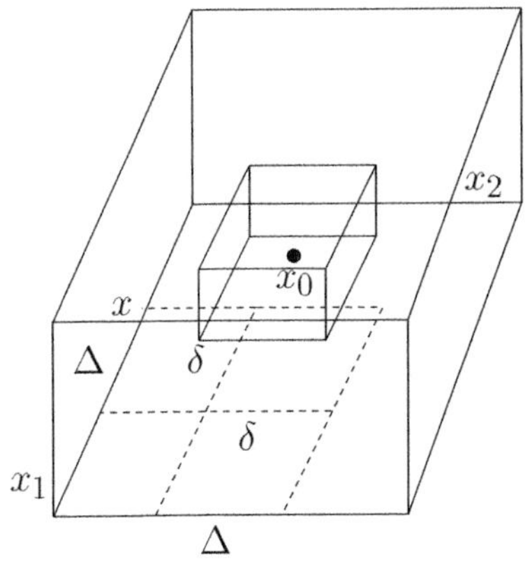

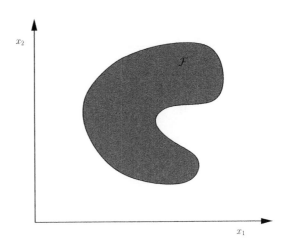

Figure 1.21 Loss of information in three dimensions.

Figure 1.22 Feasible space for two-dimensional problem.

1.3.2 A Note on Random Sampling Methods and the Improbability of Satisfying Constraints

Random sampling methods are simple ways to get quick answers to complicated problems. For unconstrained problems, by the previous analysis, if any significant percentage of a hypervolume is to be characterized (sampled), then the number of function evaluations required grows exponentially with the dimensionality of the problem.

For constrained problems, another phenomenon manifests itself as the problem size increases. Consider the feasible space of an optimization problem being described by $\mathcal{F}$, in two dimensions, as depicted in Figure 1.22.

As it is difficult with random sampling methods to get a feasible point within the feasible region $\mathcal{F}$, we try to find an outer bounding box containing the region, and sample uniformly points from within it, as shown in Figure 1.23.

If the point sampled is feasible and then improves the objective function value, then we keep it; otherwise we reject it and re-sample from the bounding box shown in Figure 1.21.

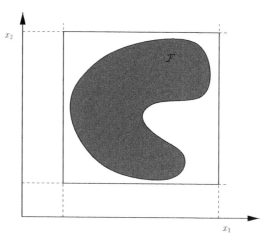

Figure 1.23 Box containing the sampling region.

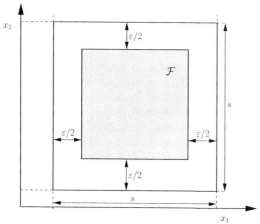

Figure 1.24 Feasible space, represented as a square, for a two-dimensional problem.

In order to observe the effect of higher dimensionalities, let us assume for simplicity that the feasible region is now given by a square, and that we sample from within an outer bounding square as in the previous case; this is shown in Figure 1.24.

According to this, the outer square has side s and the inner square (feasible region) has side $s - \varepsilon$. The probability of a random point falling in the inner square while sampling uniformly from the outer square is equal to the ratio of the two areas of the squares, or hypervolumes for many dimensions where the squares become hypercubes, which can be written as the following:

$$p(x \in \mathcal{F}) = \frac{(s - \varepsilon)^N}{s^N} = \left(1 - \frac{\varepsilon}{s}\right)^N \tag{1.2}$$

Even if ε is very small, much smaller than 1, as N the number of dimensions of the problem increases, then the probability tends to zero, *i.e.* it is improbable that a feasible point can be sampled by chance!!![4]

For example, the probability with $\varepsilon/s = 10^{-3}$ for $N = 10{,}000$ is $p = 0.0000452$. The full effect of increasing dimensions is shown in Figure 1.25.

---

[4] The implication also for higher-dimensional polyhedra is that "most" of the volume of the object is an infinitesimal distance away from the surface of the object.

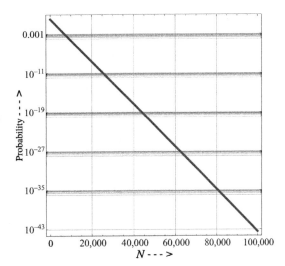

Figure 1.25 Probability of feasible sample for $\varepsilon = 10^{-3}$.

The size of the problems mentioned may not be typical for practical engineering applications; however, it is difficult in general to find tight sampling bounding boxes for complicated problems, thus the feasible region may even occupy a smaller portion of the outer bounding hypercube.

1.3.3 Efficient Algorithms for Combinatorially Complex Problems

Unless the problem is continuous and it belongs to the special class of *convex programming problems* (Chapter 7), then in general we cannot guarantee the global optimality of a solution obtained with solvers employing standard local optimization algorithms.

Special and often very successful algorithms are based on branch and bound (B&B) methodologies, both for combinatorial and continuous problems having multiple optima. See Chapter 27 on Global Optimization.

It is finally noted that there are heuristic algorithms, either employing rules-of-thumb based on engineering practice or employing some form of random sampling procedure (*Monte Carlo optimization/search methods*) to improve the values of the objective function obtained.

1.3.4 More Efficient Computers for Combinatorial Problems?

Consider a problem with N_1 variables. The question is: How much faster does a computer have to be in order to solve a problem of larger size N_2 depending on the problem complexity. For the following analysis, the time required for the evaluation is represented by CPU_1 and CPU_2.

If the complexity of the problem was *linear, i.e.* $\mathcal{O}\left(N_1\right)$, we find the following:

$$CPU_1 = \alpha \cdot N_1$$
$$CPU_2 = \alpha \cdot N_2$$

Now if $N_2 = 2 \times N_1$, the following relation is obtained:

$$\frac{CPU_2}{CPU_1} = \frac{N_2}{N_1} = 2$$

i.e. we can double the size of the problem solved in the same time on the new computer as the half-size problem.

But this is not true with exponential complexity $\mathcal{O}\left(2^{N_1}\right)$. For this case we have:

$$CPU_1 = \alpha 2^{N_1}$$

If we double the speed of the old machine, we have the following,

$$CPU_2 = 2 \cdot CPU_1$$

With double the speed, how much larger of a problem can be solved within the same time? With the new computer we have:

$$CPU_2 = \alpha \cdot 2^{N_2}$$

and using the definition of CPU_2 as a function of CPU_1 we get:

$$CPU_2 = 2CPU_1 = 2 \cdot \alpha \cdot 2^{N_1} = \alpha \cdot 2^{N_2}$$

hence,

$$2^{N_1+1} = 2^{N_2}$$

or merely

$$N_2 = N_1 + 1$$

So we can only solve a problem increased by **one** variable with double the speed if the problem has an exponential complexity $\mathcal{O}\left(2^{N_1}\right)$. Considering Moore's Law, the amount of transistors on a processor doubles every two years. Hence for combinatorial problems, having exponential complexity, we can add …one variable every two years to the maximum size of problem that can be handled!

1.4 GENERALIZED OPTIMIZATION PROBLEMS

Further to the general optimization problem given in Section 1.1, it is possible to widen the scope of optimization formulations, as in the following subsections.

1.4.1 Problems with Infinite-Dimensional Decisions: Optimal Control Problems (OCP), or Dynamic Optimization Problems (DYNOPT)

The generic formulation that was given in Equations (1.1a–1.1d), in Section 1.1, considers the case of a finite set of variables x that are to be optimally chosen.

Often, in Chemical Engineering applications we are interested not for a single variable value but for an entire *trajectory* to be optimally chosen. This is best highlighted by the following example problem:

- a catalytic tubular reactor design problem
- using a bi-functional catalyst blend along its length
- the task is to find the optimal catalyst blend
- to maximize the yield of a desired product at the outlet of the reactor

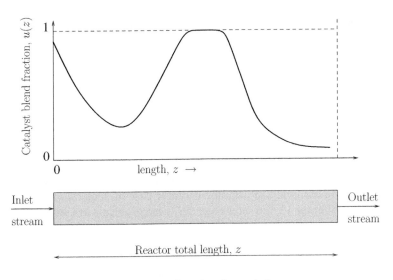

Figure 1.26 Optimal distribution of catalyst for a tubular reactor.

The reactor and the decision profile for the catalyst blend mixture are shown in Figure 1.26.

We have to determine at every point z, where $0 \leq z \leq Z$, the blending ratio function $u(z)$, ranging from 0 (catalyst A) to 1 (catalyst B).

Things to note about this problem are as follows:

- there is an infinity of points we have to determine for the blending fraction, at each point along the length of the reactor
- the blending fraction is what we call a *control function*
- the model equations are the differential balances, hence they are ODE's

A general formulation of such an optimization problem is as follows:

The objective function, evaluated at the final time t_f (which may itself be a decision parameter) is:

$$\min_{p,u(t),t_f} \phi(x(t_f), x(t_f), y(t_f), u(t_f), p) \tag{1.3a}$$

subject to the following:
   Process dynamics, described by a set of ODE's:

$$\frac{dx}{dt}(t) = f(t, x(t), y(t), u(t), p), \quad t_0 \leq t \leq t_f \tag{1.3b}$$

Coupled algebraic equations:

$$g(t, x(t), y(t), u(t), p) = 0 \tag{1.3c}$$

Initial conditions for differential process variables:

$$x(t_0) = x(t_0, p) \tag{1.3d}$$

Trajectory inequality constraints (*path constraints*), imposed at all times t:

$$h(t, x(t), y(t), u(t), p) \leq 0, \quad t_0 \leq t \leq t_f \tag{1.3e}$$

21

Simple bounds on control functions, $u(t)$, design parameters, p, and final time t_f:

$$u^L \leq u(t) \leq u^U, \quad t_0 \leq t \leq t_f \tag{1.3f}$$

$$p^L \leq p \leq p^U \tag{1.3g}$$

$$t_f^L \leq t \leq t_f^U \tag{1.3h}$$

Time-invariant design parameters are p, control functions are $u(t)$, differential state variables are $x(t)$, and algebraic state variables are $y(t)$.

Solution procedures usually involve transformation of these problems into standard optimization problems with a finite number of variables, *e.g.* using some form of discretization. We are going to examine a simple case of such an *optimal control problem* (OCP) in Chapter 13 and later consider these problems in Chapter 22.

1.4.2 Multiobjective Optimization Problem (MOO)

All the cases of optimization problems we have seen so far involve a simple objective function, subject to equality and inequality constraints.

Multiobjective optimization deals with problems with at least two objectives, as the term declares. Such a problem formulation is given by the following:

$$\min_{x} (f_1(x), f_2(x), \ldots, f_m(x))^T \tag{1.4a}$$

subject to,

$$h(x) = 0 \tag{1.4b}$$

$$g(x) \leq 0 \tag{1.4c}$$

As can be seen in the objective function in Equation (1.4a), we have to satisfy a whole vector of m-objective functions simultaneously.

These represent *conflicting objectives*, such that in the solution, improvement of any one objective can come only at the expense of worsening (enlarging) the value of the other objective(s).

Such situations can be found in many practical engineering applications, such as:

- Balancing cost minimization and risk minimization (the higher the investment in overdesigning a system the lower the risk)
- Balancing cost minimization versus environmental impact
- Maximization of both the yield and selectivity in reaction networks (*e.g.* biochemical reactor systems)

Multiobjective optimization is examined in Chapter 16.

1.4.3 Bilevel Optimization, Bilevel Programming Problems (BLP)

We first define the problem formulation directly to see how it differs from simple optimization problems.

The upper objective is:

$$\min_{x,y} F(x, y) \tag{1.5a}$$

subject to,

the upper objective constraints as follows:

$$H(x, y) = 0 \tag{1.5b}$$

$$G(x, y) \leq 0 \tag{1.5c}$$

The lower level (embedded) optimization problem, with lower level objective and lower level constraints is:

$$y \in \arg\min \{ f(x, y) : h(x, y) = 0,\ g(x, y) \leq 0 \} \tag{1.5d}$$

In simpler terms we can also write:

$$\min_{x,y} F(x, y) \tag{1.6a}$$

$$s.t.$$

$$H(x, y) = 0 \tag{1.6b}$$

$$G(x, y) \leq 0 \tag{1.6c}$$

$$\min_{y} f(x, y) \tag{1.6d}$$

$$s.t.$$

$$h(x, y) = 0 \tag{1.6e}$$

$$g(x, y) \leq 0 \tag{1.6f}$$

With this we can give a definition of a BLP as:

A bilevel programming problem contains a subset of variables that is required to be an optimal solution of a second optimization problem, *i.e.* one of the constraints of the optimization problem is itself an optimization problem (nested problem).

Application areas can be found wherever the decision-making problem involves multiple levels, each optimizing a local objective, *e.g.* when hierarchical decision making is applied in an organization.

An example of such a formulation arises in the design of new microbial strains subject to *gene knockouts* [4]. This is shown schematically in Figure 1.27.

1.4.4 Stochastic Optimization Problems (SP): Decision Making under Uncertainty

Often in real-world models there is a great deal of uncertainty, which has to be reflected by realistic process models. Modeling uncertainty in chemical processes has three sources:

1. Input variability,
2. Disturbances,
3. Parametric uncertainty.

Decision making based on such models has to take into account the uncertainty involved, so as to estimate as best as possible the expected outcomes. We assume that all the forms of uncertainty can be captured by parametric uncertainty, and hence we have the following formulations:

Objective function: optimized on average

$$\min_{x} \bar{f}(x) = E\left[f(x, \xi) \right] \tag{1.7a}$$

Genetic manipulation level, gene knockout selection

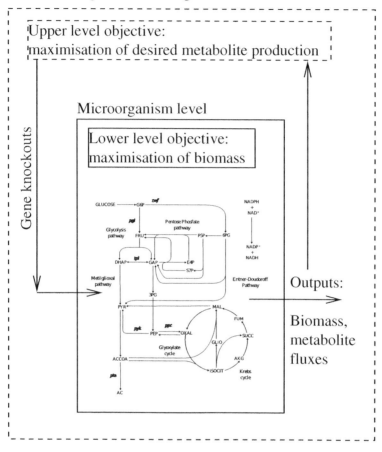

Figure 1.27 Design of new microbial strains subject to gene knockouts.

where

- $x \in \mathbb{R}^{n_x}$, is the vector of decision variables to be chosen optimally
- $\xi \in \Xi \subset \mathbb{R}^{n_\xi}$, is the vector of uncertain parameters. In fact, these are *random variables* (RV), such that there is a probability distribution function vector $P(\xi)$[5] for them.
- $\bar{f}(x) = E\left[f(x,\xi)\right] = \int_{\Xi} f(x,\xi)dP(\xi)$, is the average value definition.

Constraints:

When the uncertain parameters ξ enter the constraints then we have

$$h(x,\xi) = 0 \qquad\qquad (1.7b)$$

$$g(x,\xi) \leq 0 \qquad\qquad (1.7c)$$

---

[5] This is not a strict definition of a probability metric; the aim here is just to introduce the general concepts.

and for this case we need to define what we mean by feasibility. Often a choice of x might satisfy some realisations of the RV, ξ, while for others it might not. So we can relax strict feasibility of Equation (1.7c) by taking an average value of the constraints as follows:

$$E\left[g(x,\xi)\right] \leq 0 \tag{1.7d}$$

so that now we consider average satisfaction.

Another possibility of quantifying satisfaction of the constraints in (1.7c) is to introduce constraints on the probability of their satisfaction as follows:

$$P\left[g(x,\xi) \leq 0\right] \geq \varepsilon, \quad 0 \leq \varepsilon \leq 1 \tag{1.7e}$$

Solution procedures usually involve transformation of these problems into standard optimization problems, *e.g.* using some form of approximation of continuous probability distribution functions (*e.g. discrete samples, multiple scenarios*).

1.5 CHAPTER SUMMARY

The field of Optimization includes numerous fields that vary in terms of variables, formulations, and parameters. Variable types include the following:

- continuous
- integer
- binary
- control function

Parameters used in models can be known or unknown due to uncertainty. These parameters and variables lead to various model formulations such as:

- MP-general
- NLP
- MIP
- MOO
- BLP
- SP
- OCP

Each of these formulations is best solved with one of the following solvers:

- continuous
- mixed-integer continuous

1.6 REFERENCES

[1] Biegler, L. T., Grossmann, I. E., and Westerberg, A. W. Systematic Methods of Chemical Process Design. Prentice Hall PTR. 1997;p. 796.

[2] Méndez, C. A., Cerdá, J., Grossmann, I. E., Harjunkoski, I., and Fahl, M. "State-of-the-art review of optimization methods for short-term scheduling of batch processes." Computers & Chemical Engineering. 2006;30:p. 913–946.

[3] Smaïli, F., Vassiliadis, V. S., and Wilson, D. I. "Long-term scheduling of cleaning of heat exchanger networks: comparison of outer approximation-based solutions with a

backtracking threshold accepting algorithm." Chemical Engineering Research and Design. 2002;80:p. 561–578.

[4] Burgard, A. P., Pharkya, P., and Maranas C. D. "Optknock: A bilevel programming framework for identifying gene knockout strategies for microbial strain optimization." Biotechnology and Bioengineering. 2003;84:p. 647–657.

I.7 FURTHER READING RECOMMENDATIONS

Protein Folding Problem (Particularly Focusing on Potential Energy Objective Function Construction)

Boas, F. E. and Harbury, P. B. (2007). "Potential energy functions for protein design." Current Opinion in Structural Biology. 17:p. 199–204.

Fujitsuka, Y., Takada, S., Luthey-Schulten, Z. A., and Wolynes P. G. (2003). "Optimizing physical energy functions for protein folding." Proteins: Structure, Function, and Bioinformatics. 54:p. 88–103.

Pillardy, J., Czaplewski, C., Liwo, A., Lee, J., & Ripoll, D. R., Kazmierkiewicz, R., Oldziej, S., Wedemeyer, W. J., Gibson, K. D., Arnautova, Y. A., Saunders, J., Ye, Y.-J., and Scheraga H. A. (2001). "Recent improvements in prediction of protein structure by global optimization of a potential energy function." Proceedings of the National Academy of Sciences. 98:p. 2329–2333.

Liwo, A., Lee, J., Ripoll, D. R., & Pillardy, J., and Scheraga, H. A. (1999). "Protein structure prediction by global optimization of a potential energy function." Proceedings of the National Academy of Sciences of the United States of America. 96:p. 5482–5485.

Pardalos, P. M., Shalloway, D., and Xu, G. (1994). "Optimization methods for computing global minima of nonconvex potential energy functions." Journal of Global Optimization. 4:p. 117–133.

Lazaridis, T. and Karplus, M. (2000). "Effective energy functions for protein structure prediction." Current Opinion in Structural Biology. 10:p. 139–145.

Lee, J., Seung-Yeon, K., and Lee, J. (2004). "Design of a protein potential energy landscape by parameter optimization." American Chemical Society. 108:p. 4525–4534.

Lee, J. & Liwo, A. & Ripoll, D. R. & Pillardy, J. & Scheraga, H. A. (1999). "Calculation of protein conformation by global optimization of a potential energy function." Proteins. 37:p. 204–208.

Abagyan, R. A. and Totrov, M. (1999). "Ab initio folding of peptides by the optimal-bias Monte Carlo minimization procedure." Journal of Computational Physics. 151: p. 402–421.

Molecular Design of Small Organic Molecules

Maranas, C. D. (1996). "Optimal computer-aided molecular design: A polymer design case study." American Chemical Society. 35:p. 3403–3414.

Churi, N. and Achenie, L. E. K. (1996) "Novel mathematical programming model for computer aided molecular design." American Chemical Society. 35:p. 3788–3794.

Optimization of Biochemical Reaction Networks

Torres, N. V. and Voit, E. O. (2002). Pathway Analysis and Optimization in Metabolic Engineering. Cambridge University Press.

Process Synthesis and Scheduling by Optimization Methods

Yeomans, H. and Grossmann, I. E. (1999). "A systematic modeling framework of super-structure optimization in process synthesis." Computers & Chemical Engineering. 23: p. 709–731.

Kocis, G. R. and Grossmann, I. E. (1988). "Global optimization of nonconvex mixed-integer nonlinear programming (MINLP) problems in process synthesis." American Chemical Society. 27:p. 1407–1421.

Biegler, L. T., Grossmann, I. E., and Westerberg, A. W. (1997). Systematic Methods of Chemical Process Design. Prentice Hall PTR.

Pooling-Blending Problem

Misener, R. and Christodoulos, A. F. (2009). "Advances for the pooling problem: Modeling, global optimization, and computational studies." Applied and Computational Mathematics. 8:p. 3–22.

Greenberg, H. J. (1995). "Analyzing the pooling problem." ORSA Journal on Computing. 7:p. 205–217.

Model Predictive Control and Parameter Estimation

Huang, R, Patwardhan, S. C., and Biegler, L. T. (2009). "Multi-scenario-based robust nonlinear model predictive control with first principle models." Computer Aided Chemical Engineering. 27:p. 1293–1298.

Huang, R. and Biegler, L. T. (2009). "Robust nonlinear model predictive controller design based on multi-scenario formulation." 2009 American Control Conference. p. 2341–2342.

Huang, R., Patwardhan, S. C., and Biegler, L. T. (2009). "Robust extended Kalman filter based nonlinear model predictive control formulation." Proceedings of the 48h IEEE Conference on Decision and Control (CDC) held jointly with 2009 28th Chinese Control Conference. p. 8046–8051.

Biegler, L. T. (2009). Efficient Nonlinear Programming Algorithms for Chemical Process Control and Operations. Springer, Berlin, Heidelberg. p. 21–35.

Zavala, V. M. and Biegler, L. T. (2009). Nonlinear Programming Strategies for State Estimation and Model Predictive Control. Springer, Berlin, Heidelberg. p. 419–432.

Zavala, V. M. and Biegler, L. T. (2009). "The advanced-step NMPC controller: Optimality, stability and robustness." Automatica. 45:p. 86–93.

Routing Problems

Toth, P. and Vigo, D. (2002). "Multi-scenario-based robust nonlinear model predictive control with first principle models." Society for Industrial and Applied Mathematics. p. 367.

Laporte, G. (1992). "The vehicle routing problem: An overview of exact and approximate algorithms." European Journal of Operational Research. 59:p. 345–358.

Pataki, G. (2003). "Teaching integer programming formulations using the traveling salesman problem." SIAM Review. 45:p. 116–123.

Biegler, L. T. (2009). Efficient Nonlinear Programming Algorithms for Chemical Process Control and Operations.' Springer, Berlin, Heidelberg. p. 21–35.

Bellmore, M. and Nemhauser, G. L. (1968). "The traveling salesman problem: A survey on JSTOR." Operations Research. 16.

Optimal Control (Dynamic Optimization)

Vassiliadis, V. S., Sargent, R. W. H., and Pantelides, C. C. (1994). "Solution of a class of multistage dynamic optimization problems. 1. Problems without path constraints." Industrial & Engineering Chemistry Research. 33:p. 2111–2122.

Vassiliadis, V. S., Sargent, R. W. H., and Pantelides C. C. (1994). "Solution of a class of multistage dynamic optimization problems. 2. Problems with path constraints." Industrial & Engineering Chemistry Research. 33:p. 2123–2133.

Biegler, L. T. and Zavala V. M. (2009). "Large-scale nonlinear programming using IPOPT: An integrating framework for enterprise-wide dynamic optimization." Computers & Chemical Engineering. 33:p. 575–582.

Biegler, L. T. (2009). Efficient Nonlinear Programming Algorithms for Chemical Process Control and Operations. Springer, Berlin, Heidelberg. p. 21–35.

Cuthrell, J. E. and Biegler, L. T. (1987). "On the optimization of differential-algebraic process systems." AIChE Journal. 33:p. 1257–1270.

Multiobjective and Bilevel Optimization

Young, I. L., Pascal, F., Joulia, X. and Kim, S. D. (1999). "Multiobjective optimization in terms of economics and potential environment impact for process design and analysis in a chemical process simulator." American Chemical Society.

Marler, R.T. and Arora J. S. (2004). "Survey of multi-objective optimization methods for engineering." Structural and Multidisciplinary Optimization. 26:p. 369–395.

Young, I. L., Pascal, F., and Joulia, X. (2000). "Efficient implementation of the Normal Boundary Intersection (NBI) method on multiobjective optimization problems." American Chemical Society.

Pu-yan, N. (2005). "A note on bilevel optimization problems." International Journal of Applied Mathematical Sciences. 2:p. 31–38.

Clark, P. A. (1990). "Bilevel programming for steady-state chemical process design-II. Performance study for nondegenerate problems." Computers & Chemical Engineering. 14:p. 99–109.

Clark, P. A. and Westerberg, A. W. (1990). "Bilevel programming for steady-state chemical process design-I. Fundamentals and algorithms." Computers & Chemical Engineering. 14:p. 87–97.

Clark, P. A. and Westerberg A. W. (1983). "Optimization for design problems having more than one objective." Computers & Chemical Engineering. 7:p. 259–278.

Stochastic Optimization

Ruszczynski, A. and d Shapiro, A. (2003). "Stochastic programming models." Handbooks in OR & MS. 10:p. 1–64.

Pistikopoulos, E. N. (1995). "Uncertainty in process design and operations." Computers & Chemical Engineering. 19:p. 553–563.

Hene, T. S., Dua, V., and Pistikopoulos, E. N. (2002). "A Hybrid parametric/stochastic programming approach for mixed-integer nonlinear problems under uncertainty." Industrial & Engineering Chemistry Research. 41:p. 67–77.

Sahinidis, N. V. (2004). "Optimization under uncertainty: State-of-theart and opportunities." Computers & Chemical Engineering. 28:p. 971–983.

PART II

From General Mathematical Background to General Nonlinear Programming Problems (NLP)

2

General Concepts

2.1 "SIZE OF VECTORS"

Quite often it is necessary to measure how "big" a given vector is. For example, in solving a set of nonlinear equations we would like to set up a termination criterion that measures the error (residual) of the equations as the solution vector evolves through some algorithm. When the equations are "sufficiently" satisfied, i.e. their residual approaches zero, then we would stop the algorithm and report the solution found within that error tolerance.

This raises the question of how to measure the "length" of a vector. Such measures are called **norms of vectors** (and of matrices, but only vectors are shown here).

Given a vector (*e.g.* $\in \mathbb{R}^3$), then:

$$x = \begin{bmatrix} x_1 \\ x_2 \\ x_3 \end{bmatrix} \tag{2.1}$$

The infinity norm is:

$$\|x\|_\infty = \max_{i=1,2,3} \{|x_i|\} \tag{2.2}$$

Norm-1 is:

$$\|x\|_1 = \sum_{i=1}^{n} |x_i| \tag{2.3}$$

Norm-2 or the Euclidian norm norm (the standard Euclidean distance when dealing with 1, 2, and 3 dimensions) is:

$$\|x\|_2 = \sqrt{\sum_{i=1}^{n} x_i^2} \tag{2.4}$$

General Norm-k is:

$$\|x\|_k = \left(\sum_{i=1}^{n} |x_i|^k \right)^{\frac{1}{k}} \tag{2.5}$$

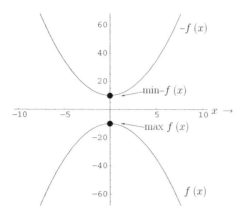

Figure 2.1 Exchange of max and min problem.

2.2 MINIMIZATION VERSUS MAXIMIZATION

In this book we shall be usually dealing with minimization problems. This does not exclude the case of maximization, as the latter problem can be cast into the former type.

Definitions for minimization are:

$$\min_x f(x); \quad x \in \mathcal{X}(\in \mathbb{R}^n) \tag{2.6}$$

$$x^* = \arg\min_x f(x) \tag{2.7}$$

Exchanging types are:

$$\max_x f(x) = -\min_x[-f(x)] \tag{2.8}$$

$$x^* = \arg\max_x f(x) \equiv \arg\min_x[-f(x)] \tag{2.9}$$

The exchange of max and min is shown in Figure 2.1.

2.3 TYPES OF EXTREMA OF A FUNCTION (STATIONARY POINTS)

The following figure outlines the general types of extrema a one-dimensional function may have. These generalize into higher dimensionality functions, as is shown in Figures 2.2–2.4.

2.3.1 Strong/Weak Extrema: A More Formal Definition

Given $\delta > 0$ and an extremum x^*, for sufficiently small δ the neighborhood around the extremum is:

$$\Omega(x) = \{x : \|x - x^*\| < \delta, x \in \mathcal{X}\} \tag{2.10}$$

Curly braces indicate a set, in this case the set of values to which x is restricted. Read the colon (:) as: "x is such that ...". $\mathcal{X}$ is the set of allowable values for the variable x. Table 2.1 summarizes the types of extrema.

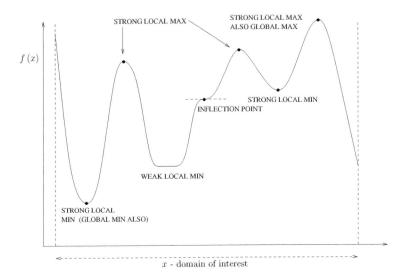

Figure 2.2 Diagram of higher dimensional function.

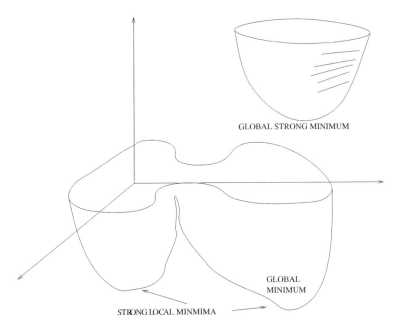

Figure 2.3 Difference between strong local and global minima.

2.4 **CONTOURS OF $f(x)$**

Contours are useful in 2-D problems ($x \in \mathbb{R}^2$).

Property: the gradient at a point (x_1, x_2) is perpendicular to the contour at that point, *i.e.* perpendicular to the tangent of the contour.

The gradient points in the direction of greatest increase of $f(x)$ at the point (x_1, x_2).

Table 2.1 Summary of types of extrema

Weak local minimum	$f(x) \geq f(x^*)$	for $x \in \Omega(x^*)$
Strong local minimum	$f(x) > f(x^*)$	for $x \in \Omega(x^*)$
Weak local maximum	$f(x) \leq f(x^*)$	for $x \in \Omega(x^*)$
Strong local maximum	$f(x) < f(x^*)$	for $x \in \Omega(x^*)$
Unique global minimum	$f(x) > f(x^*)$	for $x \in \mathcal{X}$
Global minimum	$f(x) \geq f(x^*)$	for $x \in \mathcal{X}$
Unique global maximum	$f(x) < f(x^*)$	for $x \in \mathcal{X}$
Global maximum	$f(x) \leq f(x^*)$	for $x \in \mathcal{X}$

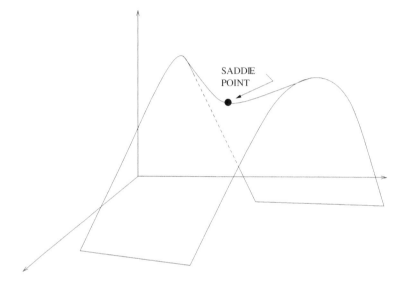

Figure 2.4 Higher dimensional function with saddle point.

Contours are found by setting function values to a constant, C, and tracing out which values of the variables, x, satisfy the relationship:

$$f(x) = C \tag{2.11}$$

Contours and properties are shown in Figures 2.5 and 2.6.

2.5   VECTORS, AND DERIVATIVES INVOLVING MATRICES AND VECTORS

Given a function of a set of variables:

$$f(x): \quad \mathbb{R}^n \mapsto \mathbb{R}^1$$

its gradient vector is a row vector with dimensions $(1 \times n)$:

$$\frac{\partial f}{\partial x} = \left(\frac{\partial f}{\partial x_1}, \frac{\partial f}{\partial x_2}, \ldots, \frac{\partial f}{\partial x_n} \right) \tag{2.12}$$

2.5 Vectors, and Derivatives Involving Matrices and Vectors

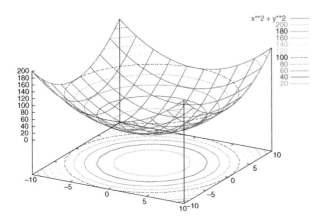

Figure 2.5 Three-dimensional plot of the function $f(x)$.

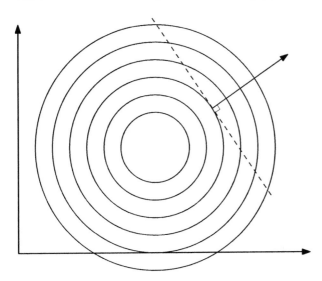

Figure 2.6 Contour plot of the function $f(x)$.

and its Hessian matrix is a matrix with dimensions $(n \times n)$:

$$\frac{\partial^2 f}{\partial x^2} = \frac{\partial}{\partial x}\left[\frac{\partial f}{\partial x}\right] = \begin{pmatrix} \frac{\partial^2 f}{\partial x_1^2} & \frac{\partial^2 f}{\partial x_1 \partial x_2} & \cdots & \frac{\partial^2 f}{\partial x_1 \partial x_n} \\ \vdots & \vdots & \vdots & \vdots \\ \frac{\partial^2 f}{\partial x_n \partial x_1} & \frac{\partial^2 f}{\partial x_n \partial x_2} & \cdots & \frac{\partial^2 f}{\partial x_n^2} \end{pmatrix} \tag{2.13}$$

Example 1: Chain Rule Application

Let $f(x) \in \mathbb{R}^1$, $x \in \mathbb{R}^n$, $y \in \mathbb{R}^m$ and x is a function of $y : x = x(y)$. Then we obtain:

$$\frac{\partial f}{\partial y} = \frac{\partial f}{\partial x}\frac{\partial x}{\partial y} \tag{2.14}$$

$$(1 \times m) = (1 \times n) \times (n \times m) \tag{2.15}$$

Example 2: Sum of Functions

$$\phi(x) = f(x) + g(x) \tag{2.16}$$

$$\frac{\partial \phi}{\partial x} = \frac{\partial f}{\partial x} + \frac{\partial g}{\partial x} \tag{2.17}$$

$$\frac{\partial^2 \phi}{\partial x^2} = \frac{\partial^2 f}{\partial x^2} + \frac{\partial^2 g}{\partial x^2} \tag{2.18}$$

Further Examples

$$x \in \mathbb{R}^3, \ y \in \mathbb{R}^2$$

$$x = \begin{bmatrix} x_1 \\ x_2 \\ x_3 \end{bmatrix}, \ y = \begin{bmatrix} y_1 \\ y_2 \end{bmatrix}$$

Given $f(x)$ and that $x \equiv x(y)$, calculate $\frac{\partial f}{\partial y}$:

$$\frac{\partial f}{\partial y} = \frac{\partial f}{\partial x} \cdot \frac{\partial x}{\partial y}$$

$$\frac{\partial f}{\partial x} = \begin{pmatrix} \frac{\partial f}{\partial x_1} & \frac{\partial f}{\partial x_2} & \frac{\partial f}{\partial x_3} \end{pmatrix} \quad (1 \times 3) \text{ row vector}$$

$$\frac{\partial x}{\partial y} = \begin{bmatrix} \frac{\partial x_1}{\partial y_1} & \frac{\partial x_1}{\partial y_2} \\ -- & -- \\ \frac{\partial x_2}{\partial y_1} & \frac{\partial x_2}{\partial y_2} \\ -- & -- \\ \frac{\partial x_3}{\partial y_1} & \frac{\partial x_3}{\partial y_2} \end{bmatrix} \quad (3 \times 2) \text{ matrix: The Jacobian of } x \text{ with respect to } y$$

$$\frac{\partial f}{\partial y} = \frac{\partial f}{\partial x} \cdot \frac{\partial x}{\partial y}$$

$$\frac{\partial f}{\partial y} \rightarrow (1 \times 2)$$

$$\frac{\partial f}{\partial x} \cdot \frac{\partial x}{\partial y} \rightarrow (1 \times 3) \times (3 \times 2)$$

$$\rightarrow (1 \times 2)$$

Hence the same dimensions are present on both sides of the equation.

Example: Inner Product of Vector Functions

Consider: $f(x), g(x) \in \mathbb{R}^m$ and $x \in \mathbb{R}^n$

$$\phi(x) = g^T(x) \cdot f(x) \tag{2.19}$$

$$\phi(x) \rightarrow (1 \times 1) \tag{2.20}$$

$$g^T(x) \cdot f(x) \rightarrow (1 \times m) \times (m \times 1) \tag{2.21}$$

$$\rightarrow (1 \times 1) \tag{2.22}$$

Find $\frac{\partial \phi(x)}{\partial x}$: row vector $(1 \times n)$.

When applying the $\frac{\partial \cdot}{\partial x}$ operator to a product, apply it to the rightmost term:

$$\frac{\partial \phi(x)}{\partial x} = g^T(x)\frac{\partial f}{\partial x} + f^T(x)\frac{\partial g}{\partial x} \tag{2.23}$$

$$\frac{\partial \phi(x)}{\partial x} \rightarrow (1 \times n) \tag{2.24}$$

$$g^T(x)\frac{\partial f}{\partial x} \rightarrow \underbrace{(1 \times m) \times (m \times n) \rightarrow (1 \times n)}_{\text{obtained by applying the operator to } f(x)} \tag{2.25}$$

$$f^T(x)\frac{\partial g}{\partial x} \rightarrow \underbrace{(1 \times m) \times (m \times n) \rightarrow (1 \times n)}_{\text{obtained by first writing}} \tag{2.26}$$

$$g^T(x) \cdot f(x) = f^T(x) \cdot g(x)$$

and then applying the operator to $g(x)$

2.6   FURTHER READING RECOMMENDATIONS

Characterisation of Extrema

Edgar, T. F., Himmelblau, D. M., and Lasdon, L. S. (2001). Optimization of Chemical Process. McGraw-Hill, Edition 2.

Derivatives Involving Vectors, Matrices, and Tensors

Marlow, W. H. (1993). Mathematics for Optimization Research. Courier Dover Publications.

3

Convexity

A very important issue in optimization is the concept of convexity of general functions and also of sets of values that function arguments take their values from.

3.1 DEFINITION OF CONVEX FUNCTION

A function is said to be convex if the chord joining any two points x_1, x_2 (generally speaking, vectors in $\mathbb{R}^n$) is always above the function for any value x on the line connecting those points, as follows.

$$\theta \in [0, 1], \quad f((1 - \theta) \cdot x_1 + \theta \cdot x_2) \leq (1 - \theta) \cdot f(x_1) + \theta \cdot f(x_2) \tag{3.1}$$

If in the relation above strict inequality $(<)$ holds (and $0 < \theta < 1$), it is known as strictly convex. The above inequality combines *linearly* two vectors in order to produce a point in the line connecting them (forming the chord in the values of $f(x)$).
Concave functions are such that:

$$\theta \in [0, 1], \quad f((1 - \theta) \cdot x_1 + \theta \cdot x_2) \geq (1 - \theta) \cdot f(x_1) + \theta \cdot f(x_2) \tag{3.2}$$

Figure 3.1 shows an example of a convex & concave function in one variable.
If $f(x)$ is a continuous function, consider what happens as one point tends to the other, *i.e.* $x_2 \rightarrow x_1$. This is shown in Figure 3.2.
Then the chord becomes tangent at x_1. Then the tangent, for a convex function is always underestimating $f(x)$. Figure 3.2 shows the tangent (in 1-D) and tangent plane (in 2-D, or higher D) as support hyperplanes of the function.
The main result for differentiable convex functions is as follows.
If a function $f(x)$ is convex, then for any points $x_1, x_2 \in \mathbb{R}^n$ the following holds:

$$f(x_2) \geq f(x_1) + \underbrace{\frac{\partial f}{\partial x}(x_1)}_{(\nabla_x f(x))^T \Big|_{x_1}} \cdot (x_2 - x_1) \tag{3.3}$$

Conversely, if Equation (3.3) holds for any points $x_1, x_2 \in \mathbb{R}^n$, then the function $f(x)$ is convex. Strong convexity of the function will be reflected in Equation (3.3) by strict inequality $(>)$ holding (also requiring that $x_1 \neq x_2$).

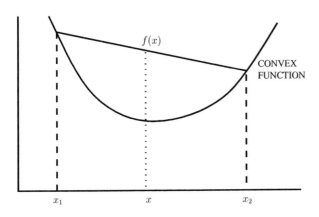

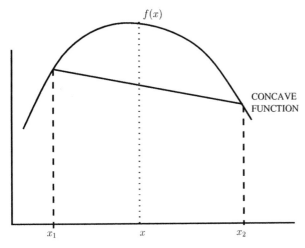

Figure 3.1 Diagram of a convex and a concave function.

Proof

At the beginning of the proof, assume that the function $f(x)$ is convex, which is defined in Equation (3.1). We consider two points, x_1 and x_2, on function $f(x)$ with a scale factor θ. Now rearrange this equation in the following way:

$$f((1 - \theta) \cdot x_1 + \theta \cdot x_2) \leq f(x_1) + \theta \cdot (f(x_2) - f(x_1)) \qquad (3.4)$$

$$f(x_1 + \theta(x_2 - x_1)) - f(x_1) \leq \theta \cdot (f(x_2) - f(x_1)) \qquad (3.5)$$

$$\frac{f(x_1 + \theta(x_2 - x_1)) - f(x_1)}{\theta} + f(x_1) \leq f(x_2) \qquad (3.6)$$

then introduce a new function:

$$g(\theta) = f(x_1 + \theta(x_2 - x_1)) \qquad (3.7)$$

Hence we now obtain:

$$\frac{g(\theta) - g(0)}{\theta} + f(x_1) \leq f(x_2) \qquad (3.8)$$

39

3 Convexity

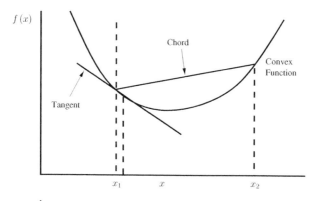

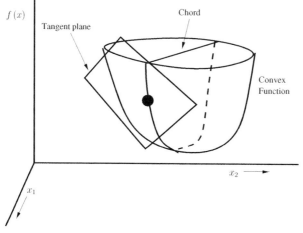

Figure 3.2 Chords of a one- and two-dimensional convex function.

Take the limit as $\theta \to 0$, which leads to:

$$\lim_{\theta \to 0} \frac{g(\theta) - g(0)}{\theta} = g'(0) \tag{3.9}$$

where:

$$g'(\theta) = \frac{\partial f}{\partial x}(x_1 + \theta(x_2 - x_1)) \cdot (x_2 - x_1) \tag{3.10}$$

$$g'(0) = \frac{\partial f}{\partial x}(x_1) \cdot (x_2 - x_1) \tag{3.11}$$

Hence we now obtain:

$$g'(0) + f(x_1) \leq f(x_2) \tag{3.12}$$

$$\frac{\partial f}{\partial x}(x_1) \cdot (x_2 - x_1) + f(x_1) \leq f(x_2) \tag{3.13}$$

The last expression is the same as Equation (3.3), hence the proof is complete.

Now that the first order sufficient condition of convexity is proven, consider the following equations:

$$f(x_2) \geq f(x_1) + \frac{\partial f}{\partial x}(x_1) \cdot (x_2 - x_1) \tag{3.14}$$

$$f(x_1) \geq f(x_2) + \frac{\partial f}{\partial x}(x_2) \cdot (x_1 - x_2) \tag{3.15}$$

Adding these two equations gives:

$$f(x_1) + f(x_2) \geq f(x_1) + f(x_2) + \left[\frac{\partial f}{\partial x}(x_1) \cdot (x_2 - x_1) + \frac{\partial f}{\partial x}(x_2) \cdot (x_1 - x_2) \right]$$

$$0 \leq (x_1 - x_2) \left[\frac{\partial f}{\partial x}(x_1) - \frac{\partial f}{\partial x}(x_2) \right]$$

Dividing the last expression by $(x_1 - x_2)^2$:

$$\frac{\frac{\partial f}{\partial x}(x_1) - \frac{\partial f}{\partial x}(x_2)}{(x_1 - x_2)} \geq 0 \tag{3.16}$$

Now as $x_1 \rightarrow x_2$ we obtain:

$$\frac{\partial^2 f(x_2)}{\partial x_2^2} \geq 0 \tag{3.17}$$

This expression can be used for a Taylor expansion of function $f(x)$ about some point x_1 as follows:

$$f(x_1) = f(x_2) + \frac{\partial f(x_2)}{\partial x_2}(x_1 - x_2) + \frac{1}{2}\frac{\partial^2 f(x_2)}{\partial x_2^2}(x_1 - x_2)^2 + H.O.T. \tag{3.18}$$

where $H.O.T.$ represents "Higher Order Terms," which can be neglected for small perturbations. Now from Equation 3.17 it can be seen that the last term in Equation 3.18 is always positive. Hence:

$$f(x_1) \geq f(x_2) + \frac{\partial f(x_2)}{\partial x_2}(x_1 - x_2) \tag{3.19}$$

which is what was obtained in Equation 3.3.

3.2 CONVEX SETS

For global optimality issues it is not sufficient that the function be convex to guarantee the (global) minimizer. It is very important that the set where the vector of variables takes its values is also a *convex set*. Otherwise, there is no guarantee of global optimality.

Simply stated, a convex set is a set such that any chord connecting any two points in the set lies entirely in the set. This is shown pictorially in Figure 3.3.

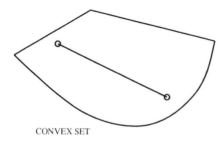

CONVEX SET

Figure 3.3 Diagram of convex and nonconvex sets.

NONCONVEX SET

3.3 FURTHER READING RECOMMENDATIONS

Properties of Convex Functions

Boyd, S. P. and Vandenberghe, L. (2004). Convex Optimization. Cambridge University Press.

3.4 EXERCISES

1. Prove convexity of the following function, using the definition of the chord presented in this chapter.

   (i) $f(x) = x^2$
   (ii) $f(x) = \exp(x)$
   (iii) $f(x) = \frac{1}{x}$, $x > 0$
   (iv) $\max_{i=1,2,\dots,m}\{f_i(x)\}$, where f_i are linear.

2. Consider a set of values $\mathcal{X} \subseteq \mathbb{R}^n$ such that it is defined by a set of inequalities $g_j(x) \le 0$, $j = 1, 2, \dots, m$:

$$x = \{x \in \mathbb{R}^n | g_j(x) \le 0, \ j = 1, 2, \dots, m\}$$

   Show that if $g_j(x)$ are convex functions in x, then set $\mathcal{X}$ is also convex.

3. Consider the transformation of a function $g(x)$ by another function h such that $f(x) = h(g(x))$, for scalar x. Prove the following two properties:

   (i) g is convex, h is convex and $h' > 0 \rightarrow f$ is convex
   (ii) g is concave, h is concave and $h' > 0 \rightarrow f$ is concave.

3.4 Exercises

4. Consider the following constrained optimization problem:

$$\min_{x \in \mathbb{R}^n} f_0(x) = \max_{i=1,2,\ldots,m_1} \{f_i(x)\}$$

subject to:

$$h_j(x) = 0, \quad j = 1, 2, \ldots, m_2$$
$$g_k(x) \leq 0, \quad k = 1, 2, \ldots, m_3$$

Show that the objective function $f_0(x)$ is convex if the individual functions $f_i(x)$, $i = 1, 2, \ldots, m_1$ are convex.

4

Quadratic Functions

The study of quadratic functions is very important in optimization theory. Not only because they are the simplest nonlinear form to begin investigating but also because any function sufficiently close to a minimizer behaves quadratically.

4.1 CONSTRUCTION OF THE MATRIX FORM OF A QUADRATIC FUNCTION BY EXAMPLE

Consider the following example in 2-D:

$$q(x) = \alpha \cdot x_1^2 + \beta \cdot x_1 \cdot x_2 + \gamma \cdot x_2^2 + \delta \cdot x_1 + \varepsilon \cdot x_2 \tag{4.1}$$

How can we write this in matrix form?

$$q(x) = x^T A x + c^T x \tag{4.2}$$

Setting $A = \begin{bmatrix} \alpha & \beta \\ 0 & \gamma \end{bmatrix}$, $c = \begin{bmatrix} \delta \\ \varepsilon \end{bmatrix}$, and $x = \begin{bmatrix} x_1 \\ x_2 \end{bmatrix}$ achieves this.

Carry out the multiplications according to Equation (4.2) and show that it results in the original quadratic given at the start.

Usually, quadratic function matrices like matrix A are written such that they are symmetric. Is there a way we can make the matrix we used symmetric (*i.e.* use another one) and still get the same result? The answer is yes, and this is how it is done. Given:

$$x^T A x \tag{4.3}$$

This is a scalar. Why? Because the original function $q(x)$ is scalar, and in any case the dimensions for the product are:

$$(1 \times n) \times (n \times n) \times (n \times 1) \tag{4.4}$$

Since the inner dimensions are correct (they match for each product term), we get from the outer dimensions that the result is (1×1), *i.e.* a scalar. Next, we do the following:

$$x^T A x = \frac{1}{2} \left[x^T A x + x^T A x \right] \tag{4.5}$$

By the property of the transpose of a product of matrices:

$$(ABC)^T = C^T B^T A^T \tag{4.6}$$

we get:

$$x^T A x = \frac{1}{2}\left[x^T A x + x^T A^T x\right] \tag{4.7}$$

$$= \frac{1}{2} x^T \left[A + A^T\right] x \tag{4.8}$$

So we may set as matrix $Q = A + A^T$, which is going to be symmetric. (If you take the transpose of it, it is just itself with a change in the order of the terms in the addition, hence symmetric). Trying it out in our example we get:

$$Q = \begin{bmatrix} \alpha & \beta \\ 0 & \gamma \end{bmatrix} + \begin{bmatrix} \alpha & 0 \\ \beta & \gamma \end{bmatrix} = \begin{bmatrix} 2\alpha & \beta \\ \beta & 2\gamma \end{bmatrix} \tag{4.9}$$

and with this, the quadratic form is written as (in standard form):

$$q(x) = \frac{1}{2} x^T Q x + c^T x \tag{4.10}$$

4.2 EIGENVALUES OF Q FOR QUADRATIC FUNCTIONS AND CONVEXITY

The eigenvectors of a matrix are vectors which when multiplied by the matrix, result in themselves being scaled by a factor which is the corresponding eigenvalue. The definition leads to:

$$Q \cdot r = \lambda \cdot r \tag{4.11}$$

Note that:

$$Q \cdot r - \lambda \cdot r = [Q - \lambda \cdot I] \cdot r = 0 \tag{4.12}$$

or to get the eigenvalues we solve (the assumption being that $r \neq 0$ and hence the matrix involved must be singular):

$$\det[Q - \lambda \cdot I] = 0 \tag{4.13}$$

which is a polynomial in λ, known as the characteristic polynomial of the matrix.
If Q is symmetric (and real) then all the λ_i are real.

1. If $\lambda_i > 0, \forall i = 1, 2, \ldots, n$ then Q is positive definite (leading to a convex quadratic function)
2. If $\lambda_i < 0, \forall i = 1, 2, \ldots, n$ then Q is negative definite (leading to a concave quadratic function)
3. If $\lambda_i = 0$, for some $i = 1, 2, \ldots, n$ then Q is singular
4. If some $\lambda_i > 0$, some $\lambda_i < 0$ then Q is indefinite

The matrix Q can be factorized into:

$$QR = R\Lambda \tag{4.14}$$

where $R = [r_1, r_2, \ldots, r_n]$ is the matrix of all eigenvectors (as columns), and

$$\Lambda = \begin{bmatrix} \lambda_1 & 0 & \cdots & 0 \\ 0 & \lambda_2 & \cdots & 0 \\ \vdots & \vdots & \ddots & \vdots \\ 0 & 0 & 0 & \lambda_n \end{bmatrix} \tag{4.15}$$

is the diagonal matrix of the eigenvalues.

For the case of a 2×2 Q matrix we have:

$$Q \cdot R = Q \cdot \begin{bmatrix} r_{11} & r_{12} \\ r_{21} & r_{22} \end{bmatrix} = \begin{bmatrix} r_{11} & r_{12} \\ r_{21} & r_{22} \end{bmatrix} \cdot \begin{bmatrix} \lambda_1 & 0 \\ 0 & \lambda_2 \end{bmatrix} = R \cdot \Lambda \tag{4.16}$$

Further, if Q is symmetric and real, implying all eigenvalues are real, then the eigenvectors (and hence matrix R) can be chosen as orthonormal:

$$R^{-1} = R^{\mathrm{T}} \tag{4.17}$$

For a quadratic $q(x) = \frac{1}{2} x^{\mathrm{T}} Q x$

$$QR = R\Lambda \implies Q = R\Lambda R^{-1} \implies Q = R\Lambda R^{\mathrm{T}} \tag{4.18}$$

The last form in the equation above is known as Canonical Form, or the spectral factorization of matrix Q. Substituting this into the assumed quadratic function we get:

$$q(x) = \frac{1}{2} x^{\mathrm{T}} R\Lambda R^{\mathrm{T}} x \tag{4.19}$$

Set in this $y = R^{\mathrm{T}} x$, and the above expression becomes:

$$q(y(x)) = \frac{1}{2} y^{\mathrm{T}} \Lambda y \tag{4.20}$$

Assuming that y is nontrivial, *i.e.* nonzero (at least one element differs from zero in the vector), then it is easy to see where the definitions of positive definiteness for $q(x)$ come in relation to the values of the eigenvalues. Explain this at home as an exercise.

Example in 3-D

$$q(y) = \frac{1}{2} (y_1 \ y_2 \ y_3) \begin{bmatrix} \lambda_1 & 0 & 0 \\ 0 & \lambda_2 & 0 \\ 0 & 0 & \lambda_3 \end{bmatrix} \begin{bmatrix} y_1 \\ y_2 \\ y_3 \end{bmatrix} \tag{4.21}$$

$$q(y) = \frac{1}{2} \left[\lambda_1 y_1^2 + \lambda_2 y_2^2 + \lambda_3 y_3^2 \right] \tag{4.22}$$

Clearly, this is always positive **iff**[1] $\lambda_i > 0$ (for $y \neq 0$) (definition of positive definiteness). If the previous holds, then the function $q(x)$ is convex everywhere.

---

[1] "if and only if"

Proofs: properties of eigenvalues and eigenvectors of symmetric real matrices

A. Orthogonality of eigenvectors

The eigenvector equation is given by:

$$Q\,v = \lambda\,v \tag{4.23}$$

Consider two different eigenvalues $\lambda_1 \neq \lambda_2$, each with their respective eigenvector v_1 and v_2:

$$Q\,v_1 = \lambda_1\,v_1 \tag{4.24}$$

$$Q\,v_2 = \lambda_2\,v_2 \tag{4.25}$$

Multiply Equation (4.24) by v_2^{T} on the left-hand side:

$$v_2^{\mathrm{T}}\,Q\,v_1 = \lambda_1\,v_2^{\mathrm{T}}\,v_1 \tag{4.26}$$

Note that matrix Q is symmetric and hence:

$$\left(v_2^{\mathrm{T}}\,Q\right) = \left(Q^{\mathrm{T}}\,v_2\right)^{\mathrm{T}} = \left(Q\,v_2\right)^{\mathrm{T}} = \lambda_2\,v_2^{\mathrm{T}} \tag{4.27}$$

Now insert the result from Equation (4.27) into Equation (4.26):

$$\lambda_2\,v_2^{\mathrm{T}}\,v_1 = \lambda_1\,v_2^{\mathrm{T}}\,v_1 \Rightarrow (\lambda_2 - \lambda_1)\,v_2^{\mathrm{T}}v_1 = 0 \tag{4.28}$$

By definition we said that $(\lambda_2 - \lambda_1) \neq 0$, hence:

$$v_2^{\mathrm{T}}\,v_1 = 0 \Rightarrow v_1 \perp v_2 \tag{4.29}$$

In the case of repeated eigenvalues, $\lambda_1 = \lambda_2$, etc., it can be shown that orthogonal eigenvectors can be chosen still (more complicated proof).

B. Product of complex numbers

Consider z_1, $z_2 \in \mathbb{C}$:

$$z_1 = a_1 + i\,b_1, \quad z_2 = a_2 + i\,b_2 \tag{4.30}$$

Their complex conjugates are:

$$z_1^* = a_1 - i\,b_1, \quad z_2^* = a_2 - i\,b_2 \tag{4.31}$$

The objective is to prove that $(z_1 z_2)^* = z_1^* z_2^*$. Multiply out the left-hand side:

$$(z_1 z_2)^* = [(a_1 + i\,b_1)\,(a_2 + i\,b_2)]^* \tag{4.32}$$

$$= [a_1 a_2 + i\,(a_1 b_2 + a_2 b_1) - b_1 b_2] \tag{4.33}$$

$$= a_1 a_2 - b_1 b_2 - i\,(a_1 b_2 + a_2 b_1) \tag{4.34}$$

$$= (a_1 - i\,b_1)\,(a_2 - i\,b_2) \tag{4.35}$$

$$= z_1^* z_2^* \tag{4.36}$$

C. If matrix Q is real and symmetric, its eigenvalues are real

For a (square) matrix: if complex eigenvalues appear, they do so in complex conjugate pairs. So λ and λ^* are both eigenvalues.

The eigenvector equation is given by:

$$Q\,v = \lambda\,v \tag{4.37}$$

Take conjugates of both sides to yield the eigenvector equation for λ^*:

$$Q\,v^* = \lambda^*\,v^* \tag{4.38}$$

Multiply both sides of Equation (4.37) on the left by $(v^*)^{\mathrm{T}}$:

$$\left(v^*\right)^{\mathrm{T}} Q\,v = \lambda\left(v^*\right)^{\mathrm{T}} v \tag{4.39}$$

As shown in Equation (4.27), we obtain:

$$\lambda^*\left(v^*\right)^{\mathrm{T}} v = \lambda\left(v^*\right)^{\mathrm{T}} v \tag{4.40}$$

Hence:

$$\left(\lambda^* - \lambda\right)\left(v^*\right)^{\mathrm{T}} v = 0 \tag{4.41}$$

Now:

$$v = \begin{bmatrix} c_1 \\ c_2 \\ \vdots \\ c_n \end{bmatrix} \tag{4.42}$$

$$v^* = \begin{bmatrix} c_1^* \\ c_2^* \\ \vdots \\ c_n^* \end{bmatrix} \tag{4.43}$$

$$\Rightarrow \left(v^*\right)^{\mathrm{T}} v = \sum_{j=1}^{n} c_j^* c_j \tag{4.44}$$

But:

$$c_j^* c_j = \left(a_j - i\,b_j\right)\left(a_j + i\,b_j\right) \tag{4.45}$$

$$= a_j^2 + b_j^2 \tag{4.46}$$

$$= \left|c_j\right|^2 \geq 0 \tag{4.47}$$

So if $v \neq 0 \Rightarrow (v^*)^{\mathrm{T}} v > 0$. Therefore from Equation (4.41) we get:

$$\lambda^* = \lambda \tag{4.48}$$

The only way for Equation (4.41) to hold is that λ must be real.

4.2.1 Summary of Convexity Analysis

1. Convexity is discussed in terms of

    1. The objective function $f(x)$
    2. The set of values for x

2. Locally, near a minimum (maximum) a function behaves as a quadratic (third-order terms in a Taylor expansion tend to zero).
3. The derivation of the properties for convexity of a function follows these steps:

    1. Chord definition of convexity (geometrical definition)
    2. From chord to tangent property
    3. Translation of tangent property via Taylor series into positive definite quadratic forms
    4. Hessian matrix
    5. Diagonalization of Hessian and eigenvalue properties

4.3 GEOMETRICAL INTERPRETATION OF EIGENVALUES AND EIGENVECTORS

Define a simple quadratic function, as follows:

$$q(x) = \frac{1}{2}(\lambda_1 x_1^2 + \lambda_2 x_2^2) = \frac{1}{2} x^{\mathrm{T}} \begin{bmatrix} \lambda_1 & 0 \\ 0 & \lambda_2 \end{bmatrix} x; \quad \lambda_1, \lambda_2 > 0 \qquad (4.49)$$

Choose a constant value c for $q(x)$ and draw the contour $q(x) = c$. The contours, generally speaking, for this function are ellipses.

Setting the two coordinates to zero in turn we get:

$$x_1 = 0 \longrightarrow x_2 = \pm\sqrt{\frac{2c}{\lambda_2}} \qquad (4.50)$$

$$x_2 = 0 \longrightarrow x_1 = \pm\sqrt{\frac{2c}{\lambda_1}} \qquad (4.51)$$

which are the intercepts with the axes. Now, the axes of the ellipse are given by:

$$D_1 = 2\sqrt{\frac{2c}{\lambda_1}} \quad \text{and} \quad D_2 = 2\sqrt{\frac{2c}{\lambda_2}}$$

This information is shown pictorially in Figure 4.1.

Scaling is reflected in relative sizes of the axes D_1 and D_2, for example if $\lambda_2 \gg \lambda_1$ then $D_2 \ll D_1$.

N.B. The $x-y$ axes are eigenvectors of Q in this example.

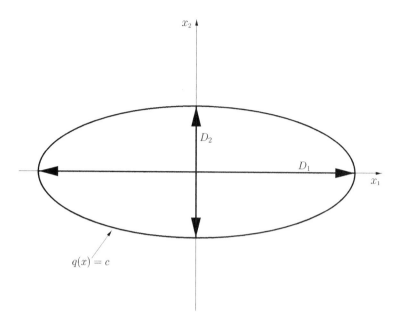

Figure 4.1 Geometric representation of eigenvectors and eigenvalues.

4.4 SOLUTION OF AN UNCONSTRAINED QUADRATIC PROGRAM (QP)

$$q(x) = \frac{1}{2}x^{\mathrm{T}}Qx + c^{\mathrm{T}}x \tag{4.52}$$

$$x^* = \arg\min_x q(x) \tag{4.53}$$

A stationary point of $q(x)$ will be a guaranteed minimum if q(x) is convex, or equivalently if Q is positive definite. For stationary points we have to solve $\frac{\partial q(x)}{\partial x} = 0_{1\times n}$.

We first need to see how to differentiate the quadratic form given in Equation (4.52).

$$\frac{\partial q(x)}{\partial x} = \frac{1}{2}\frac{\partial}{\partial x}\left(x^{\mathrm{T}}Qx\right) + \frac{\partial}{\partial x}\left(c^{\mathrm{T}}x\right) \tag{4.54}$$

Let us start with the rightmost term in the Equation (4.54) and write it out:

$$\frac{\partial}{\partial x}\left(c^{\mathrm{T}}x\right) = \frac{\partial}{\partial x}(c_1 x_1 + c_2 x_2 + \ldots + c_n x_n) = (c_1, c_2, \ldots, c_n) = c^{\mathrm{T}} \tag{4.55a}$$

From this, we learn that the gradient of a vector product (which is a scalar quantity) yields a row vector of coefficients, multiplying the vector of variables on the left (the order is important to note, as we will see next).

Let us turn our attention to the term $x^{\mathrm{T}}Qx$. We note that we differentiate such scalar forms by "removing" the rightmost vector of variables. By the logic of the previous differentiation, we could treat $(x^{\mathrm{T}}Q)$ as the vector c and apply what we know. We have to be careful though, as this term contains twice the vector of variables x with respect to which we are differentiating, so we have to apply this twice, as follows:

$$\frac{\partial}{\partial x}\left(x^{T}Qx\right) = x^{T}Q + (Qx)^{T} = x^{T}Q + x^{T}Q^{T} = 2x^{T}Q \tag{4.55b}$$

Putting the two results from Equations (4.55a) and (4.55b) into Equation (4.54) and setting equal to zero we obtain:

$$\frac{\partial q(x)}{\partial x} = x^{T}Q + c^{T} = 0_{1 \times n} \tag{4.56}$$

hence:

$$x^{T}Q + c^{T} = 0_{1 \times n} \implies Q^{T}x + c = 0_{n \times 1} \implies Qx = -c \tag{4.57}$$

Note that the matrix is symmetric, therefore $Q^{T} = Q$ was used in the last relation. Equation (4.57) requires the solution of a linear system of equations, which we either use Gaussian elimination or factorization to solve or equivalently write as:

$$x^{*} = -Q^{-1}c \tag{4.58}$$

Summary

1. Necessary condition that $q(x^{*})$ is minimum: $\left.\dfrac{\partial q(x)}{\partial x}\right|_{x=x^{*}} = 0$

2. Sufficient condition: $q(x)$ is a convex function

4.5 LEAST SQUARES FITTING AND ITS RELATION TO QP

Consider a multiple-input, single-output process such as the one shown in Figure 4.2.

For m measurements b_i, $i = 1, 2, \ldots, m$ (different settings of input vector a), to determine n unknown parameters x, with $m > n$ (more experiments than unknown parameters), we form a *linear model* such that:

$$b_i = a_i x = \sum_{j=1}^{n} a_{ij} \cdot x_j, \quad i = 1, 2, \ldots, m \tag{4.59}$$

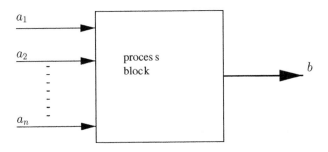

Figure 4.2 Diagram of multiple-input, single-output system.

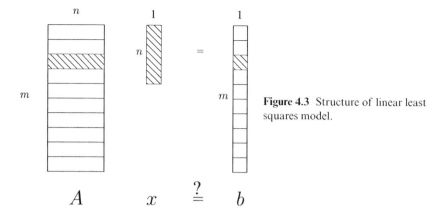

Figure 4.3 Structure of linear least squares model.

where a_i is a **row vector** of input variable values for the i-th experiment (measurement). Collecting every experiment (measurement) into a single vector equation we obtain:

$$Ax = b \tag{4.60}$$

the structure of which is shown diagrammatically in Figure 4.3.

Equality will **generally not hold** in the previous equations, as it is impossible to solve uniquely a system having more equations than variables.

Example (LSQR)

The following variables are present in the LSQR problem:

- three inputs: a_1, a_2, a_3
- one output: b
- three unknown constant parameters to fit in linear model: (x_1, x_2, x_3)

$$b\left(\underline{a}\right) = x_1 \cdot a_1 + x_2 \cdot a_2 + x_3 \cdot a_3 \tag{4.61}$$

Next, the vector measuring the error (residual) of the fitting is defined:

$$r = Ax - b \tag{4.62}$$

where

r: the error vector in $\mathbb{R}^m$
Ax: the predicted responses in $\mathbb{R}^m$
b: the measured responses in $\mathbb{R}^m$

The least squares (LSQR) objective function can be defined as the minimization of the sum of the squares of the error for each measurement, as follows:

$$\min_x q(x) = \frac{1}{2} \sum_{i=1}^{m} r_i^2 = \frac{1}{2} r^{\mathrm{T}} r \tag{4.63}$$

Substituting the definition of the error vector we obtain:

$$q(x) = \frac{1}{2} \left[\left(x^{\mathrm{T}} A^{\mathrm{T}} - b^{\mathrm{T}} \right) \left(Ax - b \right) \right] = \frac{1}{2} \left[-2b^{\mathrm{T}} Ax + x^{\mathrm{T}} A^{\mathrm{T}} Ax + b^{\mathrm{T}} b \right] \tag{4.64}$$

To find the stationary point we take the gradient and set it to zero, as follows:

$$\frac{\partial q(x)}{\partial x} = -b^T A + x^T A^T A = 0 \tag{4.65}$$

hence:

$$A^T A x = A^T b \tag{4.66}$$

Note: $A \longrightarrow (m \times n)$, $A^T A \longrightarrow (n \times m) \times (m \times n) = (n \times n)$

This is a square linear system which may be solved by an appropriate factorization method (symmetric positive definite matrix systems are solved most efficiently with Choleski factorization, *i.e.* if $A^T A$ turns out to be positive definite. If not, special techniques exist such as the singular value decomposition (SVD), which is outside the scope of the presentation here).

Equivalently, what we have obtained can be written as:

$$x^* = \left(A^T A\right)^{-1} A^T b \tag{4.67}$$

The corresponding optimal value of the residual (error) vector is:

$$r^* = A x^* - b$$
$$= \left[A \left(A^T A\right)^{-1} A^T - I_{m \times m}\right] b \tag{4.68}$$

Geometrical interpretation of LSQR solution

Look at $r = Ax - b$: the columns of matrix A define a subspace of $\mathbb{R}^m$, since they are m-dimensional vectors, as in Figure 4.4.

This space is the **range space** of matrix A: $\mathcal{R}(A)$.

The projection matrix:

$$P = A \left(A^T A\right)^{-1} A^T \tag{4.69}$$

will project the original vector b onto the vector $\hat{b}$ in the range space according to:

$$\hat{b} = A \left(A^T A\right)^{-1} A^T \cdot b \tag{4.70}$$

Pictorially, this is given in Figure 4.5.

Thus the projected point $\hat{b}$ is produced by finding $x^* = (x_1, x_2)^T$, which are the coefficients for the linear combination of vectors a_1 and a_2 in Figure 4.5.

Next we prove some basic properties for the least squares matrix $A^T A$.

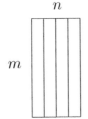

Figure 4.4 Matrix A viewed in column format.

n

m

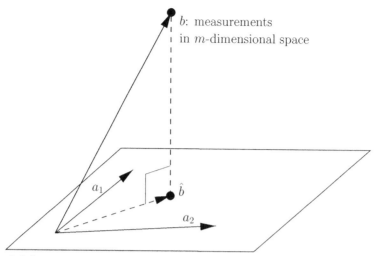

$\mathcal{R}(A)$: the set of reachable right-hand sides b

Figure 4.5 Projection (perpendicular) onto the reachable space, i.e. the range space of matrix A. The vertical vector is the minimal residual (error).

Proof: The least squares matrix

Consider the following matrix, A: $A \in \mathbb{R}^{m \times n}$, $m > n$.

This matrix appears in the LSQR objective function in a quadratic form, as follows:

$$q(x) = x^{T} A^{T} A x \tag{4.71}$$

This form is, in general, positive semi-definite because:

$$y^{T} y \geq 0$$

where $y = Ax$. $q(x)$ will be zero if there exists a vector x for which $y = Ax = 0$, other than the trivial solution $x = 0$. A nonzero solution will exist if matrix A has linearly dependent columns and $(A^{T}A)$ will not be invertible.

If not, then $Ax = 0$ holds only for $x = 0$, hence:

$$q(x) > 0, \quad \forall x \neq 0 \tag{4.72}$$

In that case, $A^{T}A$ is positive definite and invertible. This will be proven next.

Proof: Matrix with linearly independent (L.I.) columns

Consider a matrix $B \in \mathbb{R} \times k$, $n > k$.

It can be written in column form as:

$$B = (b_1, b_2, \ldots, b_k) \tag{4.73}$$

where $b_i \in \mathbb{R}^n$, $i = 1, 2, \ldots, k$.

If its columns are linearly independent, then:

$$x_1 \cdot b_1 + x_2 \cdot b_2 + \cdots + x_k \cdot b_k = 0$$

holds only for $x_i = 0$, $i = 1, 2, \ldots, k$.

Another way to write this is that:

$$B \cdot x = 0 \tag{4.74}$$

admits only $x = 0$ as its only solutions.

The subspace (set) of such solutions x is called the Nullspace of matrix B:

$$\mathcal{N}(B) = \{x | Bx = 0\} \subset \mathbb{R}^k \tag{4.75}$$

Proof: Invertibility of matrix $A^T A$ used in least squares

Consider the following matrix A: $A \in \mathbb{R}^{m \times n}$ with $m > n$ and linearly independent columns.

Since A has linearly independent columns:

$$\mathcal{N}(A) = \{0\} \tag{4.76}$$

or

$$Ax = 0 \Rightarrow x = 0 \tag{4.77}$$

is the only solution.

$A^T A$ is $(n \times m) \times (m \times n) = (n \times n)$ a square matrix.

Select a vector:

$$v \epsilon \mathcal{N}(A^T A) \Rightarrow A^T A \cdot v = 0 \tag{4.78}$$

where $v \in \mathbb{R}^n$. Multiply both sides of Equation (4.78) by v^T:

$$v^T A^T A v = 0 \tag{4.79}$$

Now define $y = Av$:

$$y^T y = \|y\|_2^2 = 0 \tag{4.80}$$

The only possibility for Equation (4.80) to hold is $y \equiv 0$ or:

$$Av = 0 \tag{4.81}$$

Equation (4.81) implies that $v \epsilon \mathcal{N}(A)$ also. So we have that if:

$$v \in \mathcal{N}(A^T A) \Rightarrow v \in \mathcal{N}(A) \tag{4.82}$$

But from Equation (4.77):

$$\mathcal{N}(A) = \{0\} \Rightarrow v \in \mathcal{N}(A) \Rightarrow v = 0 \tag{4.83}$$

Hence, any $v \in \mathcal{N}(A^T A)$ has to be the zero vector, or:

$$A^T Ax = 0 \Rightarrow 0 \tag{4.84}$$

or:

$$\mathcal{N}(A^T A) \subseteq \mathcal{N}(A) = \{0\} \tag{4.85}$$

Hence the columns of $A^T A$ are linearly independent.

Since it is a square matrix (it would have n pivot points with Gaussian elimination), it is also invertible:

$$\Rightarrow (A^T A)^{-1} \text{ exists.}$$

4.6 FURTHER READING RECOMMENDATIONS

On Properties of Quadratic Functions and Related Topics

Edgar, T. F., Himmelblau, D. M., and Lasdon, L. S. (2001). Optimization of Chemical Processes. McGraw-Hill, Edition 2.

Taha, H. A. (2010). Operations Research: An Introduction. Prentice Hall, Edition 9.

An Excellent Instructive Book on Linear Algebra

Strang, G. (2006). Linear Algebra and Its Applications. Thomson Brooks/Cole, Edition 4.

4.7 EXERCISES

1. Prove the convexity of the following functions:

   (i) $f(x) = \sum_{i=1}^{4} a_i x_i + b_i$

   (ii) $f(x) = 3(x_1 - 3)^2 + 2(x_2 - 1)^2$

   (iii) $f(x) = \exp(x_1) + x_2^2 + (x_3 - 1)^4$

   (iv) $f(x) = \sum_{i=1}^{4} x_i \ln x_i, \qquad x_i \geq 0$

   (v) $f(x) = \frac{1}{x_1} + \frac{1}{x_2^2}, \qquad x_i > 0$

2. A positive-semidefinite matrix has non-negative eigenvalues, *i.e.* some of them may be exactly zero (and may repeat). A positive semi-definite Hessian corresponds to a convex (not strictly) function. Show this and explain why.

3. Formulate the curve fitting problem of a set of pairs (x_i, z_i), $i = 1, 2, \ldots, m$, where the y variable is the response to an input variable x. The model to fit is a linear one according to which:

$$\hat{y} = a \cdot x + b$$

where $\hat{y}$ is the model predicted response and the pair (a, b) are undetermined parameters to be estimated. Use a Euclidean norm formulation (least squares) to analytically derive expressions for the unknown parameters.

5

Minimization in One Dimension

Most optimization problems found in the real world include many variables. In order to understand how the different types of optimization techniques work, it is beneficial to first consider what happens in one dimension. Once a good understanding and intuition of each technique is present, more complicated techniques in more than one dimension are considered.

5.1 BISECTION

Assume we have a domain $x \in [a, b]$ containing a single minimum. Evaluate $f(a)$, $f(x_m - \varepsilon)$, $f(x_m + \varepsilon)$, $f(b)$. This is shown pictorially in Figure 5.1.

In the example of Figure 5.1 we have:

$$f(a) > f(x_m - \varepsilon) > f(x_m + \varepsilon) < f(b)$$

Thus, we reject the left half $[a, x_m - \varepsilon]$. Continue in the new subdomain as in the initial step. Effectively, for small enough ε the method rejects half of the domain per iteration. The drawback is that it needs two function evaluations per iteration.

Can we find a more efficient domain reduction technique? It turns out that the answer to this question is an ancient one, and is examined in the following section.

5.2 GOLDEN SECTION SEARCH

The golden section search uses a ratio of proportions known to the ancient Greeks, who used it as the most aesthetically pleasing ratio in the design of temples *etc.*

The definition is:

the ratio of the smallest to the biggest part, is equal to the ratio of the biggest to the whole.

Figure 5.2 shows a diagram of the Golden Ratio.

According to Figure 5.2 and the definition, we have:

$$\frac{B}{A} = \frac{A}{A + B}$$

By setting $\xi = \dfrac{B}{A}$ we get:

$$\xi = \frac{B}{A} = \frac{1}{\dfrac{B}{A} + 1} = \frac{1}{\xi + 1}$$

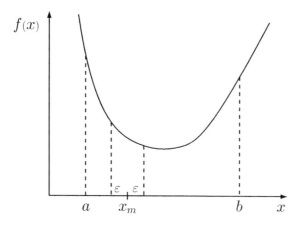

Figure 5.1 Schematic of bisection method.

Figure 5.2 Diagram of Golden Ratio.

hence we get:

$$\xi = \frac{1}{1+\xi} \quad \text{or} \quad \xi^2 + \xi - 1 = 0$$

By solving the quadratic and retaining the positive root, we finally get the golden section ratio:

$$\xi = \frac{\sqrt{5}-1}{2} = 0.618\ldots$$

This ratio, apart from being used in architecture (ancient and modern), is also found in nature, particularly in branching proportions in trees, and other natural branched structures.

5.2.1 Golden Section in 1-D Minimization

Consider the diagram in Figure 5.3. The initial domain containing a minimum is $[x_1, x_4]$, such that $L_0 = x_4 - x_1$, which is partitioned as shown below.

According to the diagram, we would reject the subinterval $[x_3, x_4]$ (verticals in the diagram are the values of the objective function, $f(x_i)$). Calculate:

$$\frac{(x_2 - x_1)}{(x_3 - x_1)} = \frac{(1-\xi)L_0}{\xi L_0} = \frac{1-\xi}{\xi}$$

Let now $L_1 = x_3 - x_1$ be the new uncertainty interval containing the minimum. But we know that $\xi^2 + \xi - 1 = 0$, hence $1 - \xi = \xi^2$, which implies that:

$$\frac{(x_2 - x_1)}{L_1} = \xi$$

Therefore, now x_2 is in L_1 what x_3 was in L_0, i.e. it is already in place as the rightmost inner point, and we need to just compute one point left to it as the innermost left point. The diagram in Figure 5.4 shows the new interval and our observations.

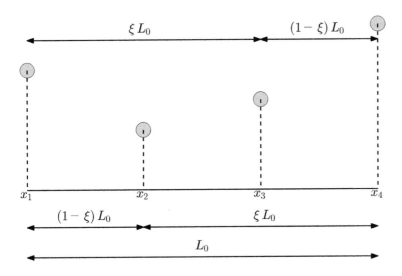

Figure 5.3 Golden Search algorithm.

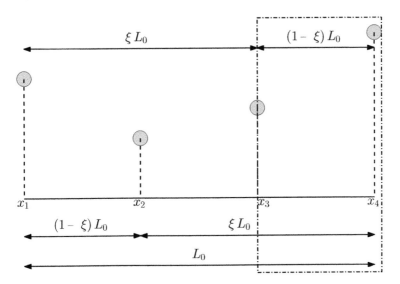

Figure 5.4 New interval for Golden Search algorithm.

Similarly, we could show the same had $[x_1, x_2]$ been the interval to reject. The point "renaming" would be within $[x_2, x_4]$, and the new point would be within $[x_3, x_4]$, as is shown for the case we examined.

In our case, we need to place a single point within $L_1 = x_3 - x_1$. Rename the points:

$$\tilde{x}_1 = x_1; \; \tilde{x}_3 = x_2; \; \tilde{x}_4 = x_3$$

and the single new point to be introduced (along with a single function evaluation) is:

$$\tilde{x}_2 = \tilde{x}_1 + (1 - \xi)L_1$$

5.2.2 Computational Cost Comparisons

Bisection

$$L_i = \frac{1}{2}L_{i-1}, \quad i = 1, 2, \ldots, K_b$$

To achieve a final interval L_f containing the minimum:

$$2^{K_b} \approx \frac{L_0}{L_f} \implies K_b = \log_2(L_0/L_f) = \ln(L_0/L_f)/\ln 2$$

Golden

$$L_i = \xi L_{i-1}; \quad L_f = L_0 \cdot \xi^{K_g} \implies K_g = \frac{\ln(L_0/L_f)}{\ln(1/\xi)}$$

Ratio

$$\frac{K_b}{K_g} = \frac{\ln(1/\xi)}{\ln 2} \approx 0.694\ldots$$

In other words, $K_g > K_b$, *i.e.* more iterations for Golden than for Bisection. However, bisection needs two function evaluations per iteration, whereas golden needs only one. The real computational cost is this, not the number of iterations alone. So:

$$F_b = 2K_b; \quad F_g = K_g$$

So the "real" computational cost is:

$$\frac{F_b}{F_g} = 2\frac{K_b}{K_g} = 2\frac{\ln(1/\xi)}{\ln 2} \approx 1.389$$

Hence, $F_b > F_g$ and therefore golden is cheaper computationally.

5.3 NEWTON'S METHOD

For Newton's method, we consider a Taylor series expansion around a given starting point $x^{(0)}$:

$$f(x + \Delta x) \approx f(x^{(0)}) + \frac{\partial f}{\partial x}\bigg|_{x^{(0)}} \cdot \Delta x + \frac{1}{2}\left(\frac{\partial^2 f}{\partial x^2}\right)\bigg|_{x^{(0)}} \cdot (\Delta x)^2 + H.O.T.$$

This represents a local quadratic approximation of the function at the point x_0. We can try to minimize this approximation with respect to Δx around x_0:

$$\frac{\partial f(x^{(0)} + \Delta x)}{\partial(\Delta x)} = \frac{\partial f}{\partial x}\bigg|_{x^{(0)}} + \left(\frac{\partial^2 f}{\partial x^2}\right)\bigg|_{x^{(0)}} \Delta x = 0$$

hence we obtain:

$$\Delta x = -\frac{\partial f}{\partial x}\bigg|_{x^{(0)}} \left(\frac{\partial^2 f}{\partial x^2}\right)^{-1}\bigg|_{x^{(0)}}$$

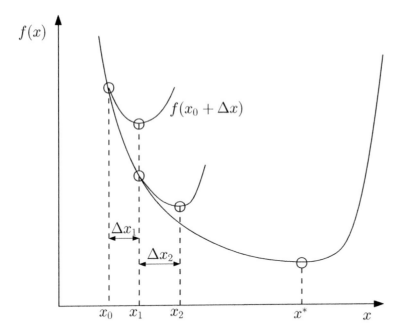

Figure 5.5 Schematic of Newton's method.

and the new point can be updated as:

$$x^{(1)} = x^{(0)} + \Delta x$$

Newton's method is the fastest if we are close enough to the minimum. For minimization, the sufficient condition is to have positive second derivative, *i.e.* $\frac{\partial^2 f}{\partial x^2} > 0$.

Newton's method is depicted in Figure 5.5 as a local quadratic approximation of the original function.

5.4 OTHER METHODS

A variety of methods exist, which mostly are based on interpolation of the function value, and/or of its derivative at various points. The interpolating polynomials are then treated as "surrogate" functions, which are minimized to yield the next iterate.

5.5 FURTHER READING RECOMMENDATIONS

More Extensive Methods on One-Dimensional Minimization

Edgar, T. F., Himmelblau, D. M., and Lasdon, L. S. (2001). Optimization of Chemical Processes. McGraw-Hill, Edition 2.

Gill, P. E., Murray, W., and Wright, M. H. (1981). Practical Optimization. Academic Press.

5.6 EXERCISES

1. Solve the following functions numerically with a) the Golden section, b) the Bisection, and c) the Newton method:

(i) $f(x) = 5x^2 + 5x - 10$, $x \in [-5,5]$
(Answer: $x^* = -0.5$, $f(x^*) = -11.25$)

(ii) $f(x) = 5x^4 - 35x^3 + 25x^2 + 155x - 150$, $x \in [-3,5]$
(Answer: $x^* = -0.9354$, $f(x^*) = -240.6$)

(iii) $f(x) = x^2 \ln x^2$, $x \in [0,2]$
(Answer: $x^* = 0.6065$, $f(x^*) = -0.3679$)

6

Unconstrained Multivariate Gradient-Based Minimization

This chapter introduces the methodology for unconstrained optimization in many variables. The solution techniques presented use the gradient to derive a direction in which to seek an update for a starting point.

6.1 MINIMUM STATIONARY POINTS

The necessary condition for x^* to be a minimum of a function $f(x)$ is the following:

$$\frac{\partial f}{\partial x}(x^*) = \nabla_x f(x^*)^T = 0 \tag{6.1}$$

Proof: First-order necessary optimality condition

For $\min_x f(x)$ such that x^* is a minimizer, then we have by definition:

$$f(x^* + \Delta x) \geq f(x^*) \tag{6.2}$$

For any Δx (see definition of local/global minimizer). Use a first-order Taylor expansion around x^* to get:

$$f(x^* + \Delta x) = f(x^*) + \nabla f(x^*)^T \cdot \Delta x + H.O.T. \tag{6.3}$$

where $H.O.T. \to 0$ as $\|\Delta x\| \to 0$.
 Let us now choose a Δx:

$$\Delta x = -\delta \cdot \nabla f(x^*) \tag{6.4}$$

for $\delta > 0$. Insert this into Equation (6.3), assuming δ is small enough so that $\|\Delta x\| \to 0$ and hence $H.O.T. \to 0$:

$$f(x^* + \Delta x) = f(x^*) - \delta \nabla f(x^*)^T \nabla f(x^*) \tag{6.5}$$

Insert this expression into Equation (6.2) to get:

$$-\delta \nabla f(x^*)^T \nabla f(x^*) \geq 0 \tag{6.6}$$

Since $\delta > 0$ and $\nabla f(x^*)^T \nabla f(x^*) = \|\nabla f(x^*)\|_2^2 \geq 0$:

$$0 \leq -\delta \nabla f(x^*)^T \nabla f(x^*) \leq 0 \tag{6.7}$$

The only way for this to hold is:

$$\nabla f(x^*) = 0 \tag{6.8}$$

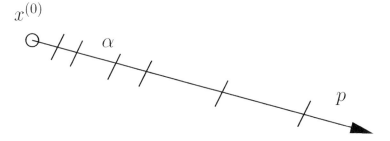

Figure 6.1 Diagram of search direction at a local point $x^{(0)}$.

Most methods demand that along a search direction, $p \in \mathbb{R}^n$, there is descent. To derive a condition for this, we look at the local point $x^{(0)} \in \mathbb{R}^n$ along with the search direction p emanating from that point. This is shown in Figure 6.1.

Let $a > 0$ be a scalar scaling parameter, and consider the first-order expansion with $\Delta x = \alpha \cdot p$:

$$f(x^{(0)} + \alpha p) = f(x^{(0)}) + \left.\frac{\partial f}{\partial x}\right|_{x^{(0)}} \alpha p \tag{6.9}$$

In order to have local descent, the new point must be improving (yielding a lower value of the objective function), as follows:

$$f(x^{(0)} + \alpha p) - f(x^{(0)}) < 0 \tag{6.10}$$

Using this condition in the previous equation we get:

$$\left.\frac{\partial f}{\partial x}\right|_{x^{(0)}} p < 0 \tag{6.11}$$

Other ways to write the gradient vector are:

$$\left.\frac{\partial f}{\partial x}\right|_{x^{(0)}} = (\nabla_x f(x))^T = g^T \tag{6.12}$$

So, in short, the descent direction condition is:

$$g^T p < 0 \tag{6.13}$$

6.2 CALCULATING DERIVATIVES

Gradient methods at least require the gradient $g(x) = \nabla_x f(x)$ and may also require the Hessian matrix (second derivatives) $H(x) = \frac{\partial^2 f(x)}{\partial x^2}$. Let us assume we have n variables. What is the cost of getting $g(x)$?

Forward Differences

This will require one base calculation to find the objective function, and then n perturbations, a variable at a time, in order to get the gradient. In other words, the order of the computational cost is:

$$\mathcal{O}(n) \tag{6.14}$$

Automatic Differentiation (AD)

If the function $f(x)$ is defined as a "computer program," or more accurately as a predefined set of algebraic steps, then automatic differentiation may be used. It works as follows:

Let the "code" that gets the value of the objective function be given as below. Variables are x1, x2, x3, intermediate variables are y1, y2, y3, and the objective is given in variable y:

```
y1 = sin(x1);
y2 = x2^2 * y1;
y3 = cos(y2) * x3
y = y1 + y2 + y3;
```

Then automatic differentiation uses inlined differentials derivation and the chain rule to augment the above code into the following:

```
y1 = sin(x1);
dy1 = cos(x1) * dx1;
y2 = x2^2 * y1;
dy2 = 2 * x2 * y1 * dx2 + x2^2 * dy1;
y3 = cos(y2) * x3
dy3 = -sin(y2) * x3 * dy2 + cos(y2) * dx3;
y = y1 + y2 + y3;
dy = dy1 + dy2 + dy3;
```

You may verify that for the objective function derivative, say with respect to the first variable x1, all you need to do is to run the above code with the values of x1, x2, x3 at the current point and set dx1 = 1, dx2 = dx3 = 0.

Automatic differentiation works either at the compilation time of a programme, or it may be done with advanced use of C++ classes. There are ready-made libraries for this, including ADOL-C, which is an example in the public domain [1].

Why are we discussing this?

Automatic Differentiation Theorem

Using the chain rule, and what is called adjoint differentiation, the cost of evaluating the gradient is **proportional** to that of the evaluation of the objective function, as follows:

$$\text{Cost}[g(x)] = K \cdot \text{Cost}[f(x)] \tag{6.15}$$

So what is new about this? The crucial thing is in the constant K. This is **at most three to five times** the cost of the objective function, **regardless of the number of variables** n.

Remember that for finite differences, using forward or backward differences, we get:

$$\text{Cost}[g(x)] = n \cdot \text{Cost}[f(x)] \tag{6.16}$$

For second derivatives, things get a bit more complicated, particularly if one wishes to also exploit sparsity for very large-scale optimization problems.

6.3 THE STEEPEST DESCENT METHOD

The steepest descent method is conceptually the easiest approach to numerical optimization. It was used by the Russian mathematician Pavel Alexeevich Neraskov as early as 1884 (Petrova and Solov'ev, 1997) and has been the foundation on which more advanced algorithms were developed.

The algorithm can be described by the following equations:

$$\min_{x} f(x) \tag{6.17}$$

$$x^{\text{new}} = x^{\text{old}} - \nabla_x f(x^{\text{old}}) \tag{6.18}$$

But, how far along the opposite direction of the gradient do we go?
Set:

$$\Delta x = -\alpha \nabla_x f(x^{\text{old}}) \tag{6.19}$$

hence

$$f^{\text{new}}(\alpha) = f\left[x^{\text{old}} - \alpha \nabla_x f(x^{\text{old}})\right] \tag{6.20}$$

Since α **is scalar**, this is now a **1-D minimization problem**. The process is called **linesearch**.

With exact linesearch:

$$\alpha^* = \arg\min_{\alpha} f^{\text{new}}(\alpha) \tag{6.21}$$

Steepest descent moves in **perpendicular steps under exact linesearch**, as is shown in Figure 6.2.

Steepest descent:

- is scale dependent,
- unless all eigenvalues are equal (quadratic case, contours become perfect concentric circles), then it will perform badly, and

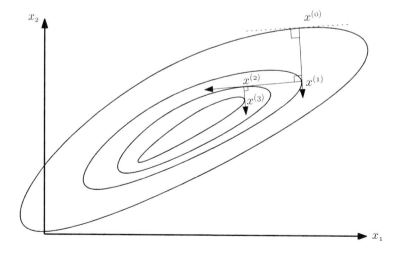

Figure 6.2 Solution trajectory of the steepest descent algorithm.

- its performance deteriorates as the **condition number** of the Hessian increases. The condition number is the ratio of maximum to minimum eigenvalue, as follows:

$$\left|\lambda_{max}/\lambda_{min}\right| \tag{6.22}$$

6.4 NEWTON'S METHOD

$$\min_{x} f(x) \tag{6.23}$$

Assume that $f(x)$ is a convex function, and seek the stationary point:

$$\left(\frac{\partial f}{\partial x}\right)^T = 0_{n \times 1} \tag{6.24}$$

Equation (6.24) defines a simultaneous system of n nonlinear equations. For a system of nonlinear equations:

$$h(x) = 0 \tag{6.25}$$

Newton's method becomes:

$$\frac{\partial h}{\partial x}\bigg|_{old} \cdot \Delta x = -h|_{old} \tag{6.26}$$

N.B. This comes from the local approximation of the system of nonlinear equations, via a Taylor expansion, by a set of hyperplanes (affine[1] functions):

$$h(x^{(0)} + \Delta x) = h(x^{(0)}) + \frac{\partial h}{\partial x}\bigg|_{x^{(0)}} \cdot \Delta x + H.O.T. \tag{6.27}$$

where H.O.T stands for Higher Order Terms.

The matrix of first derivatives $\frac{\partial h}{\partial x}\big|_{x^{(0)}}$ in the system is called the Jacobian matrix.

In the case of optimization, the system we have to solve is given in Equation (6.24). By substituting this into Equations (6.25) and (6.26), we get:

$$\frac{\partial^2 f}{\partial x^2}\bigg|_{x^{(0)}} \cdot \Delta x = -\left(\frac{\partial f}{\partial x}\right)^T\bigg|_{x^{(0)}} \tag{6.28}$$

The search direction may be thus defined as $p = \Delta x$. This system may be written compactly as:

$$H(x^{(0)}) \cdot p = -g(x^{(0)}) \tag{6.29}$$

Next, we perform linesearch in the direction p and update:

$$\Delta x = \alpha p \tag{6.30}$$

$$x^{(1)} = x^{(0)} + \Delta x = x^{(0)} + \alpha p \tag{6.31}$$

The parameter α is determined by linesearch. The choice of the search direction in Newton's method is given by:

$$p = -H^{-1}g \tag{6.32}$$

---

[1] Usually known as "linear" functions, but this is not correct.

and it is compared to that of steepest descent by:

$$p = -g \qquad (6.33)$$

The direction in Newton's method is self-scaling by the actual inverse Hessian matrix (which carries the curvature information) and is not sensitive to scaling (in theory; why not in practice also?)

6.4.1 Newton Failure Safeguarding

Even Newton's method can fail for convex functions far away from the minimizer x^* (for quadratic it solves it in one step). This is shown in Figure 6.3, where taking the entire search direction can lead to oscillation and failure.

To guarantee "global" convergence, *i.e.* ability to reach a minimum from any starting point, we need to cut the step, hence that is the reason why linesearch is introduced.

Newton may also fail if, locally, the Hessian matrix $H(x)$ is not positive definite, which leads to a non-descent direction, or even worse to no solution for the step if the Hessian becomes singular. The system is:

$$H \cdot p = -g \qquad (6.34)$$

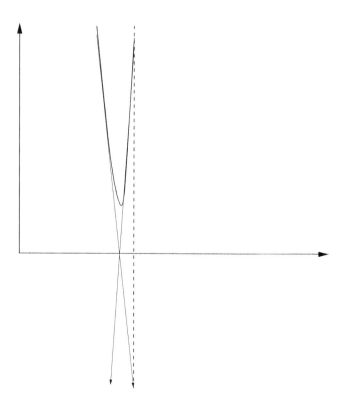

Figure 6.3 Failure of Newton's method.

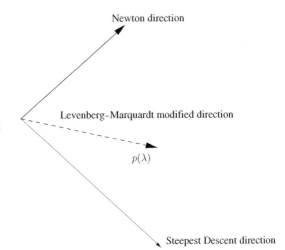

Figure 6.4 Levenberg–Marquardt search direction.

and the descent condition desired is:

$$p^T g < 0 \tag{6.35}$$

Modifications:

1. Factorization based. The Hessian matrix is factorized according to the symmetric Choleski factorization, which is modified to induce positive definiteness ($H = LDL^T$). A good book on this, and other practical optimization issues, is [2].
2. Levenberg–Marquardt scheme. Originally employed for least-squares problems, it modifies the Hessian matrix by: $[H + \lambda I] p = -g$. Different values of the scalar parameter λ are tried, $e.g.$ $10^{-3}, 10^{-2}, \ldots, 10^4, \ldots$ by trial and error until positive definiteness/descent direction is achieved.

For the Levenberg–Marquardt scheme, the following interpretation is provided:

$$\text{if} \quad \lambda = 0 \quad Hp = -g \quad \longrightarrow \quad p_{\text{Newton}} \tag{6.36}$$

$$\text{if} \quad \lambda \gg, \text{ i.e. } \lambda \to \infty \quad p \propto -g \quad \longrightarrow \quad p_{\text{Steepest Descent}} \tag{6.37}$$

For "intermediate" values of λ, the search direction is shown in Figure 6.4.

6.5 RATE OF CONVERGENCE

The Rate of Convergence represents the local rate (speed) with which an algorithm reaches a minimum when near a minimizer:

Newton – quadratic rate

$$\lim_{k \to \infty} \frac{\left\| x^{(k+1)} - x^* \right\|}{\left\| x^{(k)} - x^* \right\|^2} = K \tag{6.38}$$

where K is some constant value.

Example 6.1

Newton's method for unconstrained minimization:

$$\min f(x), \; x \in \mathbb{R}^n \tag{6.39}$$

follows a descent direction, when the function $f(x)$ is strictly convex. To prove this, answer the following sequence of steps:

(i) State the first-order condition that defines the tangent of a strictly convex function (considering its tangent at a point x_1), and prove by Taylor series expansion to second order the condition of positive definiteness of the Hessian of the function, namely that $y^T H y > 0$ for any non-zero vector $y \in \mathbb{R}^n$.

(ii) Show that a positive definite matrix is always invertible. (Hint: consider the positive definiteness property in part (i) written as $t^T (Hy) > 0$ and what this means for the inner product of the two vectors to be non-zero.)

(iii) Show that for an invertible matrix H, $\left(H^{-1}\right)^T = \left(H^T\right)^{-1}$ (Hint: use the property of the inverse of a matrix as $H^{-1}H = HH^{-1} = I$). From this show furthermore that if the matrix H is symmetric then H^{-1} is also symmetric.

(iv) Show that for a symmetric and positive definite matrix H, H^{-1} is also positive definite. (Hint: define an auxiliary vector $z = H y$ and substitute y into the definition of part (i) to prove this.)

(v) Consider now Newton's method applied to the optimization problem given in Equation (6.39). Using your result for part (iv), show that the Newton search direction is always a descent direction for a strictly convex function.

Solution

(i) The first-order condition for the tangent of a strictly convex function $f(x)$ states that the tangent at any point x_1 always underestimates the function:

$$\hat{f}(x) < f(x)$$

$$\hat{f}(x) = f(x_1) + \nabla_x f(x)\big|_{x_1}^T \cdot (x - x_1)$$

Hence:

$$f(x) > f(x_1) + \nabla_x f(x)\big|_{x_1}^T (x - x_1) \tag{6.40}$$

$$x \neq x_1$$

Expansion to second order of $f(x)$ at $x = x_1$:

$$f\underbrace{(x_1 + \Delta x)}_{x} = f(x_1) + \underbrace{\frac{\partial f}{\partial x}\bigg|_{x_1}}_{=(\nabla_x f)^T} \Delta x + \frac{1}{2}(\Delta x)^T \underbrace{\frac{\partial^2 f}{\partial x^2}}_{=H} \Delta x \tag{6.41}$$

(ignoring terms $\mathcal{O}\left(\|\Delta x\|^3\right)$ and higher).

Bring to the left-hand side the underbraced term and rewrite, as follows:

$$f(x) - \left\{ f(x_1) + \nabla_x f(x)\big|_{x_1}^T (x - x_1) \right\} = \frac{1}{2}(\Delta x)^T H \Delta x \tag{6.42}$$

But the left-hand side is identical to what we have in Equation (6.40). Hence inserting Equation (6.40) into Equation (6.42) gives:

$$\frac{1}{2} (\Delta x)^{\mathrm{T}} H \Delta x > 0$$

Δx can be substituted by any vector y and if strong convexity holds we may finally derive:

$$y^{\mathrm{T}} H y > 0 \qquad (6.43)$$

where $y \neq 0$.

(ii) A positive definite matrix is invertible. Consider the definition:

$$y^{\mathrm{T}} (Hy) > 0$$

Since this is the product of two vectors, the only case is that $y \neq y$ (by definition) and $Hy \neq 0$ for all $y \neq 0$.

This implies that the columns of H are independent, as no non-zero set of coefficients y can combine them to produce the zero vector. Hence H^{-1} exists.

(iii) The property of the inverse is:

$$H^{-1} H = H H^{-1} = I$$

Taking transposes we get:

$$H^{\mathrm{T}} \left(H^{-1} \right)^{\mathrm{T}} = \left(H^{-1} \right)^{\mathrm{T}} H^{\mathrm{T}} = I$$

This produces the invertibility property for H^{T}:

$$\left(H^{-1} \right)^{\mathrm{T}} = \left(H^{\mathrm{T}} \right)^{-1} \qquad (6.44)$$

If H is symmetric, then $H^{\mathrm{T}} = H$, which inserted into Equation (6.44) yields:

$$\left(H^{-1} \right)^{\mathrm{T}} = H^{-1} \rightarrow H^{-1} \text{ is symmetric.} \qquad (6.45)$$

(iv) A positive definite matrix implies:

$$y^{\mathrm{T}} (Hy) > 0, \ y \neq 0$$

setting $z = Hy \rightarrow y = H^{-1}z$, for which H^{-1} exists by the answer to part (ii) ($z \neq 0$ also since H is invertible). By substituting, we obtain:

$$z^{\mathrm{T}} \underbrace{\left(H^{-1} \right)^{\mathrm{T}}}_{H^{-1} \text{is symmetric}} H H^{-1} z > 0$$

$$z^{\mathrm{T}} H^{-1} \underbrace{H H^{-1}}_{=I} z > 0$$

$$z H^{-1} z > 0 \text{ for } z \neq 0$$

hence H^{-1} is also positive definite.

(v) Newton's method yields the search direction p given by:

$$H p = g$$

where $H = \left. \frac{\partial^2 f}{\partial x^2} \right|_{x^{(0)}}$ and $g = \nabla_x f(x)|_{x^{(0)}}$. Hence:

$$p = H^{-1} g$$

For a search direction to be a descent direction we must have:

$$p^T g < 0$$

By substituting Newton's search direction we get:

$$-g^T H^{-1} g^T < 0$$

Because H^{-1} is symmetric and positive definite (parts (iii) and (iv)), then $-g^T H^{-1} g^T < 0$ for any non-zero gradient vector g. Hence, with the negative sign, Newton's method always satisfies the descent direction condition.

Steepest Descent – linear rate

$$\lim_{k \to \infty} \frac{\left\| x^{(k+1)} - x^* \right\|}{\left\| x^{(k)} - x^* \right\|} < 1 \tag{6.46}$$

is generally extremely slow.

Defining the error as:

$$\varepsilon = \left\| x^{(k)} - x^* \right\| \tag{6.47}$$

Newton yields in the limit that:

$$\varepsilon^{(k+1)} = K_1 \left(\varepsilon^{(k)} \right)^2$$

and Steepest Descent yields:

$$\varepsilon^{(k+1)} = K_2 \varepsilon^{(k)} \tag{6.48}$$

For a more detailed discussion of this material please refer to [3].

6.6   THE NEWTON FAMILY OF METHODS

Newton's method has motivated the creation of Newton-like (quasi-Newton) methods. Why?

- Think of the cost of evaluating a full Hessian matrix (second derivatives of objective function).

Generally, the Hessian matrix is substituted at each step by some other approximation:

$$B \cdot p = -g \tag{6.49}$$

where B is some positive definite symmetric matrix approximating $H = \frac{\partial^2 f}{\partial x^2}$. In the literature there are many methods one can find:

- BFGS method
- Truncated Newton method (TN, suitable for many variables)
- Limited memory BFGS (LBFGS, suitable for many variables)

Good books on these topics are [3, 4].

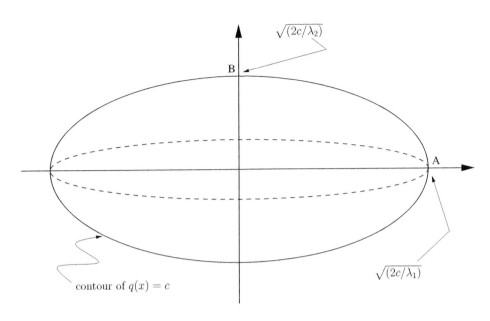

Figure 6.5 Contour plot of function $q(x) = \frac{1}{2}\left(\lambda_1 x_1^2 + \lambda_2 x_2^2\right)$. The dashed line shows $q(x)$ for $\lambda_2 \gg \lambda_1$.

6.7 OPTIMIZATION TERMINATION CRITERIA

When do we stop numerical algorithms and report a solution?

One way to measure closeness to a minimizer is to measure the gradient vector, by using a norm of it.

For example, if:

$$\|g\|_\infty = \max_{i=1,2,\dots,n} \{|g_i|\} \leq \varepsilon \tag{6.50}$$

then stop. ε is a user-specified termination tolerance.

Is this a reliable indicator of proximity to x^*?

- Reminder: section on contours of a quadratic function

  Consider $q(x) = \frac{1}{2}\left(\lambda_1 x_1^2 + \lambda_2 x_2^2\right)$ with $\lambda_1, \lambda_2 > 0$. Say $q(x) = c$, then:

$$g(x) = \nabla_x q(x) = \begin{bmatrix} \lambda_1 x_1 \\ \lambda_2 x_2 \end{bmatrix} \tag{6.51}$$

Now, consider the situation depicted in the diagram in Figure 6.5.

The gradient norm at point A has value:

$$\left\|g^{(A)}\right\|_\infty = \sqrt{2c\lambda_1} \tag{6.52}$$

while at point B it has value:

$$\left\|g^{(B)}\right\|_\infty = \sqrt{2c\lambda_2} \tag{6.53}$$

Hence, if $\lambda_2 \gg \lambda_1$, then $\left\|g^{(A)}\right\|_\infty \ll \left\|g^{(B)}\right\|_\infty$, while at both points the objective function has constant value (on the same contour):

$$q(x^{(A)}) = q(x^{(B)}) = c \qquad (6.54)$$

(exercise: verify the values for the norms at points A and B).

The inner dashed contour in Figure 6.5 shows the situation when $\lambda_2 \gg \lambda_1$. Clearly, the larger the disparity in the eigenvalue magnitudes, the larger the difference in gradient norm at the points along the two eigenvectors (which coincide with the x–y axes in our case here).

The problem with using the gradient as termination criterion is general, as functions near a minimum behave as quadratics, and we have just shown the gradient to be scale-dependent. Hence, more than one criterion is needed. Usual things to monitor from iteration to iteration along with the gradient norm are, as follows:

- Progress of objective function value,
- Progress in the norm of the solution vector itself value, and
- The norm of the search direction vector.

For more information on this issue the reader is referred to [2].

6.8 LINESEARCH METHODS

Linesearch methods deal differently with the step-size taken along the search direction derived by the central optimization algorithm (*e.g.* Newton's method). Here, a simple outline of general principles is presented.

In broad terms, there are two distinctions in linesearch methods: exact and inexact methods.

Exact

As we have seen, once the search direction p has been derived, then exact linesearch seeks a 1-D minimizer along this ray so that the exact optimal step-size value α^* is computed.

Generally, exact linesearch is almost never used in practical numerical applications, as far from the minimizer of the objective function exact search is wasteful.

Inexact

In contrast to exact linesearch, inexact methods require a sufficient decrease in the objective function over its base value at point $x^{(k)}$ in order to compute the next iterate $x^{(k+1)}$, using the current search direction $p^{(k)}$, which is a descent direction such that, $\left(g^{(k)}\right)^T p^{(k)} < 0$.

We have a 1-D minimization in terms of the step-size variable α:

$$\min_\alpha f(\alpha) = f\left(\underbrace{x^{(k)} + \alpha \cdot p^{(k)}}_{x}\right) \qquad (6.55)$$

Linearization of $f(\alpha)$ at $\alpha = 0$, which is equivalent to $x = x^{(k)}$, yields:

$$\hat{f}(\alpha) = \underbrace{f(\alpha = 0)}_{f(x^{(k)})} + \left.\frac{\partial f(\alpha)}{\partial \alpha}\right|_{\alpha=0} \cdot (\alpha - 0) \tag{6.56}$$

The partial derivative appearing on the RHS of Equation (6.56) is calculated using the chain rule:

$$\left.\frac{\partial f(\alpha)}{\partial \alpha}\right|_{\alpha=0} = \left.\frac{\partial f(x(\alpha))}{\partial x}\right|_{x(\alpha=0)} \cdot \left.\frac{\partial x}{\partial \alpha}\right|_{\alpha=0}$$

$$= \left.\frac{\partial f(x)}{\partial x}\right|_{x^{(k)}} \cdot p^{(k)}$$

$$= \left(g^{(k)}\right)^T \cdot p^{(k)} \tag{6.57}$$

Using the result in Equation (6.57) in Equation (6.56) yields the equation of the tangent of the function $f(\alpha)$ at $\alpha = 0$:

$$\hat{f}(\alpha) = f\left(x^{(k)}\right) + \alpha \cdot \left(\left(g^{(k)}\right)^T p^{(k)}\right) \tag{6.58}$$

Linesearch algorithms use the following criterion of sufficient decrease, as follows:

$$f(\alpha) = f(\underbrace{x^{(k)} + \alpha p^{(k)}}_{x^{(k+1)}}) \le \phi(\alpha) = f(x^{(k)}) + \mu\alpha(g^{(k)})^T p^{(k)} \tag{6.59}$$

where $0 < \mu \le \frac{1}{2}$, with usual choice $\mu = 10^{-4}$ or 10^{-6}.

What this condition implies is that the range of acceptable values are such that they cause sufficient decrease in the new value of the objective function. By comparing it with Equation (6.58), it can be seen that it uses a straight line with an increased slope over that of the tangent at $\alpha = 0$.

This condition is often called the *first Armijo linesearch condition*. It is shown pictorially in Figure 6.6.

A simple way to choose the step size is to first try the full step, $\alpha = 1$, which yields $f(x^{(k)} + p^{(k)})$. If the criterion in Equation (6.59) is not satisfied, then backtrack

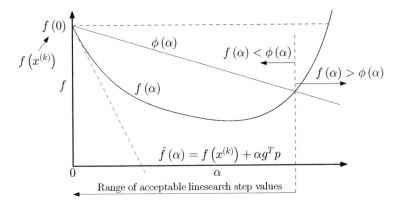

Figure 6.6 Diagram of first Armijo linesearch condition.

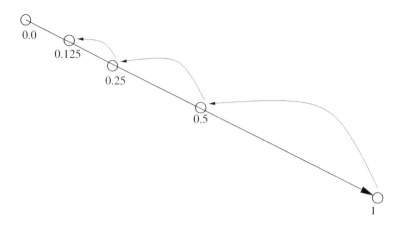

Figure 6.7 Step size determination for linesearch.

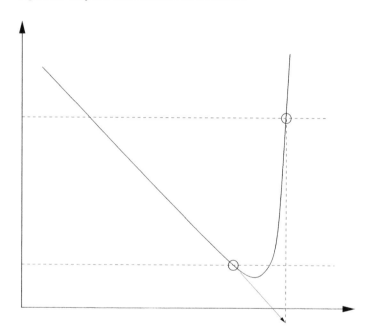

Figure 6.8 Diagram of "creeping" effect.

by setting $\alpha^{\mathrm{new}} = \alpha^{\mathrm{old}}/2$ and so on until the criterion is satisfied. This is shown in Figure 6.7.

In the example above, the final solution is:

$$x^{\mathrm{new}} = x^{\mathrm{old}} + \alpha \cdot p^k \tag{6.60}$$

or

$$x^{k+1} = x^k + \alpha \cdot p^k \tag{6.61}$$

The above scheme creates a monotonically decreasing sequence of iterates $f(x^{(k)})$ of the main optimization algorithm, such as $f(x^{(k+1)}) < f(x^{(k)})$ by Equation (6.59).

There are, however, cases where this is actually a problem. Figure 6.8 indicates the case of "creeping" iterates near a "wall" of the objective function (where it turns suddenly upwards very steeply). This will cause the linesearch method to use small steps, and many backtracking inner steps per linesearch cycle, as there will be many rejections.

A remedy is to use non-monotone linesearch, which is, however, outside the scope of this book and will not be presented here.

6.9 SUMMARY OF THE CHAPTER

Things to remember from this chapter are the following:

1. The necessary conditions for a minimum of a function $f(x)$ are given by:

$$\nabla_x f(x) = 0$$

2. For linesearch methods, it is essential to follow a descent direction to find an optimum. If the gradient of a function is given by g, and the search direction is p, a descent direction is present if:

$$g^T p < 0$$

3. Find the direction of **steepest descent** at the current point in order to find the next point of the algorithm. **Full Newton's method** fits a quadratic function at the current point, found by a Taylor expansion. This quadratic function is used to find the next point of the iteration.

4. Linesearch methods are categorized as exact and inexact: Exact linesearch methods find the minimum of the function along a certain search direction, whereas inexact linesearch methods use heuristics to choose the step length along the search direction.

5. Scaling of the underlying problem can have significant effects on the performance of linesearch methods, *e.g.* steepest descent. Newton's method, on the other hand, is not affected by scaling of the problem.

6.10 REFERENCES

[1] Walther, A. and Griewank, A. A Package for the Automatic Differentiation of Algorithms Written in C/C++. Technical report. 2010.

[2] Gill, P. E., Murray, W., and Wright, M. H. Practical Optimization. Academic Press. 1981.

[3] Luenberger, D. and Ye, Y. Linear and Nonlinear Programming. Springer. 2008.

[4] Nash, S. G. and Sofer, A. Linear and Nonlinear Programming. McGraw-Hill. 1995.

6.11 EXERCISES

1. Show that Newton's method applied to the general unconstrained minimization problem

$$\min_x f(x), \; x \in \mathbb{R}^n$$

is intensive to linear transformations of the original variables.

Hint: As a scaling scheme, use constant factors d_i, given by

$$x'_i = d_i x_i, \ i = 1, 2, \ldots, n$$

represented in matrix-vector form by

$$x' = Dx; \ x' \in \mathbb{R}^{n \times n}$$

where

$$D = \begin{bmatrix} d_1 & 0 & \cdots & 0 \\ 0 & d_2 & \cdots & 0 \\ \vdots & \vdots & \vdots & \vdots \\ 0 & 0 & \cdots & d_n \end{bmatrix}$$

is a diagonal scaling matrix. Use this information to compare the search directions for the problem with and without the scaling.

2. Minimize the surface area of a "rectangular" box (parallelpiped shape) having constant volume equal to 1 unit. Show that what you have obtained is the global optimum.

3. Consider the optimization problem in two variables given by:

$$\min_{x,y} \left(x^2 + y^2 + xy \right) \tag{6.62}$$

(i) Starting at $(1,1)$, perform two steepest descent iterations with a fixed step size $\alpha = 10^{-2}$.

(ii) The constant K is defined as

$$K = \frac{\|x^{\text{new}} - x^*\|}{\|x^{\text{old}} - x^*\|}$$

where x^{new}, x^{old} are the new and old iterates for the solution approximation, respectively, and x^* is the exact solution (minimizer). For the norms $\| \cdot \|$, use a Euclidean vector norm. By computing K for each step in part (i), show that the steepest descent method has a linear rate of convergence.

(iii) Perform the following variable transformation into Equation (6.62):

$$x' = 0.5 x$$
$$y' = 0.1 y \tag{6.63}$$

with the same starting point as in part (i), although mapped through the transformation in Equations (6.63), perform two iterations of the steepest descent method using the same fixed line search step size. Comment on your result.

4. (i) Compute the gradient and the Hessian of the following function:

$$f(x, y) = 100 \left(y - x^2 \right)^2 + (1 - x)^2$$

Calculate all of its stationary points. Classify them as local or global maxima or minima.

(ii) Among all rectangles of a fixed perimeter, which one has maximal area? Determine an extremum for the area using Lagrange multipliers. Find an alternative formulation to determine whether the extremum is a maximum.

(iii) The following optimisation problem is considered:

$$\min (2\,x_1 + 3\,x_2 - 4\,x_3)$$

subject to:

$$x_1 + x_2 \le 6$$
$$-x_1 + 2\,x_2 - 3\,x_3 = 2$$
$$x_i \ge 0,\ i = 1, 2, 3$$

Verify that the vector $x^{\mathrm{T}} = (0, 1, 0)$ is the optimal solution. Justify your findings based on the satisfaction of the necessary conditions of optimality.

5. Consider fitting a nonlinear model of a system with n inputs, a single output, and m measurements. The residuals for the system are given by:

$$r_i(x) = f\,(x,\,a_i) - b_i \tag{6.64}$$

where $r_i(x)$ is the residual, f is a general nonlinear model (function), $x \in \mathbb{R}^n$ is the set of undetermined parameters of the model, $a_i \in \mathbb{R}^n$ is the set of inputs for measurement i, and b_i is the corresponding measured output of the system.

(i) Define the least squares objective function $q(x)$ for this problem. Hence, show that the gradient of the objective function defined in Equation (6.64) is given in column format by:

$$\nabla_x q(x) = \left(\frac{\partial r(x)}{\partial x}\right)^{\mathrm{T}} r(x) = J(x)^{\mathrm{T}} r(x) \tag{6.65}$$

where $J(x)$ is the Jacobian, as indicated, of the residual vector $r(x)$.

(ii) The *Gauss–Newton* method updates a guess $x^{(k)}$ of the undetermined parameters at iteration k according to $x^{(k+1)} = x^{(k)} + \Delta x^{(k)}$, where:

$$\Delta x^{(k)} = -\left[J\left(x^{(k)}\right)^{\mathrm{T}} J\left(x^{(k)}\right) \right]^{-1} J\left(x^{(k)}\right)^{\mathrm{T}} r\left(x^{(k)}\right)$$

Show how the above result is obtained by considering Newton's method for the least squares optimization problem. Hint: differentiate the result in Equation (6.65) and drop the second derivative terms from the resulting Hessian of the least squares objective function.

(iii) The model connecting biochemical oxygen demand (BOD) with incubation time of waste water is given by:

$$\mathrm{BOD} = x_1\,(1 - \exp{(-x_2 t)})$$

After inoculation with bacteria, the load in organic material is measured as the BOD of the waste water as a function of time, in mg/l. Three measurements are presented in Table 6.1.

Table 6.1 BOD measurements

t days	BOD (mg/l)
1	8
3	18
7	20

It is desired to estimate the BOD coefficients x_1 and x_2 from the measurements in Table 6.1 using the Gauss–Newton method described in part (ii). Set up the residuals and derive the Jacobian for this problem, starting from the initial guess $x_1 = 20$, $x_2 = 0.5$. Comment on the number of sample points used in the fitting.

7

Constrained Nonlinear Programming Problems (NLP)

Constrained optimization generalizes the scope of optimizing an objective function to the general case where the variables have to obey relationships defined by a set of constraints, which can include both equality and inequality constraints.

7.1 CONVEXITY OF CONSTRAINT SET

It is not enough to have a convex objective function to guarantee a global minimum. The constraint feasible region has to also be convex. The impact of a nonconvex constraint set and the presence of local optimizers for a constrained problem is demonstrated by the diagram shown in Figure 7.1.

7.2 CONVEX PROGRAMMING PROBLEM

There is one specific type of problem that is guaranteed to have a unique global minimum, and that is defined as a convex programming problem. It can comprise the following items:

$$\min_{x} f(x) \longrightarrow \text{Convex function} \tag{7.1}$$

subject to:

$$h(x) = 0 \longrightarrow \text{Affine functions} \tag{7.2}$$

$$g(x) \leq 0 \longrightarrow \text{Convex functions} \tag{7.3}$$

The convexity of a feasible region $\mathcal{F}$ (domain) can again be defined by a connecting line (as shown in Chapter 3) such that:

$$\text{for any} \quad x_1, x_2 \in \mathcal{F} \tag{7.4}$$

Then the following point belongs also to the feasible domain:

$$x = (1 - \alpha)x_1 + \alpha x_2, \quad x \in \mathcal{F} \tag{7.5}$$

for $0 < \alpha < 1$.

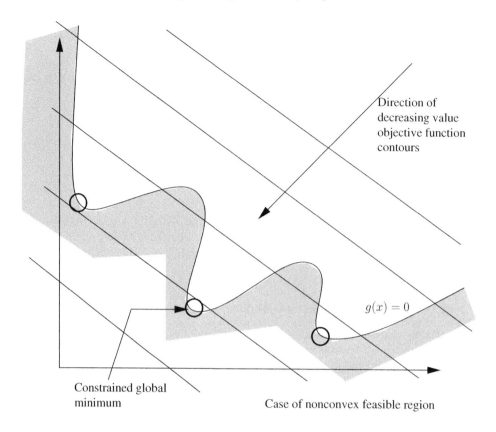

Direction of
decreasing value
objective function
contours

$g(x) = 0$

Constrained global
minimum

Case of nonconvex feasible region

Figure 7.1 Nonconvex constraint set. Points indicated with circles are value constrained local minima.

7.3 LAGRANGE MULTIPLIERS

Consider the following *equality constrained* optimization problem:

$$\min_{x} f(x) \tag{7.6}$$

subject to:

$$h(x) = 0 \tag{7.7}$$

where there are n_h equality constraints, and $x \in \mathbb{R}^n$.

We form the Lagrangian function with Lagrange multipliers[1] λ:

$$L(x, \lambda) = f(x) + \sum_{i=1}^{n_h} \lambda_i h_i(x) = f(x) + \lambda^T h(x) \tag{7.8}$$

[1] The discussion of the Lagrangian function and Lagrange multipliers is part of what is called *duality theory*. The presentation of the material in this chapter, although sufficient for the purposes of this chapter, is not the most rigorous mathematically. The Lagrangian duality theory is discussed in more detail in Chapter 19.

Take derivatives with respect to (x, λ) and set them equal to zero:

$$\frac{\partial L}{\partial x} = \frac{\partial f}{\partial x} + \sum_{i=1}^{n_h} \lambda_i \frac{\partial h_i}{\partial x} = 0 \tag{7.9}$$

$$\frac{\partial L}{\partial \lambda} = h^T(x) = 0 \tag{7.10}$$

or

$$\nabla_x L = \nabla_x f(x) + \sum_{i=1}^{n_h} \lambda_i \nabla_x h_i(x) = 0 \tag{7.11}$$

$$\nabla_\lambda L = h(x) = 0 \tag{7.12}$$

This is a system of $(n + n_h)$ simultaneous nonlinear equations in an equal number of unknowns. The equalities were set to zero, as a stationary point (minimum of the objective function) is to be determined.

If there are also n_g inequalities, an equivalent definition of the optimization problem:

$$\min_x f(x) \tag{7.13}$$

subject to:

$$h(x) = 0 \tag{7.14}$$

$$g(x) \leq 0 \tag{7.15}$$

leads to the following Lagrangian function:

$$L(x, \lambda, \mu) = f(x) + \sum_{i=1}^{n_h} \lambda_i h_i(x) + \sum_{i=1}^{n_g} \mu_i g_i(x) = f(x) + \lambda^T h(x) + \mu^T g(x) \tag{7.16}$$

7.3.1 Interpretation of Lagrange Multipliers for Equality Constraints

Consider a single equality constraint (or equivalently fix all other equalities and consider perturbations on one only equality), as follows:

$$h(x) = \varepsilon \tag{7.17}$$

The optimal solution x and Lagrange multiplier λ are dependent on the value of $\varepsilon \longrightarrow x(\varepsilon), \lambda(\varepsilon)$. The Lagrangian becomes:

$$L(x, \lambda) = f(x) + \lambda \cdot [h(x) - \varepsilon] \tag{7.18}$$

For the Lagrangian in Equation (7.18), write the optimality conditions, as follows:

$$\frac{\partial f(x(\varepsilon))}{\partial x} + \lambda(\varepsilon) \frac{\partial h(x(\varepsilon))}{\partial x} = 0 \tag{7.19a}$$

$$h(x(\varepsilon)) = \varepsilon \tag{7.19b}$$

The sensitivity of the objective function f at the solution with respect to ε is given by calculation of the partial derivative $\frac{\partial f(x(\varepsilon))}{\partial \varepsilon}$:

$$\frac{\partial f(x(\varepsilon))}{\partial \varepsilon} = \frac{\partial f(x(\varepsilon))}{\partial x} \frac{\partial x(\varepsilon)}{\partial \varepsilon} \tag{7.20}$$

Using Equation (7.19a) in the result in Equation (7.20), we obtain:

$$\frac{\partial f(x(\varepsilon))}{\partial \varepsilon} = -\lambda(\varepsilon) \frac{\partial h(x(\varepsilon))}{\partial x} \frac{\partial x(\varepsilon)}{\partial \varepsilon} \tag{7.21}$$

We now differentiate Equation (7.19b) (the equality constraint with modified right-hand side) with respect to ε to obtain:

$$\frac{\partial h(x(\varepsilon))}{\partial x} \frac{\partial x(\varepsilon)}{\partial \varepsilon} = 1 \tag{7.22}$$

Substituting the result of Equation (7.22) in Equation (7.21) yields:

$$\frac{\partial f(x(\varepsilon))}{\partial \varepsilon} = -\lambda(\varepsilon) \tag{7.23}$$

Now ε was a perturbation on the right-hand side of the equality constraint, so we may write:

$$\lambda = -\left.\frac{\partial f}{\partial h}\right|_{x^*} \tag{7.24}$$

The Lagrange multipliers thus measure the sensitivity of the objective function to a perturbation in the right-hand side of equality (active) constraints at the optimal point (minimum). They are also known as "dual variables" or "shadow prices" of a constrained mathematical programming problem. A similar analysis to what has been shown in this section can be found in [1].

7.3.2 Interpretation of Lagrange Multipliers for Inequality Constraints

Consider instead of $h(x) = 0$ the inequality $g(x) \leq 0$ *active* at the solution, *i.e.* treated as an equality constraint. The analysis of the previous section still holds and will yield:

$$\mu = -\left.\frac{\partial f}{\partial g}\right|_{x^*} \approx -\left.\frac{\Delta f}{\Delta g}\right|_{x^*} \tag{7.25}$$

Feasible Δg is: $\Delta g < 0$.

If the constraint is active, $\Delta g < 0$ makes it tighter (smaller feasible region) and hence $\Delta f \geq 0$ (worse, larger value at the minimum point). Equation (7.25) leads to the following sign restriction of the Lagrange multipliers of inequality constraints:

$$\mu \geq 0 \text{ for } g(x) \leq 0 \tag{7.26}$$

In fact, if a constraint is *inactive* the multiplier should be zero, $\mu \equiv 0$, as the constraint has no influence over the minimization and hence zero contribution. For equalities, it is finally noted that the multiplier is sign-unrestricted; it can be either negative or positive in sign.

Strictly speaking, in the presence of other constraints we should write the definitions of the Lagrange multipliers as:

$$\lambda_i = -\frac{\partial f}{\partial h_i}\bigg|_{x=x^*,h_j=\text{const.},j\neq i,g_k=\text{const.},k=1,2,...n_g} \qquad , \quad i=1,2,\ldots,n_h \qquad (7.27)$$

and:

$$\mu_i = -\frac{\partial f}{\partial g_i}\bigg|_{x=x^*,g_j=\text{const.},j\neq i,h_k=\text{const.},k=1,2,...n_h} \qquad , \quad i=1,2,\ldots,n_g \qquad (7.28)$$

where both Lagrange multipliers represent the sensitivity of the objective function with respect to the constraints.

7.4 NECESSARY CONDITIONS OF OPTIMALITY (KKT CONDITIONS)

The following conditions, known as Karush–Kuhn Tucker (KKT) conditions, are the necessary conditions of optimality for a general NLP problem:

1. Gradient of Lagrangian with respect to x must be zero

$$\nabla_x L(x^*,\lambda,\mu) = \nabla_x f(x^*) + \sum_{i=1}^{n_h}\lambda_i\nabla_x h_i(x^*) + \sum_{i=1}^{n_g}\mu_i\nabla_x g_i(x^*) = 0 \qquad (7.29)$$

2. Nonnegativity of inequality multipliers

$$\mu \geq 0 \qquad (7.30)$$

3. Feasibility of optimal point

$$g(x^*) \leq 0 \qquad (7.31)$$
$$h(x^*) = 0 \qquad (7.32)$$

4. Complementarity conditions (multiplier/active inequality constraint). They imply that either the constraint is active and the multiplier is positive, or the inequality is inactive (less than zero) and the multiplier is zero.

$$\mu_i g_i(x^*) = 0 \quad i=1,2,\ldots,n_g \qquad (7.33)$$

7.5 SUFFICIENT CONDITIONS

If a problem is a convex programming problem (see earlier section), then there is no need for the following. If, however, it is not a convex programming problem, then the following must hold at a point that satisfies the necessary (KKT) conditions:

$$p^T\frac{\partial^2 L}{\partial x^2}\bigg|_{x^*} p > 0 \qquad (7.34)$$

The stationary point will be a minimum if the above holds for any direction p satisfying all active constraints (*including equalities h_i, and active inequalities g_j*). The vector p must be such that:

$$\left(\nabla_x h_i(x^*)\right)^T p = 0, \quad i=1,2,\ldots,n_h \qquad (7.35)$$

$$\left(\nabla_x g_j(x^*)\right)^T p = 0 \quad j=1,2,\ldots,n_g, \text{ such that } g_j(x^*)=0 \qquad (7.36)$$

This comes from the linearization of the active constraints at the optimal point x^*, requiring that nearby points also satisfy the same active constraints. Set as $p = \Delta x$ and linearize the active constraints around the optimal point x^* to get:

$$\overbrace{h_i\left(x^* + p\right)}^{0} = \overbrace{h_i\left(x^*\right)}^{0} + \left(\nabla_x h_i(x^*)\right)^T p \tag{7.37}$$

$$\overbrace{g_j\left(x^* + p\right)}^{0} = \overbrace{g_j\left(x^*\right)}^{0} + \left(\nabla_x g_j(x^*)\right)^T p \tag{7.38}$$

We demand that the optimal point is feasible for the active set of constraints, so they are satisfied as equalities equal to zero. The same must hold for the perturbed point $x = x^* + p$, and it is for these points we check positive definiteness of the Lagrangian Hessian.

This says that the Hessian matrix of the Lagrangian function with respect to the primal variables x must be positive definite in the null space of the Jacobian of the active constraints.

Aside: The null space of a matrix J is the set of solution vectors p, such that:

$$J \cdot p = 0 \tag{7.39}$$

The Jacobian matrix of a set of equality constraints has as rows the gradient vectors of the constraints, so for our case $J = \left[\begin{array}{c} \frac{\partial h}{\partial x} \\ \frac{\partial g_A}{\partial x} \end{array}\right]\Bigg|_{x^*}$ (g_A is the vector of all the active inequality constraints at x^*).

The nullspace of a matrix is nonempty (apart from the trivial solution, $p = 0$) if there are linearly dependent columns in it. Since we consider problems with more variables than constraints, we are guaranteed that we will have this case, and hence through the null space Equation (7.39) we can express some of the problem variables (equal in number to the active constraints) as linear combinations of the remaining variables.

For example, if we have a total of m constraints acting as equalities at the solution and n variables, we can express m variables as linear functions of the remaining $(n - m)$ variables, which are the degrees of freedom of the problem (this assumes that the linearizations of the constraints form independent rows in the Jacobian matrix).

For the above case, the Jacobian matrix and the variables can be separated into basic and nonbasic subsets, such that the basic submatrix B is invertible to get:

$$J \cdot p = \left[\begin{array}{cc} B & N \end{array}\right] \left[\begin{array}{c} p_B \\ p_N \end{array}\right] = B \cdot p_B + N \cdot p_N = 0 \tag{7.40}$$

where $p_B \in \mathbb{R}^m$, $p_N \in \mathbb{R}^{(n-m)}$, $B \in \mathbb{R}^{m \times m}$, $N \in \mathbb{R}^{m \times (n-m)}$. By the invertibility of B, we can solve to get:

$$p_B = -B^{-1} N \, p_N \tag{7.41}$$

We can now write the original vector of variables p as:

$$p = \begin{bmatrix} p_B \\ p_N \end{bmatrix} = \begin{bmatrix} -B^{-1}N \\ I_{(n-m)} \end{bmatrix} \cdot p_N \qquad (7.42)$$

Substituting the result of Equation (7.42) in Equation (7.34), we get that in the space of the degrees of freedom we must have:

$$p_N \cdot \begin{bmatrix} -B^{-1}N \\ I_{(n-m)} \end{bmatrix}^T \cdot \left. \frac{\partial^2 L}{\partial x^2} \right|_{x^*} \cdot \begin{bmatrix} -B^{-1}N \\ I_{(n-m)} \end{bmatrix} \cdot p_N > 0 \qquad (7.43)$$

This means that the Lagrangian Hessian matrix projected to the space of the degrees of freedom, at the solution point, must be positive definite (and for example we can check that its eigenvalues are positive). This matrix, derived by Equation (7.43), is given by:

$$\hat{H} = \begin{bmatrix} -B^{-1}N \\ I_{(n-m)} \end{bmatrix}^T \cdot \left. \frac{\partial^2 L}{\partial x^2} \right|_{x^*} \cdot \begin{bmatrix} -B^{-1}N \\ I_{(n-m)} \end{bmatrix}$$

$$= \begin{bmatrix} -N^T B^{-T} & I_{(n-m)} \end{bmatrix} \cdot \left. \frac{\partial^2 L}{\partial x^2} \right|_{x^*} \cdot \begin{bmatrix} -B^{-1}N \\ I_{(n-m)} \end{bmatrix} \qquad (7.44)$$

By checking the dimensions of the product form we get that $\hat{H} \in \mathbb{R}^{(n-m)\times(n-m)}$.

7.6 DISCUSSION AND SOLUTION PROCEDURES

The previous KKT necessary conditions cannot be solved explicitly if inequalities are involved. Their value is:

1. theoretical: defining the foundation of constrained optimization and used in related theoretical analyses and proofs,
2. used to "inspire" numerical solution algorithms, and
3. used to derive measures of proximity to a potential stationary point during numerical computation (termination criteria).

In terms of numerical solution procedures there are many types of algorithms used in professional optimization packages nowadays. Some of them are:
For each of the following, explain a little and possibly give examples.

1. Active Set Methods

The active set method is applied to an optimization problem of the following form:

$$\min_x f(x)$$
$$\text{subject to :} \qquad (7.45)$$
$$c(x) \leq 0$$

where $x \in \mathbb{R}^n$, and $c(x) \in \mathbb{R}^m$. The method operates according to the following algorithmic steps:

1. Initialize the iteration counter, $k = 0$. Select a starting point for the primal variables s. Initially, set all constraints outside the active set.

2. Evaluate the constraints $c(x)$ at the current point x. Select those that are violated, if any, and append them to the active set.
3. Solve the optimization problem resulting from the objective function by treating the active set constraints as equalities. Obtain the solution of the promal variables x and of the Lagrange multipliers λ.
4. Check the Lagrange multipliers of the constraints treated as equalities. If any of the multipliers are negative, then remove the corresponding constraints from the active set.
5. Evaluate the constraints outside the active set. If there are constraints that are violated, append them to the active set.
6. If the active set has changed in steps 4 and 5, either by dropping or adding constraints, then increase the iteration counter, $k \leftarrow k + 1$, and go to step 3. Else, terminate with the optimal solution of the optimization problem of Equation (7.45).

2. Successive Linear Programming (SLP)

SLP works by linearizing all nonlinear constraints and the objective function. The linearized problem results in an LP problem, which can be solved. The solution of the LP is used as new linearization points and the procedure is repeated.

3. Successive Quadratic Programming (SQP)

SQP is based on the linearization of the constraints and approximating the objective by a quadratic function of the Lagrangian of the constrained problem [2]. This results in the solution of a sequence of QP problems

4. Generalized Reduced Gradient Methods (GRG)

GRG methods [1] operate by projecting the objective gradient onto the feasible region of the constraints, thereby proceeding toward the constrained minimum via feasible steps.

5. Penalty and Barrier Methods

More information on these methods is given in the next two chapters.

7.7   REFERENCES

[1] Edgar, T. F., Himmelblau, D. M., and Lasdon, L. S. Optimization of Chemical Processes. McGraw-Hill. 2001.

[2] Nocedal, J. and Wright, S. J. Numerical Optimization. Springer. 2006.

7.8   FURTHER READING RECOMMENDATIONS

More Advanced Material on Constrained Optimization

Biegler, L. T. (2010). Nonlinear Programming: Concepts, Algorithms, and Applications to Chemical Processes. MPS-SIAM Series on Optimization. 10.

7.9 EXERCISES

1. Solve the following problem, by guessing the active set of constraints iterating on the KKT conditions. The problem is given by:

$$\min_{x} f(x) = x_1^2 + x_2^2 - 6x_1 - 4_2$$

subject to:

$$x_1 + x_2 \leq 3$$
$$2x_1 - x_2 \leq 5$$
$$x_1 \geq 0$$
$$x2 \geq 0$$

Is the above problem guaranteed to have a global minimum?

2. The following quadratic programming (QP) problem is considered to be solved by the active set method:

$$\min_{x} z = \frac{1}{2}x_1^2 + \frac{1}{2}x_2^2 - 4x_1 - 4x_2 \qquad (7.46)$$

subject to:

$$2x_1 + x_2 - 2 \leq 0$$
$$x_1 - x_2 - 1 \leq 0$$
$$-x_1 - x_2 - 1 \leq 0$$
$$-2x_1 + x_2 - 2 \leq 0$$

(i) Find the unconstrained minimizer for the objective function in the QP of Equation (7.46).
(ii) Plot the feasible region of the constraints in the QP of Equation (7.46).
(iii) Apply the active set method to the QP of Equation (7.46), using as starting point the unconstrained minimizer found in part (i).
(iv) Is the minimizer found in part (iii) the globally optimal one?
(v) In the case of linear programming (LP) problems, what are the similarities with and the differences between the active set method and the Simplex method?

3. Consider the following QP problem:

$$\min_{x,y} f(x, y) = (x - 1)^2 + (y - 2)^2 \qquad (7.47)$$

subject to:

$$x + y \leq 0$$
$$x \geq 0$$
$$y \geq 0$$

(i) For the QP problem in Equation (7.47) enumerate all combinations of active constraints, including the unconstrained case (no constraints active).

(ii) From part (i), find all feasible constraint combination solutions and identify the n=minimum solution.

(iii) For the minimum solution found in part (ii), confirm that the KKT necessary conditions of optimality are satisfied.

(iv) Is the minimum point found in part(ii) the global optimum? Discuss.

(v) Give a geometrical interpretation of your result in part (iv).

8

Penalty and Barrier Function Methods

The key characteristic of these methods is that they "absorb" the equality and inequality constraints into a penalizing term that augments the original objective function, along with weighting parameters, to enforce increasing satisfaction of the constraints involved.

8.1 PENALTY FUNCTIONS

Consider the following optimization problem:

$$\min_x f(x) \tag{8.1}$$

subject to:

$$h(x) = 0 \tag{8.2}$$
$$g(x) \leq 0 \tag{8.3}$$

The satisfaction of the constraints may be absorbed into an *absolute value penalty* term, as follows:

$$\Phi(x) = \sum_i |h_i| + \sum_i \max\{0, g_i\} \tag{8.4}$$

or into a *quadratic penalty* term, as follows:

$$\Phi(x) = \sum_i |h_i|^2 + \sum_i \max^2\{0, g_i\} \tag{8.5}$$

Consider the penalization of the inequality constraints and use the following definitions:

$$g_+(x) = \max\{0, g_i\} \tag{8.6}$$

$$g_+^2 = (g_+)^2 \tag{8.7}$$

Figure 8.1 depicts the properties of the two penalty terms, the second of which is smoother.

Generally, with any of these penalty terms the original problem can be transformed into a sequence of unconstrained minimization problems according to:

$$\min_x P(x; M) = f(x) + M \cdot \Phi(x) \tag{8.8}$$

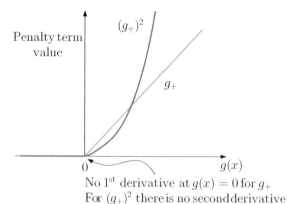

Figure 8.1 Plots of the two penalty functions g_+ and $(g_+)^2$ shown above.

Usually, one starts with a small value of M, the solution is found for the minimization problem, $x^*(M)$, and then using that as a new starting point, M is increased and so on.

These methodologies are more of theoretical value than numerical, *e.g.* used to measure constraint satisfaction during linesearch for generally constrained NLP problems (usually augmenting the Lagrangian with some penalty form to take constraints into account during the linesearch; the resulting function is known as the *merit function*).

8.1.1 Example: Quadratic Penalty for Equality Constraints

Consider the following optimization problem:

$$\min_{x,y} 2x^2 + y^2 \tag{8.9}$$

subject to:

$$x + \frac{1}{2}y^2 = 1 \tag{8.10}$$

According to the quadratic penalty scheme, we form a composite objective, which we then minimize:

$$\min_{x,y} 2x^2 + y^2 + M\left(x + \frac{1}{2}y^2 - 1\right)^2 \tag{8.11}$$

We first start with a small value of M and increase gradually until the constraint is sufficiently satisfied at the minimum obtained. Pictorially, the evolution of the optimization problem given in Equation (8.11) for the evolution of M from the value of 0.0 to 1000 is given in Figure 8.2 (read left to right, top to bottom). The last frame shows the overall evolution of the trajectory of minimizers for the subproblems.

8.1.2 Example: Quadratic Penalty for Inequality Constraints

Consider the following optimization problem:

$$\min_{x,y} x^2 + 2y^2 \tag{8.12}$$

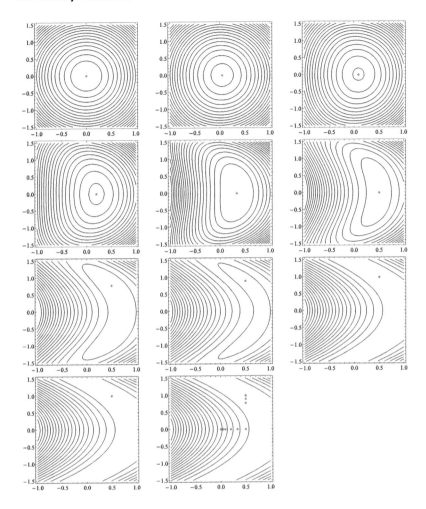

Figure 8.2 Solution trajectory with increasing values of M for equality constraints.

subject to:

$$-x + 3y + 2 \leq 0 \tag{8.13}$$

According to the quadratic penalty scheme, we form a composite objective, which we then minimize:

$$\min_{x,y} x^2 + 2y^2 + M \left(\max\{-x + 3y + 2, 0\}\right)^2 \tag{8.14}$$

We first start with a small value of M and increase gradually until the constraint is sufficiently satisfied at the minimum obtained. Pictorially, the evolution of the optimization problem given in Equation (8.14) for the evolution of M from the value of 0.0 to 1000 is given in Figure 8.3 (read left to right, top to bottom). The last frame shows the overall evolution of the trajectory of minimizers for the subproblems.

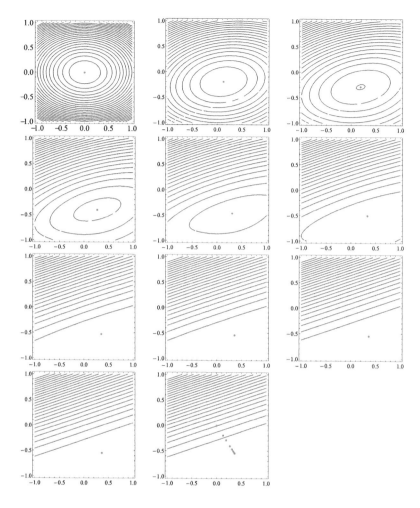

Figure 8.3 Solution trajectory with increasing values of M for the quadratic penalty.

8.2 BARRIER METHODS: LOGARITHMIC BARRIERS

For the optimization problem:

$$\min_{x} f(x) \tag{8.15}$$

subject to:

$$g(x) \leq 0 \tag{8.16}$$

Absorb the inequalities $g(x) \leq 0$ into logarithmic terms as follows:[1]

$$\min_{x} B(x; \mu) = f(x) - \mu \cdot \sum_{i=1}^{n_g} \ln\left[-g_i(x)\right] \tag{8.17}$$

---

[1] It is typical to see use of the $\log(\cdot)$ function implying the use of natural logarithms; here we explicitly use the $\ln(\cdot)$ function.

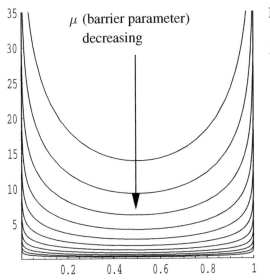

Barrier function plotted:

$$B(x, \mu) = -\mu(\ln(1 - x) + \ln(x))$$

Figure 8.4 Logarithmic barrier function with varying values of μ.

The following property holds:

$$x^*(\mu) \to x^* \text{ as } \mu \to 0 \tag{8.18}$$

The barrier terms for the bounds $x \geq 0$ and $x \leq 1$ are shown for different values of the barrier parameter μ in Figure 8.4.

The resulting algorithms are also known as *interior point methods* since they maintain always that:

$$-g(x) > 0 \tag{8.19}$$

Such algorithms are implemented similarly to penalty methods. One starts with a sufficiently large value for μ, and the corresponding unconstrained minimizer $x^*(\mu)$ is found. Then μ is *reduced* and the previous minimizer found is used as the starting point for the next unconstrained minimization. Termination is monitored by using measures on the KKT conditions of optimality (*e.g.* measure feasibility of constraints, complementarity conditions, search direction norm, progress on objective function).

8.2.1 Example of Logarithmic Barrier Method

Check to see if we can easily change the example. Consider the following optimization problem:

$$\min_{x,y} -5x + y \tag{8.20}$$

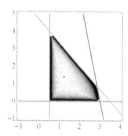

Figure 8.5 Contour plot of the feasible region with the barrier function, given in Equation (8.25).

subject to:

$$4x + 3y - 14 \leq 0 \tag{8.21}$$
$$6x + y - 18 \leq 0 \tag{8.22}$$
$$x \geq 0.5 \tag{8.23}$$
$$y \geq 0 \tag{8.24}$$

The resulting logarithmic barrier reformulation of the problem is:

$$\min_{x,y} -5x + y - \mu \Big[$$
$$\ln(-4x - 3y + 14)$$
$$+ \ln(-6x - y + 18)$$
$$+ \ln(x - 0.5)$$
$$+ \ln(y)$$
$$\Big] \tag{8.25}$$

The feasible region along with the contours of the barrier function are shown in Figure 8.5 (for a very large value of μ; reminder: the solution is approached as μ becomes very small).

Pictorially, the evolution of the optimization problem given in Equation (8.25) for the evolution of μ from the value of 100 divided by 2 each time is given in Figure 8.6 (read left to right, top to bottom). The overall evolution of the trajectory of minimizers for the subproblems is shown incrementally in the pictures.

8.3 PENALTY-MULTIPLIER METHOD (AUGMENTED LAGRANGIAN METHOD)

This section presents a special formulation of the penalty approach, according to which a penalty term is appended to the Lagrangian function of the NLP problem [3]. The resulting function is called the *augmented Lagrangian* and this in turn is minimized for the primal variables x. The method also proceeds via a step that updates Lagrange multiplier estimates at each major (outer) iteration.

Consider the equality constrained problem:

$$\min_{x} f(x) \tag{8.26}$$

subject to:

$$h_i(x) = 0, \quad i = 1, 2, \ldots, n_h, \, x \in \mathbb{R}^n \tag{8.27}$$

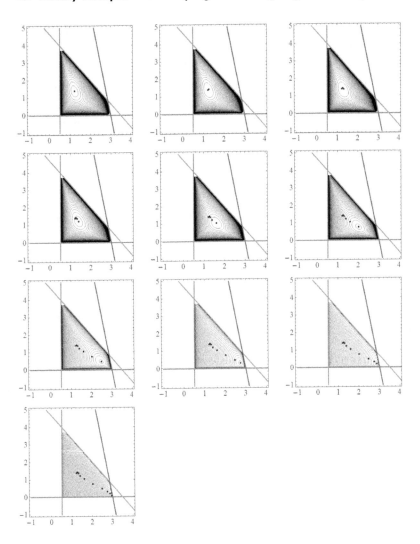

Figure 8.6 Evolution of optimal solution found with change in barrier function.

The problem is converted into an unconstrained minimization problem by the following transformed Lagrangian function:

$$\hat{L}(x;\lambda;M) = \underbrace{f(x) + \sum_{i=1}^{n_h} \lambda_i \cdot h_i(x)}_{\text{Lagrangian function}} + \underbrace{\frac{1}{2} \cdot M \cdot \sum_{i=1}^{n_h} h_i^2(x)}_{\text{quadratic penalty term}} \qquad (8.28)$$

The Augmented Lagrangian method is described by Algorithm 8.1.

Algorithm 8.1 *Augmented Lagrangian method*

1. *Initialize the values of λ and x;*
2. *Select a starting value for M;*
3. *Minimize the function $\hat{L}(x;\lambda;M)$ with respect to x;*

4. *Update the Lagrange multiplier estimates using the current minimizer obtained;*
5. *Check for convergence, if so terminate here, else do step 6;*
6. *Increase (potentially) the value of the penalty parameter M, go to step 3.*

The algorithm mentions in step 4 an update for the Lagrange multipliers. How is this to be achieved? The following derivation shows us how.

First, take the derivative (gradient) of the augmented Lagrangian with respect to the primal variables x:

$$\frac{\partial \hat{L}}{\partial x} = \frac{\partial f}{\partial x} + \sum_{i=1}^{n_h} (\lambda_i + M \cdot h_i(x)) \cdot \frac{\partial h_i}{\partial x} \tag{8.29}$$

This is followed by taking the gradient of the actual Lagrangian with respect to the primal variables x:

$$\frac{\partial L}{\partial x} = \frac{\partial f}{\partial x} + \sum_{i=1}^{n_h} \lambda_i \cdot \frac{\partial h_i}{\partial x} \tag{8.30}$$

Comparing the two, we can write the following updating scheme for the Lagrange multiplier estimates, following a minimization at the k-th iteration:

$$\lambda_i^{(k+1)} = \lambda_i^{(k)} + M \cdot h_i(x^{(k)*}), \quad i = 1, 2, \ldots, n_h \tag{8.31}$$

The method has the following interesting properties:

1. If the penalty parameter M is sufficiently large, then the function $\hat{L}$ has a local minimizer x^* (the same as of the original problem)
2. There exists a lower value $\underline{M}$ such that $x(\lambda^*; M) = x^*$ for all $M \geq \underline{M}$.
3. The sequence of $(x^{(k)*}, \lambda^{(k)}) \rightarrow (x^*, \lambda^*)$

For inequality constraints of the form $c_i(x) \leq 0$, the method is modified as for other penalty methods. Here we consider applying the same scheme as for equalities, by adding in a slack variable for each inequality constraint so that we have:

$$c_i(x) + s_i = 0, \quad i = 1, 2, \ldots, n_c \tag{8.32}$$

$$s_i \geq 0 \tag{8.33}$$

This leads to the following augmented Lagrangian (considering only the inequalities here):

$$\hat{L}(x; \mu; M) = f(x) + \sum_{i=1}^{n_c} \mu_i \cdot (c_i(x) + s_i) + \frac{1}{2} \cdot M \cdot \sum_{i=1}^{n_c} (c(x) + s_i)^2 \tag{8.34}$$

By considering minimization for s alone (treating x as fixed), it can be obtained that the augmented Lagrangian now becomes:

$$\hat{L}(x; \mu; M) = f(x) + \sum_{i=1}^{n_c} \phi(c_i(x); \mu_i; M) \tag{8.35}$$

where the function $\phi(\cdot; \cdot; \cdot)$ is defined by:

$$\phi(c_i(x); \mu_i; M) = \begin{cases} 0, & \text{if } c_i(x) \leq -\dfrac{\mu_i}{M} \\[3mm] \mu_i c_i(x) + \dfrac{1}{2} M c_i^2(x), & \text{if } c_i(x) > -\dfrac{\mu_i}{M} \end{cases} \tag{8.36}$$

The Lagrange multipliers μ_i, following a minimization at iteration k, are updated according to:

$$\mu^{(k+1)} = \begin{cases} 0, & \text{if } c_i(x^{(k)*}) \leq -\dfrac{\mu_i^{(k)}}{M} \\[3mm] \mu_i^{(k)} + M c_i(x), & \text{if } c_i(x^{(k)*}) > -\dfrac{\mu_i^{(k)}}{M} \end{cases} \tag{8.37}$$

The derivation of the case for inequalities is given in the end-of-chapter exercises.

8.4 FURTHER READING RECOMMENDATIONS

General Background on Penalty Methods and Logarithmic Barrier Methods

Edgar, T. F., Himmelblau, D. M., and Lasdon, L. S. (2001). Optimization of Chemical Processes. McGraw-Hill.

Luenberger, D. G. and Ye, Y. (2016). Linear and Nonlinear Programming. Springer.

8.5 EXERCISES

1. Solve the following problem using the augmented Lagrangian method:

$$\min_x f(x) = (x_1 - 1)^2 + (x_2 - 2)^2$$

subject to:

$$(x_1 - 1)^2 = 5x_2$$

2. Solve the following problem by using the penalty method and the augmented Lagrangian method:

$$\min_x f(x) = x_1^2 + x_2^2 - 6x_1 - 4x_2$$

subject to:

$$x_1 + x_2 \leq 3$$
$$2x_1 - x_2 \leq 5$$
$$x_1 \geq 0$$
$$x_2 \geq 0$$

Is the above problem guaranteed to have a global minimum?

3. Consider the optimization of an inequality constrained problem as follows:

$$\min_x f(x) \tag{8.38}$$

subject to:

$$c(x) \geq 0$$

where $x \in \mathbb{R}^n$, and $c(x) \in \mathbb{R}^m$.

(i) Writing the Lagrangian function as given in Equation (8.39), show that the Lagrange multipliers, $\lambda \in \mathbb{R}^m$, must be non-negative. Construct your proof by first considering perturbations of an equality constrained problem.

$$L(x, \lambda) = f(x) - \sum_{i=1}^{m} \lambda_i c_i(x) \tag{8.39}$$

(ii) The solution of the optimization problem in Equation (8.38) is approximated iteratively by solving in the kth iteration the modified barrier subproblem:

$$\min_x M\left(x, \lambda^{(k)}, \mu^{(k)}\right) = f(x) - \mu^{(k)} \sum_{i=1}^{m} \lambda_i^{(k)} \ln\left(1 + \frac{c_i(x)}{\mu^{(k)}}\right) \tag{8.40}$$

where $\lambda^{(k)}$ is an estimation of the Lagrange multiplier, μ is a positive barrier parameter, such that $\mu^{(k+1)} < \mu^{(k)}$, and $k = 1, 2, \ldots$ is the iteration counter. The optimal solution of the kth subproblem, $x^{(k),*}$, is used to update the Lagrange multiplier estimates, $\lambda^{(k+1)}$, for the next iteration. Derive this update by comparing the gradient with respect to x of Equation (8.39) with that of Equation (8.40) at the optimal point of iteration k.

9

Interior Point Methods (IPM's)

A Detailed Analysis

Interior point methods have in recent years become the method of choice for the solution of both Linear Programming (LP) problems and Nonlinear Programming (NLP) problems [1].

Originally, the use of logarithmic barriers was recognized and analyzed theoretically in the 1950s and 1960s [2], but the properties of the methods to converge in very few iterations efficiently were established much later, in the 1980s onwards. The first practical, large-scale applications for LP emerged in the 1980s, while a decade later in the 1990s applications emerged for general convex problems and, subsequently, nonlinear problems as well.

The material presented in this section uses knowledge from:

1. Lagrange multiplier theory and Lagrangians,
2. Numerical Analysis, Newton's method for nonlinear systems, and
3. Linear Algebra

9.1 NLP FORMULATIONS AND LAGRANGIANS

We have seen that the general optimization NLP problem is given by the following formulation:

$$\min_{x} f(x) \tag{9.1a}$$

subject to:

$$h(x) = 0 \tag{9.1b}$$
$$g(x) \leq 0 \tag{9.1c}$$

We have defined the Lagrange multipliers in such a way that the Lagrangian is formed by the addition of equalities, and the $\leq$ inequalities times their (positive) Lagrange multipliers:

$$L(x, \lambda, v) = f(x) + \lambda^T h(x) + v^T g(x) \tag{9.2}$$

For this, we recall that Lagrange multipliers of equalities, λ_i, are sign-free, and those of the inequalities are positive, $v_i \geq 0$.

What happens if we have $g(x) \geq 0$ inequalities?

- The obvious answer is to include them with a sign change, so that $-g(x) \leq 0$.

Thus we obtain the following Lagrangian:

$$L(x, \lambda, v) = f(x) + \lambda^T h(x) - v^T g(x) \tag{9.3}$$

We now have opposite sign definitions of the Lagrange multipliers with respect to the corresponding inequality constraints:

$$\lambda_i = -\left.\frac{\partial f}{\partial h_i}\right|_{x^*} \tag{9.4a}$$

$$v_i = \left.\frac{\partial f}{\partial g_i}\right|_{x^*} \tag{9.4b}$$

It is noted that these lead to different sign definitions for the sensitivity derivatives from what we saw in Chapter 7 on Lagrange multipliers. Please compare and understand this.

We have used the new definition of Lagrange multipliers and corresponding Lagrangians so that we derive special properties associated with the derivations to follow.

9.2 LOGARITHMIC BARRIER FUNCTIONS FOR INEQUALITY CONSTRAINED NLPS

Consider an optimization problem involving only inequality constraints:

$$\min_x f(x) \tag{9.5a}$$

subject to:

$$g_i(x) \geq 0, \quad i = 1, 2, \ldots, m \tag{9.5b}$$

We can absorb the inequality constraints into a logarithmic barrier function [3], as outlined in the previous chapter, to yield:

$$B(x; \mu) = f(x) - \mu \sum_{i=1}^{m} \ln(g_i(x)) \tag{9.6}$$

To solve the original problem given in Equations (9.5a) and (9.5b), using Equation (9.6) we proceed as follows:

1. The resulting problem is to solve the *unconstrained minimization subproblem*, given by $\min_x B(x; \mu)$, to obtain the solution $x^*(\mu)$ corresponding to the current value of the barrier parameter μ.
2. We would then reduce μ by some appropriate factor and continue from step (1), starting from the last point $x^*(\mu)$ found in the previous iteration. We continue like this until some tolerance of the KKT conditions is satisfied.

Clearly, in these two steps, the unconstrained minimization of the barrier function requires the use of Newton's method. For this, we need to calculate the gradient and the Hessian of the barrier function given in Equation (9.6), so that we can compute a search direction.

Gradient Calculation of Equation (9.6):

Show that the gradient of the barrier function is given by:

$$\nabla_x B(x;\mu) = \nabla f(x) - \mu \sum_{i=1}^{m} \frac{1}{g_i(x)} \nabla_x g_i(x) \tag{9.7}$$

Use this result to obtain an estimate of the Lagrange multipliers of $g_i(x)$ at a solution $x^*(\mu)$ of the barrier function.

Proof: Gradient Calculation of $B(x;\mu)$

$$\frac{\partial B}{\partial x}(x;\mu) = \frac{\partial f}{\partial x} - \mu \sum_{i=1}^{m} \frac{\partial}{\partial x} \left(\ln(g_i(x)) \right) \tag{9.8}$$

By the chain rule:

$$\frac{\partial}{\partial x} \left(\ln(g_i(x)) \right) = \frac{\partial}{\partial(g_i(x))} \ln(g_i(x)) \cdot \frac{\partial g_i(x)}{\partial x}$$

$$= \frac{1}{g_i(x)} \cdot \frac{\partial g_i(x)}{\partial x} \tag{9.9}$$

Putting these equations together and converting to column representations of gradient vectors $\left(\frac{\partial}{\partial x} = \nabla_x^T \cdot \right)$ we obtain:

$$\nabla_x B(x;\mu) = \nabla_x f(x) - \mu \sum_{i=1}^{m} \frac{1}{g_i(x)} \cdot \frac{\partial g_i(x)}{\partial x} \tag{9.10}$$

After a solution $x^*(\mu)$ is obtained, can we get an estimate of the Lagrange multipliers v_i for each $g_i(x^*(\mu))$?
Yes: Write the Lagrangian as:

$$L(x,v) = f(x) - \sum_{i=1}^{m} v_i\, g_i(x) \tag{9.11}$$

Compare its gradient to Equation (9.10):

$$\nabla_x L(x,v) = \nabla_x f(x) - \sum_{i=1}^{m} v_i\, \nabla_x g_i(x) \tag{9.12}$$

Thus:

$$v_i \approx \frac{\mu}{g_i(x^*(\mu))} \tag{9.13}$$

Hessian Calculation of Equation (9.6):

Show that the Hessian of the barrier function is given by:

$$\nabla_{xx}^2 B(x;\mu) = \nabla_{xx}^2 f(x) - \mu \sum_{i=1}^{m} \frac{1}{g_i(x)} \nabla_{xx}^2 g_i(x) + \mu \sum_{i=1}^{m} \frac{1}{(g_i(x))^2} \nabla_x g_i(x) \nabla_x g_i(x)^T$$

$$\tag{9.14}$$

1. Which is the Newton iteration solving the barrier optimization subproblem for fixed μ?
2. What is the limitation that must be imposed on the step size of the search direction for the iterations to work?

Proof: Hessian Calculation of $B(x; \mu)$

Start by differentiating the gradient found in Equation (9.13) once more by $\frac{\partial \cdot}{\partial x}$:

$$\nabla_{xx}^2 B(x; \mu) = \frac{\partial}{\partial x}(\nabla_x B(x; \mu))$$

$$= \underbrace{\frac{\partial}{\partial x}(\nabla_x f(x))}_{\nabla_{xx} f(x)} - \mu \sum_{i=1}^{m} \frac{\partial}{\partial x} \left(\underbrace{\frac{1}{g_i(x)}}_{\text{scalar}} \underbrace{\nabla_x g_i(x)}_{\text{vector}} \right) \qquad (9.15)$$

We have to apply the differential operator to a product form on the right-hand side. Remember we apply it to terms appearing on the right of such products, and in our case we can change the order of terms since $1/g_i(x)$ is scalar. So:

$$\frac{\partial}{\partial x}\left(\frac{1}{g_i(x)} \cdot \nabla_x g_i(x)\right) = \nabla_x g_i(x) \cdot \frac{\partial}{\partial x}\left(\frac{1}{g_i(x)}\right)$$

$$+ \frac{1}{g_i(x)} \frac{\partial}{\partial x}(\nabla_x g_i(x))$$

$$= \nabla_x g_i(x) \cdot \frac{\partial}{\partial g_i(x)}\left(\frac{1}{g_i(x)}\right)\frac{\partial g_i(x)}{\partial x}$$

$$+ \frac{1}{g_i(x)} \cdot \nabla_{xx}^2 g_i(x)$$

$$= -\frac{1}{g_i^2(x)} \cdot \nabla g_i(x) \cdot \nabla_x^T g_i(x)$$

$$+ \frac{1}{g_i(x)} \cdot \nabla_{xx}^2 g_i(x) \qquad (9.16)$$

Take this result into Equation (9.15) to obtain:

$$\nabla_{xx}^2 B(x; \mu) = \nabla_{xx}^2 f(x) - \mu \sum_{i=1}^{m} \frac{1}{g_i(x)} \cdot \nabla_{xx}^2 g_i(x)$$

$$+ \mu \sum_{i=1}^{m} \frac{1}{g_i^2(x)} \cdot \underbrace{\nabla g_i(x)}_{n \times 1} \cdot \underbrace{\nabla_x^T g_i(x)}_{1 \times n} \qquad (9.17)$$

$$\underbrace{}_{n \times n \rightarrow \text{matrix}}$$

Note: given, say, $\nabla_x g = \begin{bmatrix} g_{x_1} \\ g_{x_2} \end{bmatrix}$, $\nabla_x^T g = \begin{bmatrix} g_{x_1}, g_{x_2} \end{bmatrix}$ they produce:

$$\begin{bmatrix} g_{x_1} \\ g_{x_2} \end{bmatrix}\begin{bmatrix} g_{x_1}, g_{x_2} \end{bmatrix} = \begin{bmatrix} g_{x_1} g_{x_1} & g_{x_2} g_{x_1} \\ g_{x_1} g_{x_2} & g_{x_2} g_{x_2} \end{bmatrix} \qquad (9.18)$$

i.e. a square matrix.

The Newton equations to find a search direction p_x are:

$$\nabla_{xx}^2 B(x;\mu) \cdot p_x = -\nabla_x B(x;\mu) \tag{9.19}$$

Solving this linear system yields the search direction p_x. So:

$$x^{new} = x^{old} + \alpha \cdot p_x \tag{9.20}$$

The choice of α is crucial as we must produce x^{new} such that not only $B(x^{new},\mu)$ is reduced, but we also maintain feasibility:

$$g_i\left(x^{new}\right) > 0, \quad i = 1,2,\ldots,m \tag{9.21}$$

Also the starting point must satisfy:

$$g_i\left(x^{(0)}\right) > 0, \quad i = 1,2,\ldots,m \tag{9.22}$$

So all iterations remain feasible.

- In the case of linear constraints, feasibility is easily imposed and maintained
- In the case $g_i(x)$ are nonlinear then the primal approach is very difficult..

9.3 REFORMULATING NLP PROBLEMS INTO CANONICAL FORM

Often, in defining numerical solution procedures for optimization problems we seek to define a simpler form of NLP, which nonetheless still captures the generality of applications.

Consider the following NLP problem:

$$\min_x f(x) \tag{9.23a}$$

subject to:

$$h(x) = 0 \tag{9.23b}$$

$$x \geq 0 \tag{9.23c}$$

It is easy to show that any form of NLP, such as that in Equations (9.1a–9.1c), can be cast into the above canonical form. For this, we add deficit variables, x_d, reformulating the inequalities into the following set of constraints:

$$g(x) + x_d = 0 \tag{9.24a}$$

$$x_d \geq 0 \tag{9.24b}$$

Similarly, inequalities of the form $g(x) \geq 0$ can be reformulated with surplus variables, x_s, such that we obtain:

$$g(x) - x_s = 0 \tag{9.25a}$$

$$x_s \geq 0 \tag{9.25b}$$

We conclude that any given NLP can be cast into the canonical form of Equations (9.23a–9.23c), without any loss of generality.

The reason we are interested in this form is to facilitate the exposition of the reformulation of NLP problems in such a way that derivations are not overloaded with too many different symbols and compartments.

9.4 NEWTON'S METHOD FOR NLP WITH EQUALITY CONSTRAINTS ONLY

Consider an optimization problem involving only equality constraints, such that $x \in \mathbb{R}^n$ and $h(x) \in \mathbb{R}^m$:

$$\min_x f(x) \tag{9.26a}$$

subject to:

$$h(x) = 0 \tag{9.26b}$$

The Lagrangian for this problem is:

$$L(x, \lambda) = f(x) + \lambda^T h(x) \tag{9.27}$$

The necessary conditions of optimality are:

$$\frac{\partial L}{\partial x} = \frac{\partial f}{\partial x} + \lambda^T \frac{\partial h}{\partial x} = 0_{(1 \times n)} \tag{9.28a}$$

$$\frac{\partial L}{\partial \lambda} = h(x)^T = 0_{(1 \times m)} \tag{9.28b}$$

It is noted that the Jacobian of the equality constraints, appearing in the first equation, can be written as $\frac{\partial h}{\partial x} = J(x)$. Furthermore, we note that the necessary conditions are written in row-format, so we transpose to get the usual column-format of equations:

$$\nabla_x L(x, \lambda) = \nabla_x f(x) + J(x)^T \lambda$$

$$= \nabla_x f(x) + \sum_{i=1}^{m} \lambda_i \nabla_x h_i(x) = 0_{(n \times 1)} \tag{9.29a}$$

$$\nabla_\lambda L(x, \lambda) = h(x) = 0_{(m \times 1)} \tag{9.29b}$$

Equations (9.29a and 9.29b) comprise a nonlinear system of $(n + m)$ equations in an equal number of unknowns, the x and λ vectors combined.

To solve this system, we must derive its own Jacobian, with respect to the combined vector of variables $\begin{bmatrix} x \\ \lambda \end{bmatrix}$. The Jacobian will be a 4×4 block matrix and this is shown below[1]:

$$\begin{bmatrix} \nabla_{xx}^2 L(x, \lambda) & J(x)^T \\ J(x) & 0_{(m \times m)} \end{bmatrix} \Bigg|_{(x^{(0)}, \lambda^{(0)})} \begin{bmatrix} p_x \\ p_\lambda \end{bmatrix} = - \begin{bmatrix} \nabla_x L(x, \lambda) \\ h(x) \end{bmatrix} \Bigg|_{(x^{(0)}, \lambda^{(0)})} \tag{9.30}$$

The submatrix in the top left corner of the Jacobian is the Hessian of the Lagrangian with respect to the primal variables x:

$$\nabla_{xx}^2 L(x, \lambda) = \nabla_{xx}^2 f(x) + \sum_{i=1}^{m} \lambda_i \nabla_{xx}^2 h_i(x) \tag{9.31}$$

---

[1] Forms of the double derivative operators are: $\nabla_{xx}^2 (\cdot) \equiv \frac{\partial^2 (\cdot)}{\partial x^2}$

It is noted that the total system block Jacobian is a symmetric matrix. It is not positive definite, as it has a zero block at the lower right corner.

The vectors p_x and p_λ are the search directions derived for the primal variables x and dual variables λ, respectively. Using some appropriate linesearch scheme, which is beyond the scope of this book, a step-size α is found yielding the next set of iterates:

$$
\begin{bmatrix} x^{(1)} \\ \lambda^{(1)} \end{bmatrix} = \begin{bmatrix} x^{(0)} \\ \lambda^{(0)} \end{bmatrix} + \alpha \cdot \begin{bmatrix} p_x \\ p_\lambda \end{bmatrix} \tag{9.32}
$$

The step-size, α, is chosen so that the new point improves a combined measure of the objective function with the feasibility of the constraints.

9.5 LOGARITHMIC BARRIERS FOR NLPS WITH EQUALITIES AND BOUNDS

Here, we will turn our attention to the canonical form NLP we have defined in Equations (9.23a–9.23c), involving only equality constraints and nonnegativity bounds.

By introducing the logarithmic barrier function weighted by a barrier parameter μ, we obtain an equality-constrained problem, without inequalities (bounds):

$$
\min_x f(x) - \mu \sum_{i=1}^{n} \ln(x_i) \tag{9.33a}
$$

subject to:

$$
h(x) = 0 \tag{9.33b}
$$

Based on the properties of the barrier function method, we know that the solution of this problem will tend to the solution of the original bound-constrained problem as μ becomes very small:

$$
\lim_{\mu \to 0} x(\mu)^* = x^* \tag{9.34}
$$

A more detailed derivation with application to linesearch methods can be found in [4].

9.5.1 The Necessary Conditions of Optimality of the Barrier Problem

Based on the result for NLPs with equality constraints only, as discussed in the previous section, we can write down the necessary conditions of optimality for the problem we have here. First, we look at the corresponding Lagrangian for the problem:

$$
L(x, \lambda, v) = f(x) - \mu \sum_{i=1}^{n} \ln(x_i) + \lambda^T h(x) \tag{9.35}
$$

The necessary conditions, in column format, are thus:

$$\nabla_x L(x, \lambda) = \nabla_x f(x) - \mu X^{-1} e + J(x)^T \lambda \tag{9.36a}$$

$$= \nabla_x f(x) - \mu X^{-1} e + \sum_{i=1}^{m} \lambda_i \nabla_x h_i(x) = 0_{(n \times 1)}$$

$$\nabla_\lambda L(x, \lambda) = h(x) = 0_{(m \times 1)} \tag{9.36b}$$

where

$$X = \begin{pmatrix} x_1 & 0 & \cdots & 0 \\ 0 & x_2 & \cdots & 0 \\ \vdots & \vdots & \ddots & \vdots \\ 0 & 0 & \cdots & x_n \end{pmatrix}, \; X^{-1} = \begin{pmatrix} x_1^{-1} & 0 & \cdots & 0 \\ 0 & x_2^{-1} & \cdots & 0 \\ \vdots & \vdots & \ddots & \vdots \\ 0 & 0 & \cdots & x_n^{-1} \end{pmatrix} \tag{9.37}$$

and

$$e = \begin{pmatrix} 1 \\ 1 \\ \vdots \\ 1 \end{pmatrix} \tag{9.38}$$

with these, it can be seen that:

$$X^{-1} e = \begin{pmatrix} x_1^{-1} \\ x_2^{-1} \\ \vdots \\ x_n^{-1} \end{pmatrix} = \nabla_x \left(\sum_{i=1}^{n} \ln(x_i) \right) \tag{9.39}$$

It is noted that the only difference in the necessary conditions, with respect to those of the original problem, is in the gradient of the Lagrangian with respect to the primal variables, x, due to the presence of the logarithmic terms. It is noted that we demand that always $x > 0$ so that the inverse matrix X^{-1} is always defined.

We are next going to do some algebra on the necessary conditions to reveal the effect of the use of the logarithmic barrier function.

Introduce a new vector of variables, $v \in \mathbb{R}^n$, such that:

$$v = \mu X^{-1} e \tag{9.40}$$

Use that in the necessary conditions, Equations (9.36a–9.36b), and introduce the additional Equation (9.40) to get:

$$\nabla_x f(x) - v + J(x)^T \lambda = \nabla_x f(x) - v + \sum_{i=1}^{m} \lambda_i \nabla_x h_i(x) = 0_{(n \times 1)} \tag{9.41a}$$

$$\nabla_\lambda L(x, \lambda) = h(x) = 0_{(m \times 1)} \tag{9.41b}$$

$$v = \mu X^{-1} e \tag{9.41c}$$

We usually prefer to avoid having divisions in equalities if we can, but also to make equations neater, so we change the last equation, Equation (9.41c), to yield the following set:

$$\nabla_x f(x) - v + J(x)^T \lambda = \nabla_x f(x) - v + \sum_{i=1}^{m} \lambda_i \nabla_x h_i(x) = 0_{(n \times 1)} \tag{9.42a}$$

$$\nabla_\lambda L(x, \lambda) = h(x) = 0_{(m \times 1)} \tag{9.42b}$$

$$Xv = \mu e \tag{9.42c}$$

The last equation, Equation (9.42c), simply states that:

$$x_i v_i = \mu, \quad i = 1, 2, \ldots, n \tag{9.43}$$

Also do not forget that we have:

$$x_i > 0, \quad i = 1, 2, \ldots, n \tag{9.44}$$

which holds true for all iterations if we start from a feasible point (satisfying the bounds). Strict positivity of the primal variables x is necessary as we are using a logarithmic function, which for other values is undefined.

What does Equation (9.43) mean? What is its significance?

9.5.2 KKT Conditions of Optimality of the Canonical NLP Problem

We turn once more our attention to the canonical NLP problem we have defined in Equations (9.23a–9.23c). Let us write out its Lagrangian function:

$$L(x, \lambda, v) = f(x) - v^T x + \lambda^T h(x) \tag{9.45}$$

It can be seen that we have a contribution of the nonnegativity bounds and the equality constraints. For the bounds, being of the $\geq$ form inequalities, we adopt the subtraction of their value times the corresponding Lagrange multipliers, v.

Let us derive next the KKT conditions of optimality for this problem.

1. Lagrangian gradient with respect to the primal variables x:

$$\nabla_x f(x) - v + J(x)^T \lambda = \nabla_x f(x) - v + \sum_{i=1}^{m} \lambda_i \nabla_x h_i(x) = 0_{(n \times 1)} \tag{9.46a}$$

It is noted that this is identical to the Lagrangian gradient in Equation (9.42a), derived for the barrier problem.

2. Feasibility constraints:

$$h(x) = 0 \tag{9.46b}$$

$$x_i \geq 0, \quad i = 1, 2, \ldots, n \tag{9.46c}$$

Condition (9.46b) is identical to condition (9.42b). The nonnegativity condition in Equation (9.46c) is satisfied automatically by the barrier problem, as it is ensured in each iteration.

3. Complementarity constraints:

$$x_i v_i = 0, \quad i = 1, 2, \ldots, n \tag{9.46d}$$

It is noted that strict complementarity conditions cannot be solved directly for any NLP with inequalities, as the overall system of the KKT conditions has a singular Jacobian because of the zero right-hand side.

If we compare Equation (9.46d), with the corresponding condition in Equation (9.42c) in the barrier problem, we begin to see what the primal-dual barrier problem is doing: It relaxes the complementarity condition by μ in such a way that the bound inequality multipliers v can be determined.

The μ-relaxed complementarity conditions yield an invertible system, with a non-singular Jacobian.

4. Nonnegativity of the inequality (bounds) Lagrange multipliers:

$$v_i \geq 0, \quad i = 1, 2, \ldots, n \tag{9.46e}$$

It is noted that if $x_i > 0$, then all the bound multipliers will also be $v_i > 0$ in the barrier problem, which is the last KKT condition here.

9.5.3 The Newton Equations of the Barrier Problem

Here we will derive the Newton equations of the system defined by the necessary conditions of optimality, Equations (9.42a–9.42c), of the barrier-based reformulation of the canonical NLP problem, defined in Equations (9.33a–9.33b).

The variables appearing in the system are collected together as $(x, \lambda, v)^T$ and we differentiate the necessary conditions of optimality to derive:

$$\begin{bmatrix} \nabla_{xx}^2 L & J^T & -I \\ J & 0 & 0 \\ N & 0 & X \end{bmatrix}_{(0)} \cdot \begin{bmatrix} p_x \\ p_\lambda \\ p_v \end{bmatrix} = - \begin{bmatrix} \nabla_x f - v + J^T \lambda \\ h \\ XNe - \mu e \end{bmatrix}_{(0)} \tag{9.47}$$

The Lagrangian appearing in the left-hand side matrix of Equation (9.47) is the Lagrangian of the original canonical NLP.

The Hessian appearing in the left-hand side of the Equation (9.47), $\nabla_{xx}^2 L$, is the Hessian of the original NLP problem (with equality constraints and nonnegativity bounds only).

It is noted that the μ-relaxed complementarity conditions have been written in completely equivalent form as:

$$XNe = \mu e \tag{9.48}$$

where

$$N = \begin{bmatrix} v_1 & 0 & \cdots & 0 \\ 0 & v_2 & \cdots & 0 \\ \vdots & \vdots & \ddots & \vdots \\ 0 & 0 & \cdots & v_n \end{bmatrix} \tag{9.49}$$

Aside: The system in Equation (9.47) is non-symmetrical. So instead of solving it this way (there are special solvers for symmetric systems, exploiting sparsity), we will eliminate the last block row as follows [4]:

1. Write out the last row in matrix form (identifying the blocks from the overall system unsymmetric Jacobian):

$$N^{(0)} p_x + X^{(0)} p_v = -(X^{(0)} N^{(0)} e - \mu e) \tag{9.50}$$

Solve Equation (9.50) to find the part of the search direction with respect to the multipliers ν as a function of everything else:

$$p_\nu = -\left(X^{(0)}\right)^{-1} N^{(0)} p_x - \nu^{(0)} + \mu \left(X^{(0)}\right)^{-1} e \qquad (9.51)$$

2. Write out the first row in matrix form (identifying the blocks from the overall system non-symmetric Jacobian):

$$\nabla_{xx}^2 L^{(0)} p_x + \left(J^{(0)}\right)^T p_\lambda + \left(X^{(0)}\right)^{-1} N^{(0)} p_x + p_\nu$$
$$= -\left(\nabla f^{(0)} + \left(J^{(0)}\right)^T \lambda^{(0)} - \nu^{(0)}\right) \qquad (9.52)$$

3. Substitute into Equation (9.52) the result of Equation (9.51) to get, finally:

$$\left(\nabla_{xx}^2 L^{(0)} + \left(X^{(0)}\right)^{-1} N^{(0)}\right) p_x + \left(J^{(0)}\right)^T p_\lambda$$
$$= -\left(\nabla f^{(0)} - \mu \left(X^{(0)}\right)^{-1} e + \left(J^{(0)}\right)^T \lambda^{(0)} - \nu^{(0)}\right) \qquad (9.53)$$

Row 2 of the block Jacobian remains unchanged and we can thus replace row 1 with the result in Equation (9.53) to obtain:

$$\left[\begin{array}{cc} (\nabla_{xx}^2 L + X^{-1} N) & J^T \\ J & 0 \end{array} \right]_{(0)} \cdot \left[\begin{array}{c} p_x \\ p_\lambda \end{array} \right] = - \left[\begin{array}{c} \nabla_x f - \mu X^{-1} e + J^T \lambda \\ h \end{array} \right]_{(0)} \qquad (9.54)$$

The above system is symmetric as required. We can solve this for p_x and p_λ and then by substitution into Equation (9.51), we can obtain p_ν.

The next iteration point is found by appropriately finding a step-size α using linesearch to determine:

$$\left[\begin{array}{c} x^{(1)} \\ \lambda^{(1)} \\ \nu^{(1)} \end{array} \right] = \left[\begin{array}{c} x^{(0)} \\ \lambda^{(0)} \\ \nu^{(0)} \end{array} \right] + \alpha \cdot \left[\begin{array}{c} p_x \\ p_\lambda \\ p_\nu \end{array} \right] \qquad (9.55)$$

Again, a suitable function weighing the contribution of the objective and the constraints must be used, but this is also outside the scope of this book. Such functions are called merit functions.

It is finally noted that the step-size α must be chosen between

$$0 < \alpha < \alpha^{\max} \qquad (9.56)$$

where $\alpha^{\max}$ is the distance of iterates at point (0) to the boundary (zero) of the bounds, along the directions p. This is to guarantee that both $x^{(1)}$ and $\nu^{(1)}$ remain strictly positive. This applies to all iterations.

9.6 SUMMARY OF INTERIOR POINT METHODS

Advantages of barrier methods:

1. They deal with inequalities all at once

    1.1 Each Newton iteration progresses on all inequalities simultaneously

2. For convex problems, barrier methods exhibit polynomial complexity

    2.1 *i.e.* the number of Newton iterations is of the order $\mathcal{O}(n^K)$ where n is a characteristic size of the problem (*e.g.* number of variables) and K is a small number

    2.2 Typical case: 20–30 Newton iterations to solve a problem

3. Primal-dual methods

    3.1 Deal with equality constraints (and inequalities by conversion to equalities) directly

    3.2 Allow nonlinear equalities (and thus converted inequalities) to be violated during iterations, and they are enforced only at the solution.

9.7 REFERENCES

[1] Nocedal, J. and Wright, S. J. Numerical Optimization. Springer. 2006.
[2] Fiacco, A. V. and McCormick, G. P. Nonlinear Programming: Sequential Unconstrained Minimization Techniques. Wiley. 1968.
[3] Luenberger, D. G. and Ye, Y. Linear and Nonlinear Programming. Springer. 2016.
[4] Wächter, A. and Biegler, L. T. "On the implementation of an interior-point filter line-search algorithm for large-scale nonlinear programming." Mathematical Programming. 2006;106:p. 25.

9.8 FURTHER READING RECOMMENDATIONS

Excellent Reference Books on Interior Point Methods

Nash, S. G. and Sofer, A. (1995). Linear and Nonlinear Programming. McGraw-Hill.
Boyd, S. and Vandenberghe, L. (2004). Convex Optimization. Cambridge University Press.

PART III

Formulation and Solution of Linear Programming (LP) Problems

10

Introduction to LP Models

The third part of this book is dedicated to the formulation and automated computation of LP problems. The focus on LP formulations is explained by the following practical observations:

1. Formulating LP problems is a very good point at which to start learning how to transform problem descriptions into mathematical programming (equation-based) form.
2. LP problems find wide applications in many enterprise sectors, *e.g.* industry, commerce, finance.
3. Most real-world models, even of very large scale, are not entirely nonlinear. In fact, large-scale problems contain large linear constraint sections.
4. The applications chosen in this Part of the book are of vital importance in many situations and will almost invariably, in one form or another, be part of applied optimization problems, even nonlinear ones.

10.1 GENERAL LP PROBLEM MODEL

Linear programming solves, under certain assumptions and conditions, the problem of allocating finite resources (*e.g.* workers, materials, machines, land) between alternative and competitive with each other activities (*e.g.* product manufacture, service provision) in an optimal way.

The level of each activity is to be chosen such as to optimize the result: *e.g.* profit maximization or cost minimization of all the activities simultaneously.

Finiteness of resources also implies that there are constraints on their levels, as well as constraints on the qualities of the products produced, and there may also be other organizational or managerial constraints, depending on the model.

A general LP formulation is as follows:

$$\min_{x} \text{ or } \max_{x} z = \sum_{j=1}^{n} c_j x_j$$

$$\text{subject to:} \sum_{j=1}^{n} a_{ij} x_j \ (\leq, =, \geq) \, b_i \, i = 1, 2, \ldots, m \qquad (10.1)$$

$$x_j^L \leq x_j \leq x_j^U \, j = 1, 2, \ldots, n$$

Alternatively, we may write the problem in vector format:

$$\min_{x} \text{ or } \max_{x} z = c^T x$$

$$\text{subject to: } \underbrace{a_i}_{\text{row } i} \cdot x \, (\leq, =, \geq) \, b_i \, i = 1, 2, \ldots, m \qquad (10.2)$$

$$x^L \leq x \leq x^U$$

Finally, we can write everything in vector-matrix notation:

$$\min_{x} \text{ or } \max_{x} z = c^T x$$

$$\text{subject to:} Ax \, (\leq, =, \geq) \, bi = 1, 2, \ldots, m \qquad (10.3)$$

$$x^L \leq x \leq x^U$$

where the matrix A is:

$$A = [a_{ij}] = \begin{bmatrix} a_1 \\ a_2 \\ \vdots \\ a_m \end{bmatrix}$$

with a_i indicating the rows of matrix A (in row-vector form).

Basic Theorem of Linear Programming

If the LP problem defined in Equations (10.1) or (10.2), or (10.3), has a non-empty and bounded feasible region (defined by the constraints), then it has a unique, globally optimal solution, which is located at some vertex (more correctly, extremal point; see below immediately) of the polyhedral feasible region.

The feasible region is classified as polyhedral, since it is the result of intersecting hyperplanes (each of the constraints) in $\mathbb{R}^n$. The solution will be at some extremal point (at least one) of this polyhedral region, which clearly will be defined by the simultaneous intersection of some of the constraint hyperplanes.

It is mentioned that there may be more than one extremal point, as it may be the case that the objective function (another hyperplane for fixed value of z) is parallel/coincides with one of the faces (edges in 2-D) of the polyhedral region.

10.2 GEOMETRICAL SOLUTION AND INTERPRETATION OF LP PROBLEMS

It is instructive to observe various cases of LP formulations in two dimensions (two decision variables), as the concepts of feasible regions, vertex solutions, *etc.* become very clear.

10.2.1 Geometrical Solution of 2-D Problem, Unique Solution

Consider the following LP optimization problem:

$$\max z = x_1 + 2x_2 \qquad (10.4a)$$

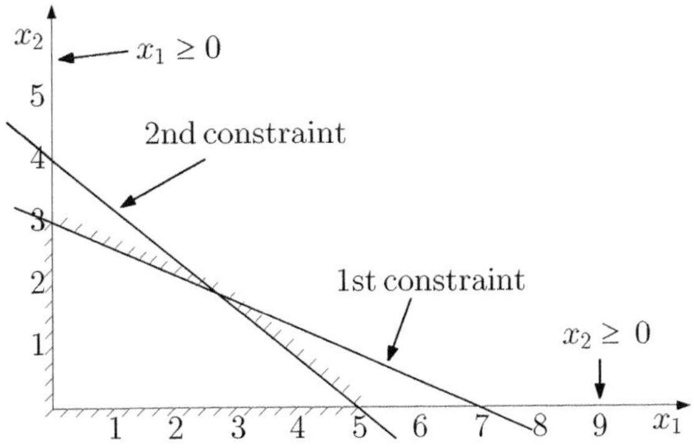

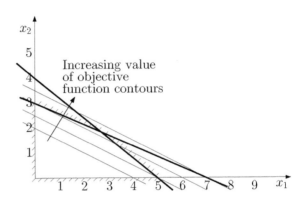

Figure 10.1 Feasible region of the LP problem given in Equation (10.4).

Figure 10.2 Feasible region with objective function for the LP problem given in Equation (10.4).

subject to:

$$3x_1 + 7x_2 \leq 21 \tag{10.4b}$$
$$4x_1 + 5x_2 \leq 20 \tag{10.4c}$$
$$x_1 \geq 0 \tag{10.4d}$$
$$x_2 \geq 0 \tag{10.4e}$$

The feasible region and contours of the objective are depicted in Figures 10.1 and 10.2.

By having recognized the active constraints, we can solve them simultaneously to obtain:

$$x_1^* = \frac{35}{13} = 2.69$$

$$x_2^* = \frac{24}{13} = 1.85$$

The objective function value at the optimal point is:

$$z^* = \frac{83}{13} = 6.38$$

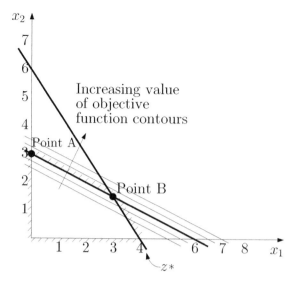

Figure 10.3 Feasible region and objective function for the LP problem in Equation (10.5).

10.2.2 Geometrical Solution in 2-D Continued, Infinity of Solutions

Consider the same LP optimization problem with a modified objective function this time:

$$\max z = x_1 + 2x_2 \tag{10.5a}$$

subject to:

$$3x_1 + 2x_2 \le 12 \tag{10.5b}$$
$$2x_1 + 4x_2 \le 12 \tag{10.5c}$$
$$x_1 \ge 0 \tag{10.5d}$$
$$x_2 \ge 0 \tag{10.5e}$$

The feasible region is the same as before, and the contour lines of the new objective are superimposed on it as shown in Figure 10.3.

It is clear that there are two vertices that maximize the objective function this time:

Vertex A: found by the intersection of the second inequality constraint with the bound $x_1 \ge 0$:

$$x_1^* = 0$$
$$x_2^* = 3$$

Vertex B: found by the intersection of the two inequality constraints:

$$x_1^* = 3$$
$$x_2^* = \frac{3}{2}$$

Both yield the same objective function value:

$$z^* = 6$$

10.2 Geometrical Solution and Interpretation of LP Problems

Any linear combination of the two solutions is also a feasible solution, which yields the same objective function. The linear combination of vertices A and B is given by:

$$x^* = \alpha \underbrace{\begin{bmatrix} 0 \\ 3 \end{bmatrix}}_{\text{vertex A}} + (1 - \alpha) \underbrace{\begin{bmatrix} 3 \\ \frac{3}{2} \end{bmatrix}}_{\text{vertex B}}$$

$$0 \leq \alpha \leq 1$$

10.2.3 Infinite Solution, Unbounded Region

Consider the following LP optimization problem:

$$\max z = 7x_1 + 6x_2 \tag{10.6a}$$

subject to:

$$-3x_1 + 8x_2 \leq 16 \tag{10.6b}$$

$$5x_1 + 7x_2 \geq 35 \tag{10.6c}$$

$$x_1 \geq 0 \tag{10.6d}$$

$$x_2 \geq 0 \tag{10.6e}$$

The feasible region and the objective contours are shown in Figure 10.4.

The objective function can increase to infinity, since the feasible region is unbounded above, and also the variable values tend to infinity as well in the same direction.

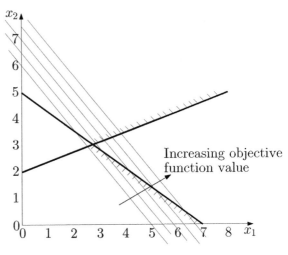

Figure 10.4 Feasible region for the LP problem shown in Equation (10.6).

Increasing objective function value

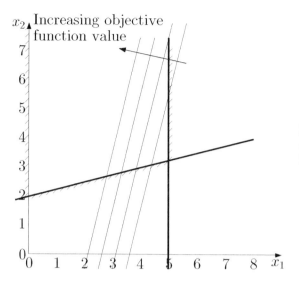

Figure 10.5 Feasible region and objective function for the LP problem in Equation (10.7).

10.2.4 Finite Solution (for Some Variable Values), Unbounded Region

Consider the following optimization problem:

$$\max z = -3x_1 + 2x_2 \tag{10.7a}$$

subject to:

$$x_1 \leq 5 \tag{10.7b}$$

$$-x_1 + 4x_2 \geq 8 \tag{10.7c}$$

$$x_1 \geq 0 \tag{10.7d}$$

$$x_2 \geq 0 \tag{10.7e}$$

The feasible region and objective contours are shown in Figure 10.5.

10.2.5 Finite Optimal Solution of Objective, with Infinite Variable Values

Consider the following LP formulation:

$$\max z = 3x_1 - 4x_2 \tag{10.8a}$$

subject to:

$$-x_1 + 4x_2 \leq 8 \tag{10.8b}$$

$$6x_1 - 8x_2 \geq \frac{4}{3} \tag{10.8c}$$

$$x_1 \geq 0 \tag{10.8d}$$

$$x_2 \geq 0 \tag{10.8e}$$

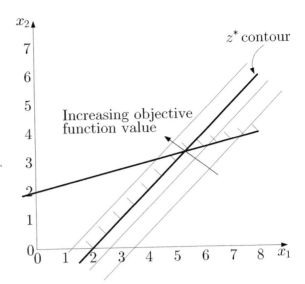

Figure 10.6 Feasible region and objective function for the LP problem shown in Equation (10.8).

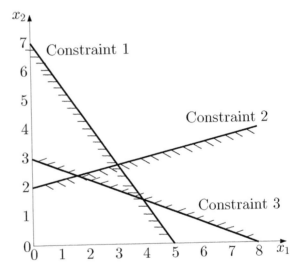

Figure 10.7 Feasible region for the infeasible LP problem shown in Equation (10.9).

The corresponding feasible region and objective function contours are plotted in Figure 10.6.

The objective function value has a *finite value* (why?), while there is no definite value for the variable values (why?)

10.2.6 No Feasible Solution: Inconsistent Constraints

The case of an empty feasible region is demonstrated by the following optimization problem:

$$\max z = 2x_1 + x_2 \tag{10.9a}$$

subject to:

$$7x_1 + 5x_2 \geq 35 \tag{10.9b}$$

$$x_1 + 4x_2 \geq 8 \tag{10.9c}$$

$$-3x_1 + 8x_2 \leq 24 \tag{10.9d}$$

$$x_1 \geq 0 \tag{10.9e}$$

$$x_2 \geq 0 \tag{10.9f}$$

The feasible region and the objective function contours are shown in Figure 10.7.

10.3 FURTHER READING RECOMMENDATIONS

More Detailed Discussion of Linear Programming

Nash, S. G. and Sofer, A. (1995). Linear and Nonlinear Programming. McGraw-Hill.
Vanderbei, R. J. (2014). Linear Programming. Springer.
Luenberger, D. G. and Ye, Y. (2016). Linear and Nonlinear Programming. Springer.
Taha, H. (2006). Operations Research. Prentice Hall.

11

Numerical Solution of LP Problems Using the Simplex Method

The simplex algorithm for linear programming was created in 1947 by the American mathematician George Dantzig. It is considered one of the top ten algorithms of the twentieth century [1].

Although nowadays Interior Point methods are very popular for the solution of LP problems, the simplex method is still used in several cases.

From an educational point of view, it offers a means to further the understanding of LP optimization and a simple way to solve small LP problems by hand.

11.1 INTRODUCTION

Here we will consider the solution of an LP defined as:

$$\min_x f = c^T x \tag{11.1a}$$

subject to:

$$Ax = b \tag{11.1b}$$

$$x \geq 0 \tag{11.1c}$$

where $x \in \mathbb{R}^n$, and there are m equalities such that $A \in \mathbb{R}^{m \times n}$, and $m < n$.

11.2 RECTANGULAR SYSTEMS OF LINEAR EQUATIONS

We consider a linear system of equations in m equations and n unknowns:

$$Ax = b \tag{11.2}$$

also subject to $x \geq 0$. It is also assumed that $b > 0$. This is known as the *canonical form of an LP*. We can cast any problem posed into this form:

1. $Ax \leq b, \quad x \geq 0 \longrightarrow Ax + y = b, \quad x \geq 0, y \geq 0$, *i.e.* with the introduction of deficit variables.
2. $Ax \geq b, \quad x \geq 0 \longrightarrow Ax - y = b, \quad x \geq 0, y \geq 0$, *i.e.* with the introduction of surplus variables.
3. x without sign restriction $\longrightarrow x = x_1 - x_2, \quad x_1 \geq 0, x_2 \geq 0$.

We further assume that the system $Ax = b$ is in a form such that we can read off immediately a *basic solution*. This leads to the following form, also called the *simplex tableau*:

x_1	x_2	$\cdots$	x_p	$\cdots$	x_m	x_{m+1}	x_{m+2}	$\cdots$	x_q	$\cdots$	x_n	b
1	0	$\cdots$	0	$\cdots$	0	$a_{1,m+1}$	$a_{1,m+2}$	$\cdots$	$a_{1,q}$	$\cdots$	$a_{1,n}$	b_1
0	1	$\cdots$	0	$\cdots$	0	$a_{2,m+1}$	$a_{2,m+2}$	$\cdots$	$a_{2,q}$	$\cdots$	$a_{2,n}$	b_2
$\vdots$	$\vdots$	$\vdots$	$\vdots$	$\vdots$	$\vdots$	$\vdots$	$\vdots$	$\vdots$	$\vdots$	$\vdots$	$\vdots$	$\vdots$
0	0	$\cdots$	1	$\cdots$	0	$a_{p,m+1}$	$a_{p,m+2}$	$\cdots$	$a_{p,q}$	$\cdots$	$a_{p,n}$	b_p
$\vdots$	$\vdots$	$\vdots$	$\vdots$	$\vdots$	$\vdots$	$\vdots$	$\vdots$	$\vdots$	$\vdots$	$\vdots$	$\vdots$	$\vdots$
0	0	$\cdots$	0	$\cdots$	1	$a_{m,m+1}$	$a_{m,m+2}$	$\cdots$	$a_{m,q}$	$\cdots$	$a_{m,n}$	b_m

From this tableau we can see that if we set the nonbasic variables equal to zero:

$$x_{m+1} = 0$$
$$x_{m+2} = 0 \tag{11.3}$$
$$\vdots$$
$$x_n = 0$$

then we can solve for the basic variables trivially:

$$x_1 = b_1$$
$$x_2 = b_2 \tag{11.4}$$
$$\vdots$$
$$x_m = b_m$$

What if we cannot find in $Ax = b$ an immediate basic solution, such as the one shown in the simplex tableau above? Then consider the introduction of artificial variables $y_1, y_2, \ldots, y_m$ (all $y_i \geq 0$, $i = 1, 2, \ldots, m$) such that:

$$Ax + y = b \tag{11.5}$$

and set as nonbasic to zero all the x's:

$$x_1 = x_2 = \ldots = x_n = 0$$

and solve for the artificial variables:

$$y_1 = b_1$$
$$y_2 = b_2 \tag{11.6}$$
$$\vdots$$
$$y_m = b_m$$

We will deal with artificial variables later. But the point to be taken from here is that we can always find a basic solution from any rectangular system (provided it is well-posed and does not have linearly dependent rows).

11.3 RECTANGULAR SYSTEMS AND THE OBJECTIVE FUNCTION OF THE LP PROBLEM

Consider appending the objective function, $f = \sum_{i=1}^{n} c_i x_i$, as an extra row (an extra equation) to the system we have defined in the previous simplex tableau. We thus obtain:

x_1	x_2	$\ldots$	x_p	$\ldots$	x_m	x_{m+1}	x_{m+2}	$\ldots$	x_q	$\ldots$	x_n	f	b
1	0	$\ldots$	0	$\ldots$	0	$a_{1,m+1}$	$a_{1,m+2}$	$\ldots$	$a_{1,q}$	$\ldots$	$a_{1,n}$	0	b_1
0	1	$\ldots$	0	$\ldots$	0	$a_{2,m+1}$	$a_{2,m+2}$	$\ldots$	$a_{2,q}$	$\ldots$	$a_{2,n}$	0	b_2
$\vdots$	$\vdots$	$\vdots$	$\vdots$	$\vdots$	$\vdots$	$\vdots$	$\vdots$	$\vdots$	$\vdots$	$\vdots$	$\vdots$	$\vdots$	$\vdots$
0	0	$\ldots$	1	$\ldots$	0	$a_{p,m+1}$	$a_{p,m+2}$	$\ldots$	$a_{p,q}$	$\ldots$	$a_{p,n}$	0	b_p
$\vdots$	$\vdots$	$\vdots$	$\vdots$	$\vdots$	$\vdots$	$\vdots$	$\vdots$	$\vdots$	$\vdots$	$\vdots$	$\vdots$	$\vdots$	$\vdots$
0	0	$\ldots$	0	$\ldots$	1	$a_{m,m+1}$	$a_{m,m+2}$	$\ldots$	$a_{m,q}$	$\ldots$	$a_{m,n}$	0	b_m
c_1	c_2	$\ldots$	c_p	$\ldots$	c_m	c_{m+1}	c_{m+2}	$\ldots$	c_q	$\ldots$	c_n	-1	0

We note that we have also introduced an extra variable column for the objective function variable f. It is not necessary to keep it there, so dropping that extra column we get the following simplex tableau:

x_1	x_2	$\ldots$	x_p	$\ldots$	x_m	x_{m+1}	x_{m+2}	$\ldots$	x_q	$\ldots$	x_n	b
1	0	$\ldots$	0	$\ldots$	0	$a_{1,m+1}$	$a_{1,m+2}$	$\ldots$	$a_{1,q}$	$\ldots$	$a_{1,n}$	b_1
0	1	$\ldots$	0	$\ldots$	0	$a_{2,m+1}$	$a_{2,m+2}$	$\ldots$	$a_{2,q}$	$\ldots$	$a_{2,n}$	b_2
$\vdots$	$\vdots$	$\vdots$	$\vdots$	$\vdots$	$\vdots$	$\vdots$	$\vdots$	$\vdots$	$\vdots$	$\vdots$	$\vdots$	$\vdots$
0	0	$\ldots$	1	$\ldots$	0	$a_{p,m+1}$	$a_{p,m+2}$	$\ldots$	$a_{p,q}$	$\ldots$	$a_{p,n}$	b_p
$\vdots$	$\vdots$	$\vdots$	$\vdots$	$\vdots$	$\vdots$	$\vdots$	$\vdots$	$\vdots$	$\vdots$	$\vdots$	$\vdots$	$\vdots$
0	0	$\ldots$	0	$\ldots$	1	$a_{m,m+1}$	$a_{m,m+2}$	$\ldots$	$a_{m,q}$	$\ldots$	$a_{m,n}$	b_m
c_1	c_2	$\ldots$	c_p	$\ldots$	c_m	c_{m+1}	c_{m+2}	$\ldots$	c_q	$\ldots$	c_n	0

11.4 THE SIMPLEX METHOD

Now, let us pivot the basic variables into the last (added) row so that we obtain:

x_1	x_2	$\cdots$	x_p	$\cdots$	x_m	x_{m+1}	x_{m+2}	$\cdots$	x_q	$\cdots$	x_n	b
1	0	$\cdots$	0	$\cdots$	0	$a_{1,m+1}$	$a_{1,m+2}$	$\cdots$	$a_{1,q}$	$\cdots$	$a_{1,n}$	b_1
0	1	$\cdots$	0	$\cdots$	0	$a_{2,m+1}$	$a_{2,m+2}$	$\cdots$	$a_{2,q}$	$\cdots$	$a_{2,n}$	b_2
$\vdots$	$\vdots$	$\vdots$	$\vdots$	$\vdots$	$\vdots$	$\vdots$	$\vdots$	$\vdots$	$\vdots$	$\vdots$	$\vdots$	$\vdots$
0	0	$\cdots$	1	$\cdots$	0	$a_{p,m+1}$	$a_{p,m+2}$	$\cdots$	$a_{p,q}$	$\cdots$	$a_{p,n}$	b_p
$\vdots$	$\vdots$	$\vdots$	$\vdots$	$\vdots$	$\vdots$	$\vdots$	$\vdots$	$\vdots$	$\vdots$	$\vdots$	$\vdots$	$\vdots$
0	0	$\cdots$	0	$\cdots$	1	$a_{m,m+1}$	$a_{m,m+2}$	$\cdots$	$a_{m,q}$	$\cdots$	$a_{m,n}$	b_m
0	0	$\cdots$	0	$\cdots$	0	$\bar{c}_{m+1}$	$\bar{c}_{m+2}$	$\cdots$	$\bar{c}_q$	$\cdots$	$\bar{c}_n$	$-f_0$

where clearly:

$$f_0 = \sum_{i=1}^{n} c_i x_i = \sum_{i=1}^{n} c_i b_i$$

What do the coefficients $\bar{c}_j$ signify for the nonbasic variables?

- They are reduced-cost coefficients, indicating the unit reduction if < 0, or unit increase if > 0, for a nonbasic variable becoming basic.
- We normally choose a negative coefficient to make basic its corresponding nonbasic variable, say if $\bar{c}_q < 0$ then x_q will become basic.
- If all $\bar{c}_j > 0$, $j = m + 1, m + 2, \ldots, n$, then we have the optimal solution: No nonbasic variable entering the basis can improve the objective function value.
- The feasible basic solutions of the system of equations $Ax = b$ (i.e. which also satisfy $x \geq 0$) are the vertices of the feasible region of the constraints.[1] Thus the number of basic variables has to be kept constant at m and that of the nonbasic variables to $(n - m)$.
- Thus when choosing a new variable to become basic, another one which is basic must be chosen to become nonbasic. In this manner, we move on the surface of the polytope[2] of the constraints, from one vertex to an adjacent vertex.

To make a basic variable into nonbasic (i.e. to exit the basis), we look into column q of the tableau.

Having decided on variable x_q to become basic, we hold all other nonbasic variables to zero, and observe the effect it has on the basic variables as we increase it:

---

[1] This implies that the maximum number of feasible vertex solutions is equal to $C_m^n = \dfrac{n!}{(n-m)!\,m!}$

[2] A polytope is a finite region in n dimensions enclosed by a finite number of hyperplanes [2]

$$x_1 = b_1 - a_{1q}x_q$$
$$x_2 = b_2 - a_{2q}x_q$$
$$\vdots$$
$$x_p = b_p - a_{pq}x_q \tag{11.7}$$
$$\vdots$$
$$x_m = b_m - a_{mq}x_q$$

where we have to remember that we have demanded that all $b_i > 0$.

The effect on the objective is similarly given by:

$$f = f_0 + \bar{c}_q x_q \tag{11.8}$$

where we recall that we have that $\bar{c}_q < 0$.

From Equation (11.8), we see that we would like to increase x_q from its zero value to as large as possible a value. But how large can we make it?

- In Equations (11.7), as $x_q \uparrow$ some variables will become negative if $a_{iq} \geq 0$.
- If all $a_{iq} \leq 0$, then x_q can increase to infinity, thus the problem is unbounded (pathological case).
- Thus we check the values of $a_{iq} > 0$ and x_q will be chosen as the smallest value that causes some basic variable to become zero:

$$x_q = \min_{\substack{i \\ a_{iq} > 0}} \frac{b_i}{a_{iq}} \tag{11.9}$$

In other words, we form all the ratios $\frac{b_i}{a_{iq}}$ for which $a_{iq} > 0$, and select the smallest ratio. Suppose this happens for the element in row p, *i.e.* the minimum ratio is achieved for element a_{pq}. Pictorially, the operations are:

x_1	$\cdots$	x_p	$\cdots$	x_m	x_{m+1}	x_{m+2}	$\cdots$	x_q	$\cdots$	x_n	b	ratios
1	$\cdots$	0	$\cdots$	0	$a_{1,m+1}$	$a_{1,m+2}$	$\cdots$	$a_{1,q}$	$\cdots$	$a_{1,n}$	b_1	$b_1/a_{1,q}$
0	$\cdots$	0	$\cdots$	0	$a_{2,m+1}$	$a_{2,m+2}$	$\cdots$	$a_{2,q}$	$\cdots$	$a_{2,n}$	b_2	$b_2/a_{2,q}$
$\vdots$	$\vdots$	$\vdots$	$\vdots$	$\vdots$	$\vdots$	$\vdots$	$\vdots$	$\vdots$	$\vdots$	$\vdots$	$\vdots$	$\vdots$
0	$\cdots$	1	$\cdots$	0	$a_{p,m+1}$	$a_{p,m+2}$	$\cdots$	$a_{p,q}$	$\cdots$	$a_{p,n}$	b_p	$b_p/a_{p,q}$
$\vdots$	$\vdots$	$\vdots$	$\vdots$	$\vdots$	$\vdots$	$\vdots$	$\vdots$	$\vdots$	$\vdots$	$\vdots$	$\vdots$	$\vdots$
0	$\cdots$	0	$\cdots$	1	$a_{m,m+1}$	$a_{m,m+2}$	$\cdots$	$a_{m,q}$	$\cdots$	$a_{m,n}$	b_m	$b_m/a_{m,q}$
0	$\cdots$	0	$\cdots$	0	$\bar{c}_{m+1}$	$\bar{c}_{m+2}$	$\cdots$	$\bar{c}_q$	$\cdots$	$\bar{c}_n$	$-f_0$	

- Element a_{pq} now becomes a pivot. We first normalize row p by a_{pq}, and then we eliminate row p from all other rows, including the last row of the objective function cost coefficients.

- With these operations, variable x_q becomes basic, and variable x_p becomes nonbasic.
- We continue with these operations, *identifying a variable to enter and a variable to exit the basis*, until all the reduced cost coefficients become positive.

This is the simplex method, which is theoretically proven to converge to the optimal solution. Effectively, the simplex method operates by changing the basic solution from vertex to vertex of the constraint polytope, always decreasing the objective function value.

11.5 REVISITING THE USE OF ARTIFICIAL VARIABLES FOR AN INITIAL BASIC SOLUTION

For an initial basis, it was mentioned that we can use the introduction of artificial variables in each row. The resulting problem is thus first to pivot out all the artificial variables and get a basis involving only the natural variables of the problem.

This is the so-called two-phase algorithm. Phase I is the following problem:

$$\min_{x,y} \sum_{i=1}^{m} y_i$$

subject to:

$$Ax + y = b$$
$$x \geq 0$$
$$y \geq 10$$

With pivoting, and feasible $Ax = b$, we get a basic solution for x and all $y_i = 0$. The resulting system will be of the form $\hat{A}x = \hat{b}$, which is equivalent to the original one (why?). We then solve the original problem with the new matrix and right-hand side, constituting Phase II of the method:

$$\min_{x} \sum_{i=1}^{n} c_i x_i$$

subject to:

$$\hat{A}x = \hat{b}$$
$$x \geq 0$$

11.6 NUMERICAL EXAMPLES OF THE SIMPLEX METHOD

In this section we will explore the simplex method using several small LP problems.

Example 11.1

Solve the following LP problem with the Simplex method:

$$\min z = -x_1 - 2x_2$$

subject to:

$$3x_1 + 7x_2 \leq 21$$
$$4x_1 + 5x_2 \leq 20$$
$$x_1 \geq 0$$
$$x_2 \geq 0$$

Solution

Since no basic solution can be obtained for this problem straight away, artificial variables have to be introduced to give the following problem:

$$\min z = y_1 + y_2$$

subject to:

$$3x_1 + 7x_2 + s_1 + y_1 = 21$$
$$4x_1 + 5x_2 + s_2 + y_2 = 20$$
$$x_i, s_i, y_i \geq 0, \ i = 1, 2$$

where s_i and y_i are the slack and artificial variables, respectively.
This problem is represented in the following way within a Simplex tableau:

x_1	x_2	s_1	s_2	y_1	y_2	b
3	7	1	0	1	0	21
4	5	0	1	0	1	20
0	0	0	0	1	1	0

The first iteration of Phase-I involves removing the artificial variables from the last row in the following way:

x_1	x_2	s_1	s_2	y_1	y_2	b
3	7	1	0	1	0	21
4	5	0	1	0	1	20
−7	−12	−1	−1	0	0	−41

Next, the most negative cost coefficient is chosen and, within that column, the variable with the smallest b/a ratio is chosen (in this case the value in the first row shown in bold). This variable is taken into the basic set in the following manner:

x_1	x_2	s_1	s_2	y_1	y_2	b
0.429	1	0.143	0	0.143	0	3
1.86	0	−0.714	1	−0.714	1	2.69
−1.86	0	0.714	−1	1.71	0	−5

The same procedure is repeated now to give the following tableau:

x_1	x_2	s_1	s_2	y_1	y_2	b
0	1	0.308	−0.231	0.308	−0.231	1.85
1	0	−0.385	0.539	−0.385	0.539	2.69
0	0	0	0	1	1	0

This concludes Phase-I of the Simplex algorithm. A basic solution not involving artificial variables y_1 and y_2 was found such that Phase-II can now be applied. After introducing the original objective function of our problem, the initial tableau for Phase-II is:

x_1	x_2	s_1	s_2	b
0	1	0.308	−0.231	1.85
1	0	−0.385	0.539	2.69
−1	−2	0	0	0

This tableau shows that the basic solution that was found in Phase-I is the optimal solution to the LP problem in this example. The final step of the Simplex algorithm is to eliminate the negative cost coefficients to obtain the objective function value, shown in bold:

x_1	x_2	s_1	s_2	b
0	1	0.308	−0.231	1.85
1	0	−0.385	0.539	2.69
0	0	0.231	0.0769	**6.38**

Hence, the final solution is given by:

$$x_1^* = 2.69$$
$$x_2^* = 1.85$$
$$z^* = -6.38$$

11.7 REFERENCES

[1] Dongarra, J. and Sullivan, F. "The top 10 Algorithms," Computing in Science & Engineering. 2000;1:p. 22–23.
[2] Coxeter, H. S. M. Regular Polytopes. Dover Publication. 1973.

11.8 FURTHER READING RECOMMENDATIONS

Linear Programming and the Simplex Method

Nash, S. G. and Sofer, A. (1995). Linear and Nonlinear Programming. McGraw-Hill.
Leunberger, D. and Ye, Y. (2008). Linear and Nonlinear Programming. Springer.
Taha, H. (2006). Operations Research. Prentice Hall.

11.9 EXERCISES

1. Consider the following LP problem:

$$\min f = 2x_1 + x_2$$

subject to:

$$3x_1 + 2x_2 \leq 10$$
$$2x_1 + 3x_2 \geq 3$$
$$2x_1 + 4x_2 \leq 5$$

$$x_i \geq 0, \; i = 1, 2$$

(i) Reformulate the problem into the canonical form required by the simplex method.

(ii) Explain why it is not straightforward from the formulation produced in part (i) to identify directly an initial basic feasible solution. Introduce artificial variables, and use the simplex method to identify an initial basic feasible solution.

(iii) Use the simplex method to solve the problem as reformulated in part (i), starting from the initial basic solution found in part (ii). Explain which inequalities are active. (Solution: $f = 1$)

2. Solve the following LP problem using the Simplex algorithm:

$$\max f = 5x_1 + 3x_2 + 4x_3 + 2x_4$$

subject to:

$$3x_1 + 2x_2 + 4x_3 + x_4 \leq 7$$
$$3x_1 + x_2 + 2x_3 + x_4 \geq 1$$
$$-x_1 + x_2 - 4x_3 + 2x_4 \leq -5$$

$$x_i \geq 0, \; i = 1, 2, 3, 4$$

(Solution: $f = 9$)

3. Solve the following LP problem using the Simplex algorithm:

$$\max f = 4x_1 + x_2 + 2x_3$$

subject to:

$$2x_1 + x_2 \leq 7$$
$$x_1 + 5x_3 \leq 6$$
$$2x_2 + 3x_3 \leq 15$$
$$x_1 + 2x_2 + x_3 \leq 8$$
$$x_1 + 3x_2 \leq 2$$

$$x_i \geq 0, \; i = 1, 2, 3$$

How many artificial variables are necessary to find a solution to this problem? (Solution: $f = 9.6$)

4. Solve the following LP problem using the Simplex algorithm:

$$\max f = 10x_1 + 5x_2 + 7x_3 + ex_4 + 2x_5$$

subject to:

$$2x_1 + 4x_2 + 2x_4 + 3x_5 \geq 4$$
$$2x_1 + x_3 + 2x_4 \geq 1$$
$$x_1 + 2x_2 + 5x_3 + 2x_4 \geq 6$$

$$x_i \geq 0, \ i = 1, 2, 3$$

(Solution: $f = 8.8$)

12

A Sampler of LP Problem Formulations

In this chapter, our attention is focused on formulations of LP problem models. Here, we begin by looking at a selection of small case studies where the description of a problem is given and then the translation into an LP mathematical programming formulation is derived.

The cases examined in the next sections are typical applications of LP in specific domains of interest and hence the titles indicate the area where it is applied. Wider areas of application of LP optimization are examined in dedicated chapters that follow.

12.1 PRODUCT MIX PROBLEM

A company has surplus production capacity and considers directing this for the production of one or more of three new products, 1, 2, and 3. The available machine capacity of the company is given in Table 12.1.

The number of machine hours required for each unit of the three different new products is given in Table 12.2.

The sales department of the company estimates that the sales capability for products 1 and 2 exceeds the amounts that can be produced, while for product 3 it is estimated at 15 units per week. The profit per product unit is predicted at 10, 15, and 12 thousand pounds, respectively for products 1, 2, and 3.

With this data formulate the LP model for the determination (optimal mix) of the number of units of each product that the company must produce in order to maximize its profits. Assume that product units produced are continuous variables.

Product Mix Problem Formulation

Variables x_1, x_2, x_3 are present, subject to operating constraints:

$$4x_1 + 3x_2 + 6x_3 \leq 100$$
$$2x_1 + x_2 \leq 150$$
$$4x_1 + 3x_3 \leq 80$$

and variable bounds:

$$x_3 \leq 15$$
$$x_1, x_2, x_3 \geq 0$$

Table 12.1 Production capacities for each machine

Machine Type	Production capacity (hours per week)
A	120
B	70
C	130

Table 12.2 Machine hours required for each product by each machine

Machine type	Product 1	Product 2	Product 3
A	1	–	3
B	6	4	1
C	2	3	–

Table 12.3 Vitamin content of each food type in mg, and cost of each food type in pence

Vitamin	Litre milk	kg meat	Dozen eggs
A	0.3	650	3
C	0	15	0
D	0.02	0.003	0.03
Cost	70	500	120

For this problem, the objective function is given by:

$$\max z = 10x_1 + 15x_2 + 12x_3$$

12.2 DIET PROBLEM

A classical problem where LP is applied is the "diet problem." The aim here is to calculate the amounts of certain dietary items that must be consumed in order to satisfy daily dietary prescribed demands, while minimizing cost.

Assume that three food types, milk, meat, and eggs, are considered. Further assume that the targets to satisfy include the daily intake of vitamins A, C, and D. The amount of each of the vitamins contained (mg) in each of the three food types is given according to Table 12.3 (in corresponding units for each food type) along with the cost of each unit (in pence).

The minimum required daily intake of each vitamin in mg is 0.5, 80, 0.008 for each of the vitamins A, C, and D, respectively. With this data formulate and solve the resulting LP problem.

Diet Problem Formulation

The following variables are present:

$$x_{milk}, \; x_{meat}, \; x_{eggs}$$
$$\underbrace{\text{litres}} \quad \underbrace{\text{kg}} \quad \underbrace{\text{dozens}}$$

subject to the minimum dietary requirements:

$$A: 0.2x_{milk} + 700x_{meat} + 2x_{eggs} \qquad \leq 0.6$$

$$C: 15x_{meat} \qquad \leq 75$$

$$D: 0.01x_{milk} + 0.0025x_{meat} + 0.025x_{eggs} \qquad \leq 0.005$$

and variable bounds:

$$x_{milk}, x_{meat}, x_{eggs} \geq 0$$

The objective function is given by:

$$\max z = 70x_{milk} + 600x_{meat} + 100x_{eggs}$$

12.3 BLENDING PROBLEM

We consider here an oil refinery blending problem, in simplified form.

The refinery produces three types of fuel: A, B, and C, with the blending of four different distillate products, 1–4. The availability of raw distillate products (daily amounts in barrels) along with their costs ($/barrel) is given in Table 12.4.

For the blending, it is necessary to maintain certain proportions (%) between the raw distillate products blended for the production of the three fuels. The proportions must be within given limits, which are provided in Table 12.5, along with the sale price ($/barrel) of each type of fuel. Maximize the profit by formulating an appropriate LP model and solving it.

Table 12.4 Raw distillate product supply in barrel and unit cost in $/barrel

Raw material	Available units	Unit cost
1	1,500	3
2	2,000	2
3	1,800	5
4	3,000	6

Table 12.5 Fuel composition constraint and selling price in $ per unit

Fuel	Proportions in % raw material content	Sale price per unit
A	No more than 35 of 1	4.00
	No less than 25 of 2	
	No more than 45 of 3	
B	No more than 55 of 1	6.00
	No less than 15 of 2	
C	No more than 55 of 1	5.00

Blending Problem Formulation

The following variables are used:

$$x_{ij} : \underbrace{\text{raw } i}_{\{1,2,3,4\}} , \underbrace{\text{fuel } j}_{\{A, B, C\}}$$

x_i : total raw i used

x_j : total fuel j produced

$$\Rightarrow 12 + 4 + 3 = 19 \text{ variables}$$

The production levels of fuels are:

$$x_A = x_{1A} + x_{2A} + x_{3A} + x_A$$
$$x_B = x_{1B} + x_{2B} + x_{3B} + x_B$$
$$x_C = x_{1C} + x_{2C} + x_{3C} + x_C$$

The raw materials used are:

$$x_1 = x_{1A} + x_{1B} + x_{1C}$$
$$x_2 = x_{2A} + x_{2B} + x_{2C}$$
$$x_3 = x_{3A} + x_{3B} + x_{3C}$$
$$x_4 = x_{4A} + x_{4B} + x_{4C}$$

The bounds on the raw materials are:

$$x_1 \leq 1000, \ x_2 \leq 3000, \ x_3 \leq 2000, \ x_4 \leq 1500$$

The blending recipe constraints are given by:

$$x_{1A} \leq 0.25x_A, x_{2A} \leq 0.35x_A, x_{3A} \leq 0.55x_A$$
$$x_{1B} \leq 0.45x_B, x_{2B} \leq 0.2x_B$$
$$x_{1C} \leq 0.6x_C$$

Overall bounds are given by:

$$x_{ij}, x_i, x_j \geq 0$$

The objective function is given by:

$$\max z = 6.0x_A + 4.0x_B + 3.0x_C$$
$$- 2x_1 - 5x_2 - 6x_3 - 3x_4$$

12.4 TRANSPORTATION PROBLEM

Three production facilities, A, B, and C, belonging to the same manufacturer, are located in geographically different locations and produce the same single product.

The customers buying the product are located in five different cities.

The three production facilities produce 200, 400, and 300 units of the product, while the customers have ordered 150, 120, 180, 230, and 200 units.

The transportation costs, in consistent units, from each production facility to each customer are provided in Table 12.6.

Find the amounts of products sent from each production facility to each customer that minimize the total transportation cost the manufacturer will have to incur.

12.4 Transportation Problem

Table 12.6 Costs of buying one unit of product for each customer from each individual production facility

Facility	Customers					
	1	2	3	4	5	6
A	12	10	15	8	10	13
B	14	17	19	12	14	20
C	21	13.5	18	12	16	19.5
D	20	14	17	19.5	12	20

Transportation Problem Formulation

$$\underbrace{x_{ij}}_{\text{Flow } i}, \underbrace{\overbrace{c_{ij}}^{}}_{\text{to } j}$$

$$\underbrace{\qquad\qquad}_{\text{Cost per unit } i \text{ to } j}$$

Total production capacity $= 900 >$ Total demand $= 880$
Variables:

$$x_{ij} : i = \{A, B, C\} \text{ facilities}$$
$$j = \{1, 2, 3, 4, 5\} \text{ customers}$$
$$\rightarrow 15 \text{ variables}$$

Production upper bounds:

$$\sum_{j=1}^{5} x_{ij} \leq U_j, \, i = \{A, B, C\}$$

$$U_j = \{200, 400, 300\}$$

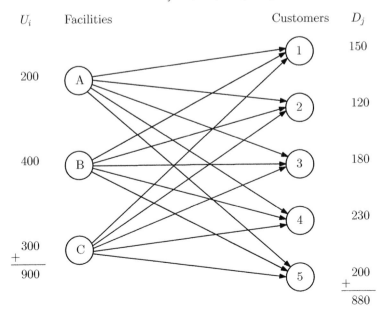

Customer demand satisfaction:

$$\sum_{i=A}^{C} x_{ij} \leq D_j, \ j = \{1, 2, 3, 4, 5\}$$

$$D_j = \{150, 120, 180, 230, 200\}$$

Objective function:

$$\min_{x_{ij}} \sum_{i=A}^{C} \sum_{j=1}^{5} c_{ij} \times x_{ij}$$

12.5 LINEAR ODE OPTIMAL CONTROL PROBLEM (OCP)

What is an optimal control problem (OCP)? It is best described by an example.[1] Consider a car that starts from rest at distance 0 units. The control variable in the system is the acceleration, which can take the values: $-2 \leq \gamma \leq +1$, *i.e.* we can control the vehicle by stepping on the gas or using the break (the units are all assumed to be consistent). The aim is to arrive at distance 300 units (straight line, say) from the start at minimal time and with zero velocity there.
The problem can be formulated as:

$$\min_{\gamma(\cdot), t_f} t_f$$

subject to
system dynamics:

$$\frac{dx(t)}{dt} = v(t)$$

$$\frac{dv(t)}{dt} = \gamma(t)$$

control variable bounds:

$$-2 \leq \gamma(t) \leq +1$$

time:

$$0 \leq t \leq t_f$$

initial conditions:

$$x(0) = 0$$

$$v(0) = 0$$

final conditions:

$$x(t_f) = 300$$

$$v(t_f) = 0$$

---

[1] A more detailed analysis of OCPs is carried out in Chapter 20.

Report the optimal control (acceleration) profile for all times, and produce appropriate plots to demonstrate the evolution of the state variables involved in the formulation.

Optimal control problems follow the general lines of the simple problem given here (car problem) and most practical approaches consider their transformation into standard optimization problems via discretization.

The simplest form we will consider here is the transformation via backward differences of all the time derivatives into finite form. This yields a set of ordinary algebraic equations, and the overall problem becomes a large optimization problem.

Example of Dynamic Variable Discretization

Using five finite elements γ_1, γ_2, γ_3, γ_4, and γ_5, the profile of $\gamma(t)$ shown in Figure 12.1 is obtained.

The number of discretization steps N is kept constant. The length of each step is defined by:

$$h = \frac{t_f - t_0}{N}$$

In our example $t_0 = 0$, hence:

$$h = \frac{t_f}{N}$$

This enables the discretization of all dynamic variables over the finite elements (finite differences).

Since the right-hand sides of the ODEs in the problem are linear, and if the objective is linear, the resulting optimization problem would be expected to be an LP.

The discretization may be carried over N equal-sized finite elements (intervals) such that we have:

$$\min_{x,v,\gamma,t_f} t_f$$

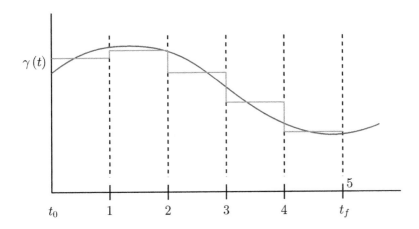

Figure 12.1 Discretization of a function $\gamma(t)$ into five piecewise constant values.

$$x^{(i)} - x^{(i-1)} = v^{(i)}\frac{t_f}{N}$$

$$v^{(i)} - v^{(i-1)} = \gamma^{(i)}\frac{t_f}{N}$$

$$-2 \le \gamma^{(i)} \le +1$$

$$x^{(0)} = 0$$

$$v^{(0)} = 0$$

$$x^{(N)} = 300$$

$$v^{(N)} = 0$$

There is, however, a problem: The equations are not linear, since the final time is not fixed, and this yields an NLP problem. Here we are going to do something unorthodox: fix t_f, and change the objective to the minimization of a slack variable measuring the satisfaction of the final time constraints. The new model we have is converted to:

$$\phi(t_f) = \min_{x,v,\gamma,\varepsilon} \varepsilon$$

subject to:

$$x^{(i)} - x^{(i-1)} = v^{(i)}\frac{t_f}{N}$$

$$v^{(i)} - v^{(i-1)} = \gamma^{(i)}\frac{t_f}{N}$$

$$-2 \le \gamma^{(i)} \le +1$$

$$i = 1, 2, \ldots, N$$

$$x^{(0)} = 0$$

$$v^{(0)} = 0$$

$$-\varepsilon \le x^{(N)} - 300 \le +\varepsilon$$

$$-\varepsilon \le v^{(N)} \le +\varepsilon$$

$$\varepsilon \ge 0$$

What do the last two constraints mean and where did ε come from? They are the relaxation of absolute values $|x^{(N)} - 300|$, $|v^{(N)}|$:

$$\left|x^{(N)} - 300\right| \le \varepsilon$$

$$\left|v^{(N)}\right| \le \varepsilon$$

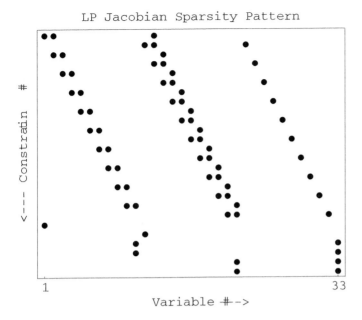

Figure 12.2 Graphical representation of the constraints for the discretized problem.

Rather than minimizing with respect to final time t_f, the deviation of the final time constraints ε are reduced until ideally a value of $\varepsilon = 0$ is achieved, *i.e.* the constraints are satisfied.

Since the value of t_f is kept constant, the smallest value of t_f that still satisfies the constraints will give the optimum. What is the meaning of the function $\phi(t_f)$? The function measuring satisfaction of the final time constraints, as a function of the final time, t_f.

Discretized problems, using finite-difference or finite elements, result in *structured matrices (banded)*, and the matrix of the constraints of the discretized problem shown is given in Figure 12.2 (for $N = 10$ elements).

We then scan for the *minimal* value of t_f that yields the optimal value $\phi(t_f) = 0$. In other words, we solve the one-dimensional optimization problem of finding the smallest t_f according to the nested $\phi(t_f)$ problem:

$$\min_{t_f} t_f$$

subject to:

$$\phi(t_f) = 0$$

By trying various values of t_f, for $N = 1000$ elements, the curve in Figure 12.3 is constructed.

In Figure 12.3 the satisfaction of the constraint $\varphi(t_F)$ is seen with respect to the value of t_F. As t_F increases, it is seen that the value of $\varphi(t_F)$ decreases up until $t_F = 30$. This means that the final constraint of $\phi(t_F) = 0$ is only achieved when the final time $t_f \geq 30$. No feasible solution is obtained for smaller values of t_f.

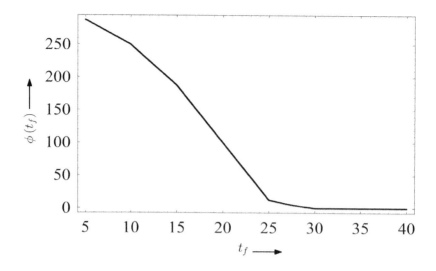

Figure 12.3 The value of the final time constraint $\phi\left(t_f\right)$ with respect to the value of t_f.

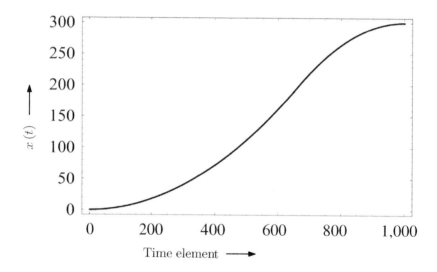

Figure 12.4 The distance profile with respect to time for the optimal solution.

In Figure 12.4 the evolution of distance with each time element, equivalent to time because constant step sizes are used, is seen. A continuous profile of distance with time is obtained as expected. Furthermore, at the final time element the distance of 300 is reached as specified by the constraints. For the optimal value of t_f we have found, Figures 12.5 and 12.6 provide the evolution of the velocity and acceleration (control) profiles, respectively, as functions of time from 0 to t_f (plotted against their value in each sequential time-element, for 1,000 elements used in the discretization).

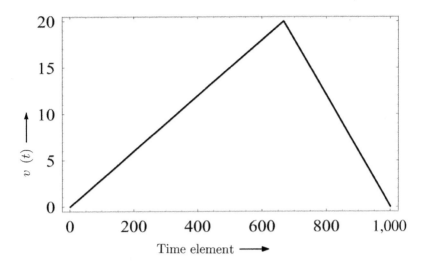

Figure 12.5 The speed profile with respect to time for the optimal solution.

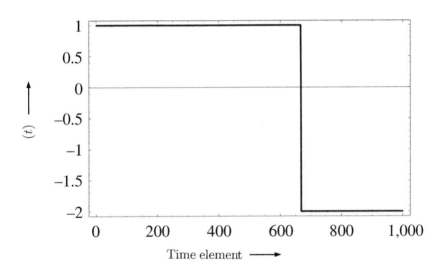

Figure 12.6 The acceleration profile with respect to time for the optimal solution.

12.6 FURTHER READING RECOMMENDATIONS

Transformation of Optimal Control Problems (Dynamic Optimization Problems) into Standard NLPs

Biegler, L. T. (2010). Nonlinear Programming: Concepts, Algorithms, and Applications to Chemical Processes. MPS-SIAM Series on Optimization.

Vassiliadis, V. S. (1993). 'Computational Solution of Dynamic Optimization Problems with General Differential-Algebraic Constraints.' PhD thesis, University of London (Imperial College London).

12.7 EXERCISES

1. The following table of unit transportation costs and supply/demands depicts a transportation problem:

	Facility	\multicolumn{6}{c}{Destinations}	
		1	2	3	4	5	6	s_i
	A	2	4	2	5	M	6	40
	B	4	6	M	3	7	8	55
Sources	C	5	4	2	12	8	11	90
	D	4	6	13	8	7	2	30
	d_j	10	20	40	30	40	25	–

where m is a very large number.

   (i) Is this problem a balanced transportation problem?
   (ii) What are the variables of this problem?
   (iii) Write the mathematical programming problem model.
   (iv) Solve the problem with four different settings for M: 1, 10, 100, 1000.
   (v) What is the meaning of M?
   (vi) Solve the problem numerically and report the feasible optimal solution found, according to the answer in item (ii).
   (vii) Reformulate the model such that the use of M is obviated. Solve again, and compare the results to the previous solutions found.

2. A company is planning its production schedule for the next 9 months, for one of its products. Table 12.7 shows for the individual months the demands per month (in terms of orders already placed), the unit production costs, and the inventory costs for carrying one unit of product from one month over to the next. At present, for month 1 there is stock held in inventory of 150 units, and after the 9 months it is not desired to have any stock left in inventory for month 10. The production capacity is fixed for each month at 100.

   (i) Formulate the corresponding optimization problem minimizing the total cost of the production plan.
   (ii) Find the optimal solution, compute the inventory for each month, and evaluate the objective function.
   (iii) Calculate the Lagrange multipliers corresponding to the constraints of the optimization problem in part (i) evaluated at the solution point found in part (ii).

Table 12.7 Information for problem 2

	\multicolumn{9}{c}{Months}								
	1	2	3	4	5	6	7	8	9
Demand	40	110	110	60	200	150	110	80	140
Unit production cost, £	1.5	1.5	2	2.5	2	3	2.5	2	1
Unit inventory cost, £	0.5	0.6	0.5	0.7	0.8	0.5	0.6	0.6	0.7

Table 12.8 Machine utilization data

	Machine utilization (hours/unit of product)		
Machine	Product 1	Product 2	Product 3
A	–	3	5
B	3	–	8
C	4	2	–
D	5	2	1

Table 12.9 Machine availability data

	Machine availability (hours/week)	
Machine	Case 1	Case 2
A	250	350
B	120	160
C	80	60
D	70	130

(iv) If you were to invest in expanding the production capacity of a single month, which month would you select and why? Base your analysis on the Lagrange multipliers found in part (iii).

3. A company wants to optimize the weekly production level of three products (1, 2, 3). The products need to be processed on four machines (A, B, C, D). The processing hours for each product unit on each machine are given in Table 12.8. The availability of each machine is given in Table 12.9; the current case of machine availability, and the expanded mode of operation with increased machine availability. The product prices are 15, 10, and 7 thousand pounds per unit product, for products 1, 2, and 3, respectively.

   (i) Write the optimization model for the determination of the production levels, maximizing the sales revenue of the three products for one week of production.

  (ii) For Case I, determine which constraints of the model derived in part (i) are active, and their corresponding Lagrange multipliers.

 (iii) Based on sensitivity analysis of the optimal solution for Case I, predict what the optimal sales revenue will be for the expanded machine availability case, Case II.

 (iv) Compute the actual optimal objective function value for the optimal solution for Case II, and compare it to the predicted value in part (iii). Explain your findings.

13

Regression Revisited

Using LP to Fit Linear Models

Model estimation is very important in engineering and science applications, as well as in financial activities. This chapter presents three alternative ways that a linear model can be fitted over a set of experimental data, or measurements of an input-output system.

13.1 L_2 / EUCLIDEAN NORM FITTING (LEAST SQUARES)

Chapter 4 presented the case of least squares fitting as follows: For m measurements b_i, $i = 1, 2, \ldots, m$ (different settings of input row vector a_i), determine n unknown parameters x, with $m > n$ (more experiments than unknown parameters), for a *linear model* such that:

$$b_i = a_i x = \sum_{j=1}^{n} a_{ij} \cdot x_j, \quad i = 1, 2, \ldots, m$$

It was then discussed that equality will generally not hold, as it is impossible to solve uniquely a system having more equations than variables. Next, the vector measuring the error (residual) of the fitting is defined:

$$r = Ax - b$$

where r is the error vector in $\mathbb{R}^m$, Ax is the predicted responses in $\mathbb{R}^m$, and b is the measured responses in $\mathbb{R}^m$.

Least squares fitting is the attempt to minimize the sum of all squared errors by choosing the optimal tuning values for the unknown parameters x:

$$\min_x q(x) = \frac{1}{2} \sum_{i=1}^{m} r_i^2 = \frac{1}{2} r^T r \tag{13.1}$$

subject to:

$$r = Ax - b \tag{13.2}$$

It is noted here that this problem is an equality-constrained quadratic programming (QP) problem (with diagonal Hessian matrix). As an optimization problem it is further noted that bounds on the variables x as well as more general constraints may be implemented.

In any case, for the optimization problem as it stands we have seen that it is possible to obtain an analytic solution (when the matrix $A^T A$ is invertible) according to:

$$x^* = (A^T A)^{-1} A^T b$$

which is the standard linear model least squares solution.

The objective function in Equation (13.1) can be written as the minimization of the square of norm-2 (l_2) (Euclidean distance), i.e.

$$\min_x \frac{1}{2} \|r\|_2^2$$

subject to the definition of the residual vector in Equations (13.2).

Is it possible to consider other norms then? What would the formulation look like and how can these problems be solved?

13.2 L_1 NORM FITTING

Here we may use norm-1 (l_1) as an objective function:

$$\min_x \|r\|_1 = \sum_{i=1}^{m} |r_i| \tag{13.3}$$

subject to:

$$r_i = a_i x - b_i \tag{13.4}$$

$$i = 1, 2, \ldots, m$$

The objective function in Equation (13.3) contains absolute values, and hence it is clearly non-differentiable. Is this a problem though? Consider re-writing the problem as the following LP:

$$\min_x = \sum_{i=1}^{m} \varepsilon_i \tag{13.5}$$

subject to:

$$-\varepsilon_i \leq a_i x - b_i \leq +\varepsilon_i \tag{13.6}$$

$$\varepsilon_i \geq 0 \tag{13.7}$$

$$i = 1, 2, \ldots, m$$

The individual residuals are relaxed with a dedicated slack variable for each instead of minimizing $\|r\|_1$ directly. The values of ε_i are decreased with every iteration. The optimization is only valid as long as the constraints shown in Equation (13.6) are satisfied. The smallest value for $\sum_{i=1}^{m} \varepsilon_i$ for which the constraints are still satisfied gives the optimal solution. The new and initial formulations hence are equivalent.

13.3 L_∞ NORM FITTING

Here, we may use the infinity norm (l_∞) as an objective function:

$$\min_x = \|r\|_\infty = \max_{i=1,2,\dots,m} \{|r_i|\} \tag{13.8}$$

subject to:

$$r_i = a_i x - b_i \tag{13.9}$$

$$i = 1, 2, \dots, m$$

The objective function in Equation (13.8) contains absolute values and selects the largest of them from a list, hence it is also clearly non-differentiable. Again, consider re-writing the problem as the following LP:

$$\min_x = \varepsilon \tag{13.10}$$

subject to:

$$-\varepsilon \le a_i x - b_i \le +\varepsilon \tag{13.11}$$

$$i = 1, 2, \dots, m$$

$$\varepsilon \ge 0$$

The new formulation has similar reasoning to the l_1 norm fitting: Instead of minimizing $\|r\|_\infty$ directly, the values of ε are decreased with every iteration. The optimization is only valid as long as the constraints shown in Equation (13.11) are satisfied.

The smallest value for ε for which the constraints are still satisfied gives the optimal solution.

13.4 APPLICATION: ANTOINE VAPOR PRESSURE CORRELATION FITTING

We consider the vapor pressure data for chlorobenzene as tabulated in Table 13.1 in pairs of temperature ($^\circ$C) and pressure (kPa) measurements.

The Antoine correlation is given by

$$P(T) = \exp\left(A - \frac{B}{T + C}\right)$$

To turn this into a linear model the following transformation steps are used:

$$\ln P = A - \frac{B}{T + C}$$

$$T \ln P + (C) \ln P = T(A) + (AC) - (B)$$

or:

$$(A)T + (AC) - (B) - (C) \ln P = T \ln P$$

13.4 Application: Antoine Vapor Pressure Correlation Fitting

Table 13.1 Vapor pressure vs Temperature data for chlorobenzene

Temperature (°C)	Pressure (kPa)	Temperature (°C)	Pressure (kPa)
0	0.316	55	6.748
5	0.458	60	8.672
10	0.639	65	10.47
15	0.867	70	12.87
20	1.170	75	15.66
25	1.487	80	17.67
30	2.048	85	22.96
35	2.682	90	27.60
40	3.477	95	30.58
45	4.359	100	36.22
50	5.409		

Table 13.2 The Antoine parameters found using each individual fitting parameter

Fitting	A	B	C
L^2	12.8893	2748.03	195.762
L^1	12.6063	2616.42	190.170
L^∞	13.6971	3131.79	210.498

Terms in parentheses indicate unknown coefficients (compound, where applicable). Note that there appear to be four groups to compute. But this would be wrong (see classroom demonstration). The compounding has to be done as follows:

$$\underbrace{(A)}_{x_1} \underbrace{T}_{a_1} + \underbrace{(AC - B)}_{x_2} \cdot \underbrace{1}_{a_2} - \underbrace{(C)}_{x_3} \underbrace{\ln P}_{a_3} = \underbrace{T \ln P}_{b}$$

The underbraces indicate the notation we have used previously for parameters, inputs, and outputs of a system for least squares problems.

The various types of fitting yield the coefficients as shown in Table 13.2.

To verify the quality of the fittings, we proceed to plot the data points with the three fittings calculated. The vapor pressure versus temperature plot is given in Figure 13.1.

The three curves seem to satisfy the original data points very accurately, and are almost indistinguishable from each other.

Further Comments on the l_2, l_1, and l_∞ Fitting Methods

l_2 (least squares): penalizes large errors more

For small errors, *i.e.* smaller than 1, squaring the residual leads to a smaller error value. If, on the other hand, errors are larger than 1, squaring the error increases the residual. Hence, large errors are penalized more if l_2 is used.

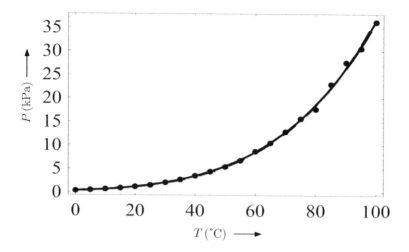

Figure 13.1 Vapor pressure vs temperature. The solid line represents a fitting according to the Antoine equation.

l_1: penalizes more "equally," can be used to ignore large outliers in the data

If using l_1, every error is treated in the same way. This also means that very large errors are not penalized more than other errors. Hence, large outliers can be ignored when using the l_1 fitting method.

l_∞: penalizes worst-case scenario

If using l_∞, only the largest error is considered. This fitting method is useful if the regression model is not supposed to have any large outliers. On the other hand, not a lot of emphasis is put on the smaller errors.

13.5 FURTHER READING RECOMMENDATIONS

More Detailed Discussion on the Use of LP Formulations for Regression

Vanderbei, R. J. (2014). Linear Programming. Springer.

14

Network Flow Problems

Many linear programming problems (and in fact many nonlinear programming problems) have an underlying structure that can be recognized as the transportation of material through a "network" (either literal or as an abstraction):

- to meet demands at some nodes,
- while other nodes provide the supply

Such problems are generally termed *network flow problems.*

Before proceeding to examine specific standardized types of network optimization problems, we need to go through a few definitions.

14.1 NETWORK, ARCS, GRAPHS

A network is a collection of nodes and arcs connecting them.

- we denote as $\mathcal{N}$ the set of nodes and
- as $\mathcal{A}$ the set of arcs.

The collection of the two sets put together comprises a *graph*, and in the case where the arcs are *directed*, we call it a *digraph* (most of our problems here will require directed arcs to indicate direction of "flow"). The network (graph) is represented generally as:

$$G(\mathcal{N}, \mathcal{A})$$

We assume that all problems we study have directed arcs, and for an arc connecting node i with node j we write the corresponding arc as the ordered pair:

$$(i, j)$$

For a given network, this allows us to define the set of arcs as follows:

$$\mathcal{A} = \{(i, j)\}$$

and this set is clearly a subset of all possible arcs connecting all the nodes in set $\mathcal{N}$:

$$\mathcal{A} \subset \{(i, j) : i, j \in \mathcal{N}, i \neq j\}$$

In typical cases of networks, the set of arcs contains far fewer than the maximum number that would yield a fully connected graph. Most real applications usually contain connections between neighboring nodes only, thus the set of arcs is not full. A good introduction to graphs for the purposes of network flow problems is given in [2].

It is noted that for most of the special cases presented in this chapter there are either specialized algorithms that can solve them or a special implementation of the simplex LP solution method that is very efficient on network problems. In this book the emphasis is on understanding the underlying structured model pattern these problems lead to, rather than how to solve them individually in the most efficient way.

14.2 MINIMUM COST NETWORK FLOW PROBLEM

This is the simplest problem in the class of network problems. To specify this problem, we need to indicate the supply (or demand) for material at each of the nodes of the network.

For $i \in \mathcal{N}$, we augment the data of the network $G(\mathcal{N}, \mathcal{A})$ with the supply of material b_i at each node $i \in \mathcal{N}$. By convention, supply of material is taken to be a positive number, whereas to indicate demand the number will be negative.

We further assume that we have a balanced network, where total supply equals total demand, in other words the following holds:

$$\sum_i^m b_i = 0$$

(for networks where we do not know these in advance, fixed demands or supplies can be defined more generally by setting supply/demands to upper/lower bounds of availability, *etc.*).

- To balance each node we need to consider what flows out of a node, what flows into it, and what is supplied and write appropriate material balance constraints (Out - In = Supply).
- To have a cost minimization we have to have costs associated with each individual arc (cost/unit of flow), c_{ij}.

The overall network problem can be written as the following LP problem:
Objective function:

$$\min_{x_{ij}} \sum_{\substack{i=1}}^{m} \sum_{\substack{j=1 \\ (i,j) \in \mathcal{A}}}^{m} c_{ij} x_{ij}$$

subject to:
Material balance equations for each node k:

$$\sum_{\substack{j: \\ (k,j) \in \mathcal{A}}} x_{kj} - \sum_{\substack{i: \\ (i,k) \in \mathcal{A}}} x_{ik} = b_k, \quad k = 1, 2, \ldots, m$$

Positivity lower bounds on flow (we do not want flows to reverse directions!):

$$x_{ij} \geq 0, \quad \forall (i, j) \in \mathcal{A}$$

Figure 14.1 shows pictorially the structures used in the material balances.

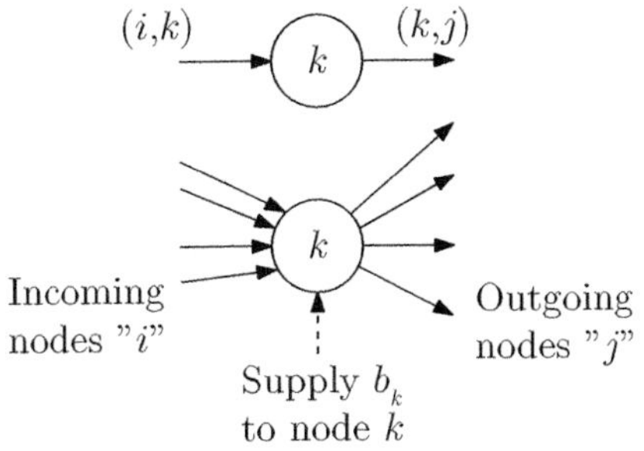

Figure 14.1 Material balance across a node k.

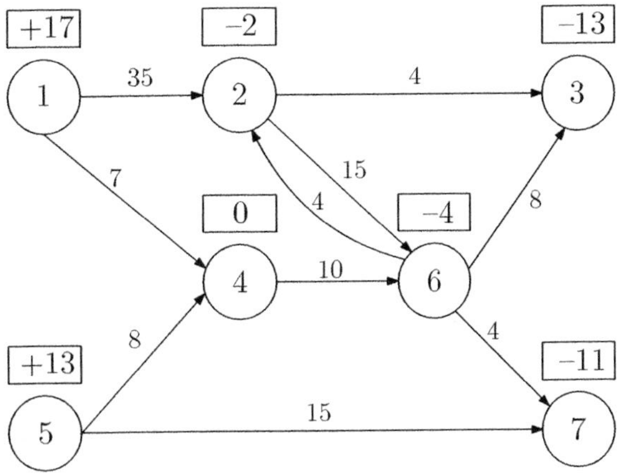

Figure 14.2 Diagram of a sample cost network flow problem.

14.2.1 Example of Minimum Cost Network Flow Problem

Consider the example shown in Figure 14.2:

- the numbers above the arcs indicate the associated cost,
- the numbers in the circles are the node numbers (names),
- and the numbers in small squares next to the node indicate the supply of material at each node (either supply, or if negative then it will be demand).

The optimal solution is given in Table 14.1. It is interesting to observe that the resulting solution from running the problem with the LP solver yielded an all-integer solution. Further examples of network flow problems can be found in [1].

14.3 INTEGRALITY OF SOLUTION THEOREM

Any network flow problem with integer data (*i.e.* right hand side of source terms) has a fully integer optimal solution (integer-valued arc flows).

This is very important, as in many realistic applications the material allocation through the arcs (or equivalently whatever the network signifies) needs to take on only integer values. For example, it would not make sense to send 3.14 airplanes, with 151.6 seats, with 1.9254 pilots ... *etc.*!

It is noted that there is a special formulation of LP problems (and NLP problems) for integrality to be included. Formulation and solution of integer linear programming problems is a very extensive field, with many real-world applications.

Table 14.1 Optimal variable values for the cost network flow problem

Index	Variable Name	Optimal Value
1	$x_{1,2}$	0
2	$x_{1,4}$	17
3	$x_{2,3}$	0
4	$x_{2,6}$	0
5	$x_{4,6}$	30
6	$x_{5,4}$	13
7	$x_{5,7}$	0
8	$x_{6,2}$	2
9	$x_{6,3}$	13
10	$x_{6,7}$	11

However, the formulations also require specialized solvers, e.g. Branch and Bound methods. The solution of such problems is often computationally very demanding, as it relies on the solution of a large number of LP sub-problems. Integer programming problems are considered in Chapter 18.

14.4 CAPACITATED MINIMUM COST NETWORK FLOW PROBLEM

This is exactly the same formulation as for the minimum cost network flow problem. The only difference is that now the arcs have an upper bound on the flow rate of material they will admit through them. The overall LP formulation becomes:

Objective function:

$$\min_{x_{ij}} \sum_{\substack{i=1}}^{m} \sum_{\substack{j=1 \\ (i,j)\in\mathcal{A}}}^{m} c_{ij} x_{ij}$$

subject to:

Material balance equations for each node k:

$$\sum_{\substack{j: \\ (k,j)\in\mathcal{A}}} x_{kj} - \sum_{\substack{i: \\ (i,k)\in\mathcal{A}}} x_{ik} = b_k, \quad k = 1, 2, \ldots, m$$

Bounds on flow:

$$0 \leq x_{ij} \leq x_{ij}^{U}, \quad \forall (i, j) \in \mathcal{A}$$

14.5 SHORTEST PATH (OR MINIMUM DISTANCE) PROBLEM

We will formulate the problem as implied directly: finding the path through a network that minimizes the distance from a given starting node to a given ending node.[1]

---

[1] where we assume we have a single source, "0"-th node, and a single sink, "m"-th node, and where these do not exist we may for simplicity's sake introduce them as dummy nodes

Objective function:

$$\min_{x_{ij}} \sum_{i=0}^{m} \sum_{\substack{j: \\ (i,j) \in \mathcal{A}}} d_{ij} x_{ij}$$

subject to:
Source node balance:

$$\sum_{\substack{j: \\ (0,j) \in \mathcal{A}}} x_{0j} = 1$$

Intermediate node balances:

$$\sum_{\substack{i: \\ (i,k) \in \mathcal{A}}} x_{ik} = \sum_{\substack{j: \\ (k,j) \in \mathcal{A}}} x_{kj}, \quad k = 1, 2, \ldots, m$$

Sink (destination) node balance:

$$\sum_{\substack{i: \\ (i,m) \in \mathcal{A}}} x_{im} = 1$$

Lower bound on arc flows:

$$x_{ij} \geq 0, \quad \forall (i, j) \in \mathcal{A}$$

14.6 TRANSPORTATION PROBLEM

The transportation problem seeks to determine the optimal (minimal cost) transportation schedule of a *single commodity* from a number of sources to a number of destinations (suppliers to customers for example). The information to set up such a problem model contains:

1. The supply at each source (either a fixed amount or an upper level in the capacity of the source).
2. The demand at each destination (either a fixed amount or a set of bounds in the affinity at the destination).
3. The transportation cost of the commodity from sources to destinations.
4. The connectivity information, showing which sources and destinations are connected, or if a particular connection is not possible (*e.g.* when a direct connection between a source and a destination does not exist).
5. The underlying assumption is that the transportation cost on all routes is proportional to the number of units transported.

The transportation problem is again a network flow problem, and the general case is shown in Figure 14.3.

The graph of the problem has two distinct sets of nodes (sources and destinations) with connections only from one set to the other. Such a graph is called a *bipartite graph.*

We have already solved a transportation problem model in Chapter 12.

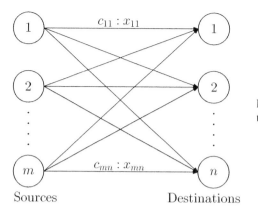

Figure 14.3 Diagram of a sample transportation problem.

Sources   Destinations

The general model formulation for a transportation problem is as follows:
Objective function:

$$\min \sum_{i=1}^{m} \sum_{j=1}^{n} c_{ij} x_{ij}$$

subject to:
Capacities at sources:

$$\sum_{j=1}^{n} x_{ij} \leq \underbrace{s_i^U}_{\text{supply upper bound}}, \quad i = 1, 2, \ldots, m$$

Affinity at destinations:

$$\underbrace{d_j^L}_{\text{demand lower bound}} \leq \sum_{i=1}^{m} x_{ij} \leq \underbrace{d_j^U}_{\text{demand upper bound}}, \quad j = 1, 2, \ldots, n$$

Bounds on flows:

$$x_{ij} \geq 0, \quad i = 1, 2, \ldots, m, \ j = 1, 2, \ldots, n$$

To be feasible this model, must have the property that the total maximum supply available must meet or exceed the minimal total affinity (demand) at the destinations:

$$\sum_{i=1}^{m} s_i^U \geq \sum_{j=1}^{n} d_j^L$$

If we have a situation where equalities hold for the supply and demand balances, and that the total demand equals the total supply, we call the problem a *balanced transportation model*. The model for it is as follows:
Objective function:

$$\min \sum_{i=1}^{m} \sum_{j=1}^{n} c_{ij} x_{ij}$$

subject to:
Capacities at sources:

$$\sum_{j=1}^{n} x_{ij} = s_i, \quad i = 1, 2, \ldots, m$$

Affinity at destinations:

$$\sum_{i=1}^{m} x_{ij} = d_j, \quad j = 1, 2, \ldots, n$$

Bounds on flows:

$$x_{ij} \geq 0, \quad i = 1, 2, \ldots, m, \ j = 1, 2, \ldots, n$$

For this to be feasible, it must have the property that the total maximum supply available must equal the minimal total affinity (demand) at the destinations, as already mentioned:

$$\sum_{i=1}^{m} s_i = \sum_{j=1}^{n} d_j$$

14.7 TRANSSHIPMENT PROBLEM

- In the transportation model we have assumed that a direct route between a source and a destination exists, and that this is the only connectivity provided.
- In the transshipment model it is possible for supply to pass through another source, transiently. This is shown in Figure 14.4.

An alternative transshipment problem involves intermediate distribution centers, or depots, which form a layer between sources and destinations, as shown in Figure 14.5. One can formulate again the problem as in the transportation problem case, with either a general formulation or with one specialized for the case of a balanced system. We will not formulate the model here, as it is repetitive of what we have already seen. It is noted that the second transshipment problem shown is like having two transportation problems in sequence, front-to-back.

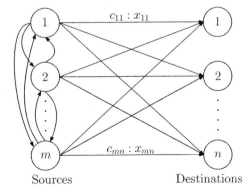

Figure 14.4 Diagram of a network for a transshipment problem.

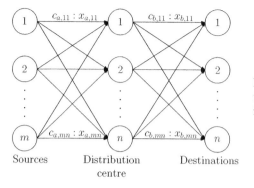

Sources   Distribution   Destinations
centre

Figure 14.5 Diagram of a transshipment problem with intermediate distribution centers.

14.8 THE ASSIGNMENT PROBLEM

The assignment problem may be considered a special case of the balanced transportation problem. It deals with the situation where we have to assign n jobs (or workers) to n machines. Jobs may be considered sources and machines as destinations. The supply for each job is 1, and the demand at each destination is 1, precisely. The costs of "transporting" jobs to machines are the costs of assigning job i to machine j, c_{ij}. The flow (assignment) variables are as follows:

$$x_{ij} = \begin{cases} 0, & \text{if the } i\text{-th job is not assigned to the } j\text{-th machine} \\ 1, & \text{if the } i\text{-th job is assigned to the } j\text{-th machine} \end{cases}$$

The mathematical programming model is given by:
Objective function:

$$\min \sum_{i=1}^{n} \sum_{j=1}^{n} c_{ij} x_{ij}$$

subject to:
Job balance:

$$\sum_{\text{machines } j=1}^{n} x_{ij} = 1, \quad \text{jobs } i = 1, 2, \ldots, n$$

Machine balance:

$$\sum_{\text{jobs } i=1}^{n} x_{ij} = 1, \quad \text{machines } j = 1, 2, \ldots, n$$

Bounds on assignments:

$$x_{ij} \geq 0; \quad i = 1, 2, \ldots, n; \quad j = 1, 2, \ldots, n$$

Integrality of the solution is guaranteed.

14.8.1 Example of Assignment Problem

Consider the cost data given in Table 14.2.
The optimal assignments are reported in Table 14.3.

Table 14.2 Hours required by each machine for each individual job

		Machine		
		1	2	3
	1	4	9	12
Job	2	15	11	10
	3	16	7	12

Table 14.3 Optimal solution for each variable for the assignment problem given in Table 14.2

Index	Variable Name	Optimal Value
1	$x_{1,1}$	1
2	$x_{1,2}$	0
3	$x_{1,3}$	0
4	$x_{2,1}$	0
5	$x_{2,2}$	0
6	$x_{2,3}$	1
7	$x_{3,1}$	0
8	$x_{3,2}$	1
9	$x_{3,3}$	0

14.9 REFERENCES

[1] Williams, H. P. Model Building in Mathematical Programming. Wiley. 1999.
[2] Vanderbei, R. J. Linear Programming: Foundations and Extensions. Kluwer's International Series. 2007.

14.10 FURTHER READING RECOMMENDATIONS

On Network Flow Problems the Following References Are Very Informative:

Taha, H. (2006) Operations Research. Prentice Hall.

14.11 EXERCISES

1. A company wants to optimize its production plan and its distribution schedule simultaneously. N_S plants (sources) are each producing N_P products, which are distributed to N_D depots (destinations). The production and distribution are planned on a monthly basis, over a horizon of N_T months. The production costs, per unit of product, are available as $c_{s,p,t}^{(1)}$ for $s = 1, 2, \ldots, N_S$, $p = 1, 2, \ldots, N_P$, $t = 1, 2, \ldots, N_T$. The transportation costs, assuming full connectivity of plants to depots, are also given as $c_{s,d,p,t}^{(2)}$ for $s = 1, 2, \ldots, N_S$, $d = 1, 2, \ldots, N_D$, $p = 1, 2, \ldots, N_P$, $t = 1, 2, \ldots, N_T$. The maximum production capacity of each plant

for one month for each product is constant over time and is given as $c_{s,p}^{(3)}$, for $s = 1, 2, \ldots, N_S$, $p = 1, 2, \ldots, N_P$. The demands at each depot are fixed and must be satisfied exactly and are given as $c_{d,p,t}^{(4)}$, for $d = 1, 2, \ldots, N_D$, $p = 1, 2, \ldots, N_P$, $t = 1, 2, \ldots, N_T$.

(i) Derive a single LP model that minimizes the total cost over the scheduling horizon. Use the minimum number of variables and constraints. Show that the overall LP problem is separable, clearly formulating the resulting independent LP subproblems.

(ii) How big are the overall LP problem and the decomposed subproblems in part (i) in the case that there are 10 plants, 100 depots, 1,000 products, and a 12-month horizon? Define the problem size in terms of variables and constraints. How many subproblems must be solved in the case of the separable formulation derived in part (i)?

(iii) For the optimization problem to be feasible, what are the relationship conditions between production capacity and demand for each product p at each point in time?

15

LP and Sensitivity Analysis, in Brief

In this chapter, a review of the interpretation of Lagrange multipliers for LP problems will be presented.

Although sensitivity analysis in LP problems covers extensive ground, we will restrict our attention to interpretation studies of active constraints.

It is noted that the Lagrange multiplier sensitivity analysis presented here can in fact be used for any NLP optimization problem as well in a similar fashion.

15.1 THE VALUE OF LAGRANGE MULTIPLIERS

As a reminder (see Chapter 7) consider the following optimization problem:

$$\min_x f(x)$$

subject to:

$h(x) = 0$ equality constraints with associated Lagrange multipliers λ

$g(x) \leq 0$ inequality constraints with associated Lagrange multipliers μ

then the following hold:

$$\lambda = -\left.\frac{\partial f}{\partial h}\right|_{x^*}$$

$$\mu = -\left.\frac{\partial f}{\partial g}\right|_{x^*} \tag{15.1}$$

Thus, Lagrange multipliers measure the sensitivity of the objective function to changes in the right-hand side of the constraints at the optimal point.

15.2 LAGRANGE MULTIPLIERS IN LP

For convenience, let us consider an LP in canonical form:

$$\min_x z = c^T x \tag{15.2a}$$

subject to:

$$Ax = b \tag{15.2b}$$

$$x \geq 0 \tag{15.2c}$$

where $x, c \in \mathbb{R}^n$, $b \in \mathbb{R}^m$, and $A \in \mathbb{R}^{m \times n}$.

The solution of any LP problem is a vertex solution. Vertex solutions of the canonical problem in Equations (15.2a–15.2c) correspond to separation of the variables into two groups: basic variables, $x_B \geq 0 \in \mathbb{R}^m$, which are solved for from Equation (15.2b), and nonbasic $x_N \in \mathbb{R}^{(n-m)}$, which are set to zero, $x_N = 0$

Using this partitioning, such that $x = \begin{bmatrix} x_B \\ x_N \end{bmatrix}$, we can write Equation (15.2b) by also partitioning matrix A:

$$Ax = \begin{bmatrix} B & N \end{bmatrix} \cdot \begin{bmatrix} x_B \\ x_N \end{bmatrix} = b$$

where $B \in \mathbb{R}^{m \times m}$, and $N \in \mathbb{R}^{m \times (n-m)}$. Hence we can write:

$$Bx_B + Nx_N = b \tag{15.3}$$

Similarly, the objective function can be written in partitioned form as:

$$z = c_B^T x_B + c_N^T x_N \tag{15.4}$$

We note that the matrix B corresponding to the basic variables x_B is invertible (to constitute a basis), and that we set the nonbasic variables to zero, $x_N = 0$. So from Equation (15.3) we get overall:

$$x_B = B^{-1} b \tag{15.5}$$

The objective function value (assuming x_B is optimal) will be given by:

$$z = c_B^T B^{-1} b \tag{15.6}$$

Now suppose that:

1. We perturb the right-hand side of the constraints by Δb, so that

$$\hat{b} = b + \Delta b \tag{15.7}$$

2. and that the active constraints remain the same, *hence the partitions x_B and x_N remain the same.*

Corresponding to this perturbation the solution changes to:

$$x_B = B^{-1} \hat{b}$$

and the new objective becomes:

$$\hat{z} = c_B^T B^{-1} \hat{b}$$
$$= c_B^T B^{-1} (b + \Delta b)$$
$$= z + c_B^T B^{-1} \Delta b$$

This gives us the change in the objective, $\Delta z = \hat{z} - z$, subject to changes in the right-hand side of the constraints, *assuming the same basic set of variables:*

$$\Delta z = c_B^T B^{-1} \Delta b \tag{15.8}$$

Considering only changes in one constraint at a time, and constraint $j = 1$, $2, \ldots, m$, we get:

$$\Delta z = c_B^T B^{-1} e_j \Delta b_j$$

where $e_j = \begin{pmatrix} 0, & 0, \cdots, & \underbrace{1}_{j\text{th location}} &, \cdots, & 0 \end{pmatrix}^T$, *i.e.* the jth unit vector. Taking the ratio

with Δb_j of both sides of the above equation we get[1]:

$$\frac{\Delta z}{\Delta b_j} = c_B^T B^{-1} e_j \tag{15.9}$$

Now, clearly we can take the limit as $\Delta b_j \to 0$, and the ratio will still be valid (as it is constant for any Δb_j, provided the basis remains the same). But now what we have is the derivative of the objective function value at the optimal point with respect to changes in the right-hand side of constraint j, so:

$$\left. \frac{\partial z}{\partial b_j} \right|_{x^*} = c_B^T B^{-1} e_j, \quad j = 1, 2, \ldots, m$$

But, by the definition of the Lagrange multipliers in Equation (15.1), what we have proven for LP is the following:

$$\lambda_j = -c_B^T B^{-1} e_j, \quad j = 1, 2, \ldots, m \tag{15.10}$$

Conclusion: Important Properties to Note for Lagrange Multipliers in LP

1. They are independent of the right-hand side of the constraints they correspond to.
2. They remain *constant* if the active set of constraints (set of basic variables) remains the same.

15.3 EXAMPLE OF SENSITIVITY ANALYSIS

Consider the following optimization problem, involving three variables:
Objective function:

$$\min_{x_1, x_2, x_3} z = 10.0x_1 + 40.0x_2 + 5.0x_3$$

Constraint 1 (equality), resource A usage:

$$1.0x_1 + 2.0x_2 + 2.5x_3 = 1200.0$$

Constraint 2 (inequality), resource B availability:

$$0.1x_1 + 5.0x_2 + 1.0x_3 \leq 200.0$$

Constraint 3 (inequality), resource C availability:

$$0.1x_1 + 0.1x_2 + 0.1x_3 \leq 1000.0$$

Lower bounds:

$$x_1 \geq 0, \quad x_2 \geq 0, \quad x_3 \geq 0$$

---

[1] note that $B^{-1} e_j$ is equal to $(B^{-1})_j$, *i.e.* the jth column of the inverse of the basis matrix.

Upper bounds:

$$x_1 \le 1000, \quad x_2 \le 5000, \quad x_3 \le 150$$

The problem represents a production planning model in which:

- x_i are the production levels of three products, $i = 1, 2, 3$
- constraint 1 is imposing the use of an exact amount of resource A, represented by its right-hand side of 1,200 units
- constraint 2 is imposing an upper bound on the use of resource B, which is 200 units
- constraint 3 is imposing an upper bound on the use of resource C, which is 1,000 units
- the lower bounds indicate the feasibility of the problem in that production levels must be nonnegative
- the upper bounds indicate the maximum production capability of the system in terms of product units (or, an upper level in demand for the three products)

15.3.1 The Solution for the Example LP Problem

The optimal solution of the problem given in the previous section was obtained with a numerical solver and it is:

- optimal production levels: $x_1^* = 933.33, \quad x_2^* = 0, \quad x_3^* = 106.67$
- optimal objective function value: $z^* = 9866.67$
- active constraints: constraint 1 (always active as it is an equality), constraint 2, lower bound for x_2

The particular solver used did not provide any Lagrange multiplier information. To get it we must carry out the following calculations:

- Define Lagrange multipliers
  - for constraints: $\lambda_1^C, \ \lambda_2^C, \ \lambda_3^C$
  - for lower bounds: $\lambda_1^L, \ \lambda_2^L, \ \lambda_3^L$
  - for upper bounds: $\lambda_1^U, \ \lambda_2^U, \ \lambda_3^U$
  - only λ_1^C is unrestricted in sign (equality constraint), the others have to be nonnegative

- Define the Lagrangian function

$$
\begin{aligned}
L(x, \lambda) = {}& 10.0x_1 + 40.0x_2 + 5.0x_3 \\
& + \lambda_1^C [1.0x_1 + 2.0x_2 + 2.5x_3 - 1200] \\
& + \lambda_2^C [0.1x_1 + 5.0x_2 + 1.0x_3 - 200] \\
& + \lambda_3^C [0.1x_1 + 0.1x_2 + 0.1x_3 - 1000] \\
& - \lambda_1^L x_1 - \lambda_2^L x_2 - \lambda_3^L x_3 \\
& + \lambda_1^U (x_1 - 1000) \\
& + \lambda_2^U (x_2 - 5000) \\
& + \lambda_3^U (x_3 - 150)
\end{aligned}
\tag{15.11}
$$

- Finding the Lagrange multipliers for active constraints

  - set the following multipliers to zero, as they correspond to inactive constraints (and bounds):

    $$\lambda_3^C = \lambda_1^L = \lambda_3^L = \lambda_1^U = \lambda_2^U = \lambda_3^U = 0$$

  - find the reduced Lagrangian by substituting the above into Equation (15.11), so that it involves only the active constraints (acting as equalities at the solution)

    $$
    \begin{aligned}
    L(x, \lambda) = {} & 10x_1 + 40x_2 + 5x_3 \\
    & + \lambda_1^C [1.0x_1 + 2.0x_2 + 2.5x_3 - 1200] \\
    & + \lambda_2^C [0.1x_1 + 5.0x_2 + 1.0x_3 - 200] \\
    & - \lambda_2^L x_2
    \end{aligned}
    $$

  - differentiate the above expression for $L(x, \lambda)$ with respect to the x_i variables, and set equal to zero to find the first set of necessary conditions of optimality, involving the Lagrange multipliers[2]

    $$
    \begin{bmatrix} \dfrac{\partial L}{\partial x_1} \\[2mm] \dfrac{\partial L}{\partial x_2} \\[2mm] \dfrac{\partial L}{\partial x_3} \end{bmatrix}
    =
    \begin{bmatrix} 10 + +1.0\lambda_1^C + 0.1\lambda_2^C \\[2mm] 40 + 2.0\lambda_1^C + 5.0\lambda_2^C - \lambda_2^L \\[2mm] 5 + 2.5\lambda_1^C + 1.0\lambda_2^C \end{bmatrix}
    =
    \begin{bmatrix} 0 \\[2mm] 0 \\[2mm] 0 \end{bmatrix}
    $$

  - solution of the above linear system yields the following values for the Lagrange multipliers corresponding to the active constraints

    $$\lambda_1^C = -12.67, \quad \lambda_2^C = +26.67, \quad \lambda_2^L = +148.0$$

15.3.2 Sensitivity Analysis of the Solution

Using the values of the Lagrange multipliers from the previous section we can engage in some analysis of the solution and find ways to improve it.

By looking at the value of the multipliers for the two active constraints in our problem, we can see that $|\lambda_1^C| < |\lambda_2^C|$ so it would be more cost-effective to improve the resource limitation imposed by constraint 2, as a change in it will have more impact on the objective function value.

For example, focusing on constraint 2, and denoting its right-hand side by b_2, we see that:

$$\left. \frac{\partial z}{\partial b_2} \right|_{x^*} = -\lambda_2^C = -26.67$$

We might thus wonder how we can reduce the objective by modifying our production facility to allow more of resource B to become available based on:

$$\Delta z = -26.67 \cdot \Delta b_2 \tag{15.12}$$

---

[2] The other set would be obtained by differentiation with respect to the λ variables, but that would yield back the original constraints involved, and for them we already have the primal solution x^* from the solver

Solution with a Small Perturbation

By trying out an increase of $\Delta b_2 = +20$ units of resource B, we predict a decrease in the objective of $\Delta z = -533.33$ using Equation (15.12), or a new value of the objective equal to $\hat{z} = 9333.33$.

To confirm this we set up the problem with the new bound for constraint 2, *i.e.* $b_2 = 220$ and solve the problem again with our solver. The new solution is:

- optimal production levels: $x_1^* = 866.67$, $x_2^* = 0$, $x_3^* = 133.3$
- optimal objective function value: $z^* = 9333.33$ (as predicted by our sensitivity analysis)
- active constraints: constraint 1 (always active as it is an equality), constraint 2, lower bound for x_2 (the same active constraints as in the nominal case)

By carrying out the analysis in the steps of 15.3.1, we find the following values for the Lagrange multipliers of the active constraints:

$$\lambda_1^C = -12.67, \quad \lambda_2^C = +26.67, \quad \lambda_2^L = +148.0$$

i.e. they are exactly identical as they were in the nominal case. This should be expected by the results found in Section 15.2.

Solution with a Large Perturbation

Let us try out a bigger increase with $\Delta b_2 = +50$ units of resource B, for which we predict a decrease in the objective of $\Delta z = -1333.33$ using Equation (15.12), or a new value of the objective equal to $\hat{z} = 8533.33$.

To confirm this we set up the problem with the new bound for constraint 2, *i.e.* $b_2 = 250$ and solve the problem again with our solver. The new solution is:

- optimal production levels: $x_1^* = 825.0$, $x_2^* = 0$, $x_3^* = 150.0$
- optimal objective function value: $z^* = 9000.0$ (higher than our prediction)
- active constraints: constraint 1 (always active as it is an equality), lower bound for x_2, upper bound for x_3 (different active constraints from the nominal case)

Clearly, our sensitivity analysis is no longer valid for $\Delta b_2 = +50$, as this leads to a different set of active constraints at the new solution. To find the new Lagrange multipliers we set all multipliers of inactive constraints to zero, *i.e.* $\lambda_2^C = \lambda_3^C = \lambda_1^L = \lambda_3^L = \lambda_1^U = \lambda_2^U = 0$, and solve for λ_1^C, λ_2^L, λ_3^U (using the procedure shown in 15.3.1) to obtain:

$$\lambda_1^C = -10.0, \quad \lambda_2^L = +20.0, \quad \lambda_3^U = +20.0$$

15.4 SUMMARY OF CHAPTER

Sensitivity analysis based on Lagrange multipliers for Linear Programming:

1. is not just local, but applies accurately for right-hand side perturbations provided the set of active constraints does not change at the new optimal solution
2. can be used to find out how to modify a process to improve the objective by changing appropriately the right-hand side of constraints (as in our example in this chapter by making available more of a resource)

3. plays a very important role in the economic evaluation of processes and retrofitting according to technical constraints

15.5 FURTHER READING RECOMMENDATIONS

LP and Parametric Programming

Pistikopoulos, E. N., Georgiadis, M. C., and Dua, V. (2007). Multi-Parametric Programming. Wiley.

16

Multiobjective Optimization

16.1 PROBLEM STATEMENT

Back in the late nineteenth century the engineer and economist Vilfrido Pareto formulated the resource allocation optimization problem. This problem dealt with the paradigm of having scarce resources which had to be distributed to meet the needs of different groups of people. And it turned out to actually be a multiobjective optimization problem, as there were several objective functions (one for each group of people), which had conflicting criteria (people had different preferences of resource investment). This is agreed to be the first multiobjective optimization problem recorded in the literature. Pareto was also one of the first to apply rigorous mathematical techniques to optimize more than one conflicting objective, which meant achieving the optimum for one objective required some compromise from the other objectives.

Multiobjective optimization problems arise naturally in chemical processes. Typical examples are the trade-offs between capital cost and operating cost, selectivity and conversion, profit and environmental impact, and safety cost, all of which are common examples in industrial practice.

In recent decades, chemical engineering has seen a rise in the solution of these problems, given that computational power now allows us to solve more challenging optimization problems.

Formally, multiobjective optimization (MOO) refers to problems where a system has many criteria that must be optimized. It is also known as multicriteria optimization, referring specifically to finding the decision variable values that correspond to and provide the optimum of more than one objective. Hence, MOO involves special methods, which are able to consider more than one objective and analyze the results obtained.

To aid in presentation, let us consider a special case of multiobjective problems, a biobjective optimization problem, defined as follows:

$$\min_{x} f_1(x) \tag{16.1a}$$

$$\min_{x} f_2(x) \tag{16.1b}$$

subject to:

$$g(x) \leq 0 \tag{16.1c}$$

$$h(x) = 0 \tag{16.1d}$$

$$x^L \leq x \leq x^U \tag{16.1e}$$

where $x \in \mathbb{R}^n$, $f_1 : \mathbb{R}^n \to \mathbb{R}$, $f_2 : \mathbb{R}^n \to \mathbb{R}$, $g : \mathbb{R}^n \to \mathbb{R}^{m_{inq}}$, $h : \mathbb{R}^n \to \mathbb{R}^{m_{eq}}$.

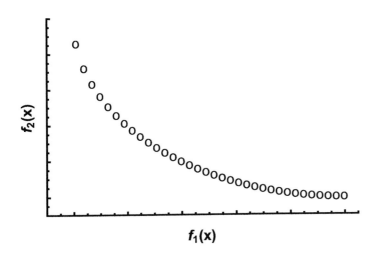

Figure 16.1 Pareto frontier.

In problems such as the biobjective problem stated in the previous paragraphs, when two objectives (*e.g.* $f_1(x)$, $f_2(x)$) are conflicting there will be several optimal solutions to the MOO problem. These solutions are known as Pareto-optimal. To characterize these different Pareto-optimal solutions we define a frontier (meaning a set of points), which can also aid in the visualization of the problem at hand. This is done as follows:

If we were to use a scalar $0 \leq \alpha \leq 1$ to combine the objective functions linearly,[1] *e.g.* $F(x, \alpha) = \alpha f_2(x) + (1 - \alpha) f_1(x)$, and plot the optimal solutions of $F(x, \alpha)$ for several values of α, we would obtain what is called a *Pareto frontier*, as shown in Figure 16.1.

Figure 16.1 shows on the vertical axis the function value of objective function $f_2(x)$, and on the x-axis the function value for objective function $f_1(x)$. On one hand, the top-left point corresponds to $\alpha = 1$, and hence this is the optimal solution for the problem where only $f_2(x)$ is taken into account. On the other hand, the bottom-right point corresponds to $\alpha = 0$, where only $f_1(x)$ is considered. For all the other points, it can be observed that a decrease in one of the functions results in an increase in the other, and hence one cannot express preference for any of the computed points simply by the mathematical formulation.

Optimal solutions in a multiojective optimization problem (such as those shown in Figure 16.1) are also referred to as non-dominated, non-inferior, efficient, or simply Pareto-optimal solutions.

A generalization of the biobjective problem is a multiobjective optimization problem, and can be defined as follows:

$$\min_{x} \quad [f_1(x), f_2(x), \ldots, f_K(x)] \tag{16.2a}$$

---

[1] The method of combining linearly the objectives is called sclarization. The method, although appealing for its simplicity, will not be able to resolve sections of the Pareto front if the problem is not convex.

subject to:

$$g(x) \leq 0 \tag{16.2b}$$

$$h(x) = 0 \tag{16.2c}$$

$$x^L \leq x \leq x^U \tag{16.2d}$$

where in this case the optimization problem is comprised of k different optimization criteria.

Similar to the case of single-objective optimization, a definition of what an *optimal solution* is must be defined from a mathematical standpoint for multiobjective optimization, and the following section will define the criteria that optimal solutions must meet in these problems.

16.2 PARETO OPTIMALITY THEORY

In this section we present the optimality definitions relevant to multiobjective optimization. It is important to note that continuous multiobjective optimization problems have an infinite number of Pareto-optimal solutions. Furthermore, problems arising from chemical engineering systems may have a Pareto set (consisting of the Pareto-optimal solutions) which could be nonconvex and disconnected.

Another important remark on Pareto optimality theory is that locally Pareto-optimal solutions have no practical relevance (unless this is also a global optimum), because they are located in the interior of the feasible objective region (*i.e.* it is possible to improve at least one objective function value without decreasing another). Whereas globally Pareto-optimal solutions are always located on the boundary of the feasible objective region (*i.e.* no objective function can be further improved without decreasing another). Multiobjective optimization problems may be convex if the feasible objective region is convex or if the feasible region is convex and the objective functions are quasiconvex with at least one beind a strictly quasiconvex function. In this chapter we will describe general nonconvex multiobjective problems unless stated otherwise.

In the context of multiobjective optimization it is useful to define a *decision maker* (DM). A DM is one or more individuals entrusted with the task of selecting one of the Pareto-optimal solutions for implementation based on their experience and other considerations not included in the multiobjective optimization problem.

Let us start by defining the *ideal*, *nadir* and *utopian* objective vectors, which are used both for theoretical and practical purposes. We define the set of feasible x as:

$$x = \left\{ x \in \mathbb{R}^n \mid h(x) = 0, \ g(x) \leq 0, \ x^L \leq x \leq x^U \right\}$$

Ideal Objective Vector

This vector contains the optimal objective values if they were to be optimized individually (*i.e.* the best possible value for each objective function). This is mathematically defined as follows:

$$f_i^{\text{ideal}} = \inf_{x \in X} \ f_i(x), \quad i = 1, 2, \ldots, K$$

It is known as an *ideal* objective vector because it will be unreachable (unless your objectives are aligned); however, it has several practical uses, such as defining a reference point to which one would like to get as close as possible.

Nadir Objective Vector

The nadir objective vector contains the worst (*i.e.* greatest) objective function values in the Pareto frontier. In other words, it contains the upper bounds of the objectives in the Pareto frontier. This can be defined by

$$f_i^{\text{nad}} = \sup_{x \in X \text{ is Pareto-optimal}} f_i(x), \quad i = 1, 2, \dots, K$$

Unfortunately, no efficient method exists to calculate the nadir objective vector, as in general the Pareto set contains an infinite number of solutions.

Utopian Objective Vector

For numerical reasons it is, in many cases, advantageous to define an objective vector that is just ε worse than the ideal objective vector. This is the case of the utopian objective vector, defined as follows

$$f_i^{\text{uto}} = f_i^{\text{ideal}} - \varepsilon, \quad i = 1, 2, \dots, K$$

where ε is a small positive constant.

The nadir and ideal points are schematically presented in Figure 16.2.

Even if solutions appear in the Pareto frontier, they can be distinguished depending on their quality with respect to other Pareto points, and can be classified as efficient, weak, strict, proper, among other definitions. These categories of Pareto-optimal solutions are introduced in the following sections.

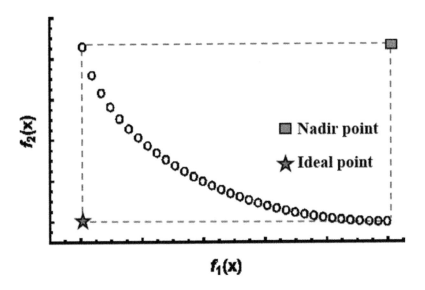

Figure 16.2 Nadir and ideal points.

16.2.1 Pareto-Optimal (Efficient) Solutions

Let us define an equivalent multiobjective optimization problem as in Equations (16.2a–16.2d) as:

$$\min_{x \in \mathcal{X}} \left[f_1(x), f_2(x), \ldots, f_K(x) \right] \tag{16.3}$$

Let us next define the image of the feasible set $\mathcal{X}$ under the mapping of the objective function vector f to be $\mathcal{Y} = f(\mathcal{X})$. We now define Pareto-optimal solutions more formally.

Definition A feasible solution $\hat{x} \in \mathcal{X}$ is called Pareto-optimal (also called *efficient*), if there is no other $x \in \mathcal{X}$ such that $f(x) \leq f(\hat{x})$. If $\hat{x}$ is Pareto-optimal, $f(\hat{x})$ is known as a non-dominated point. If x^1, $x^2 \in \mathcal{X}$ and $f(x^1) \leq f(x^2)$ we say x^1 dominates x^2 and $f(x^1)$ dominates $f(x^2)$. The set of all Pareto-optimal (efficient) solutions $\hat{x} \in \mathcal{X}$ is denoted $\mathcal{X}_E$ and known as the efficient set. The set of all non-dominated points $\hat{y} = f(\hat{x}) \in \mathcal{Y}$, where $\hat{x} \in \mathcal{X}_E$, is denoted $\mathcal{Y}_N$ and known as the non-dominated set.

This defines $\hat{x}$ as Pareto-optimal if no other feasible vector of decision variables, $x \in \mathcal{X}$, exists, which would decrease some criterion without causing a simultaneous increase in at least one other criterion. In practice, any x which is not Pareto-optimal cannot represent a most preferred alternative for a DM, because there exists at least one other feasible solution $\hat{x} \in \mathcal{X}$ for which $f_i(\hat{x}) \leq f_i(x)$ for all $i = 1, 2, \ldots, K$, where strict inequality holds at least once ($f_j(\hat{x}) < f_j(x)$ for some $j \in \{1, 2, \ldots, K\}$), and hence $\hat{x}$ should be clearly preferred over x.

16.2.2 Weakly and Strictly Pareto-Optimal Solutions

A strictly *Pareto-optimal solution* is such that no other point has an objective function $f_i(x)$ for all $i = 1, 2, \ldots, K$ with a better function value. On the other hand, a weakly Pareto-optimal solution is such that there is no other single point that has a better objective function $f_i(x)$ for all $i = 1, 2, \ldots, K$. These are formally defined as follows:

Definition A feasible solution $\hat{x} \in \mathcal{X}$ is known as strictly Pareto-optimal (strictly efficient) if there is no $x \in \mathcal{X}$, $x \neq \hat{x}$ such that $f_i(x) \leq f_i(\hat{x})$ for all $i = 1, 2, \ldots, K$. A strictly efficient set is denoted $\mathcal{X}_{sE}$.

The above definition states that when a strictly efficient Pareto-optimal is compared with any other point, it will have at least one objective function value which is better than the other point.

Definition A feasible solution $\hat{x} \in \mathcal{X}$ is known as weakly Pareto-optimal (weakly efficient) if there is no $x \in \mathcal{X}$ that $f_i(x) < f_i(\hat{x})$ for all $i = 1, 2, \ldots, K$. The point $y = f(\hat{x})$ is then known as weakly non-dominated. A weakly efficient and a non-dominated set are denoted $\mathcal{X}_{wE}$ and $\mathcal{Y}_{wN}$, respectively.

The above definition states that no point can have a better objective function value for all objectives when compared with a weakly efficient Pareto-optimal point.

From the above definitions it is possible to state:

$$\mathcal{Y}_N \subset \mathcal{Y}_{wN}$$

and

$$\mathcal{X}_{sE} \subset \mathcal{X} \subset \mathcal{X}_{wE}$$

It should be noted that for sets $\mathcal{Y} \subset \mathbb{R}^K$ there is no such concept for strict non-dominance. By definition, strict efficiency prohibits solutions x^1, $x^2 \in \mathcal{X}$ with $f(x^1) = f(x^2)$, *i.e.* strict efficiency is the multicriterion analog of unique optimal solutions in scalar optimization.

16.2.3 Proper Pareto-Optimal Solutions

Another optimality concept in multiobjective optimization is that of *properly Pareto-optimal*. This is defined as follows.

Definition A decision vector $\hat{x} \in \mathcal{X}$ is properly Pareto-optimal (in the sense of Geoffrion [3]) if it is Pareto-optimal and if a real number M exists such that for each $f_i(x)$ and each $x \in \mathcal{X}$ satisfying $f_i(x) < f_i(\hat{x})$ at least one $f_j(\hat{x})$ exists such that $f_j(\hat{x}) < f_j(x)$ and:

$$\frac{f_i(\hat{x}) - f_i(x)}{f_i(x) - f_i(\hat{x})} \leq M$$

Intuitively, a properly Pareto-optimal solution follows that there is at least one pair of objectives for which a finite decrease in one objective is possible only at the expense of some reasonable increase in the other objective.

For the interested reader, further theoretical properties derived from the multiobjective (or multicriterion) optimization domain can be found in [3].

16.3 SOLUTION PROCEDURES GENERATING PARETO POINTS

Before presenting different solution procedures, let us first classify solution methods according to the input needed from the *decision maker* (DM) to get an optimal solution. As mentioned earlier, a DM can be thought of as an agent who needs to choose between the different Pareto-optimal solutions; therefore solution methods are often classified according to the role of a DM in the solution process.

The **neutral-preference method** is the approach where the aim is to find some neutral-compromise solution without any additional preference information. This means that instead of having a particular preference for some objectives over others, some assumptions are made about what a "reasonable" compromise is. This is the only approach where the DM is not assumed to take part of the solution process.

A priori methods are those where the DM states preference information and the solution process finds a Pareto-optimal solution satisfying them as much as possible. This is a straightforward approach but the difficulty is that the DM must know the possibilities and limitations of the problem beforehand as well as the impact of the defined preferences.

In *a posteriori* methods Pareto-optimal solutions are first generated and then the DM selects the most preferred among them. This approach gives more flexibility to the DM; however, if more than two objectives are present, this advantage might be diminished as it may be difficult to visualize and analyze the solutions computed. Furthermore, computing the set of Pareto-optimal solutions may be computationally expensive, and in most cases evolutionary multiobjective optimization algorithms are used. However, evolutionary algorithms have the drawback that the real Pareto-optimal set might not be reached. This means that the solutions produced are non-dominated in the current population, but not necessarily actually Pareto-optimal (*e.g.* if the execution time is limited).

Interactive methods are iterative solution frameworks where the DM exchanges information with the solution process. After each iteration, some information is given to the DM and in turn the DM specifies new preferences (or may maintain the same ones) given this new information. What is noteworthy is that the DM can adjust preferences between each iteration and at the same time learn about the interdependencies in the problem [1].

It should be noted that the above is one of the classifications of multiobjective optimization; however, there are other possible classifications depending on the focus of the author *e.g.* [20, 5].

Test Case

In this section different approaches to address multiobjective optimization problems will be presented, and to aid in understanding, a small example problem will be used.

Let us consider a problem where we have two conflicting objectives and one decision variable. This problem, schematically presented in Figure 16.3, is the following:

$$\min_{x} x^2 - 4x + 5 \tag{16.4a}$$

$$\max_{x} \sqrt{x+1} \tag{16.4b}$$

subject to:

$$-1 \le x \le 10 \tag{16.4c}$$

which can be reformulated as follows:

$$\min_{x} \left[x^2 - 4x + 5, \ -\sqrt{x+1} \right] \tag{16.4d}$$

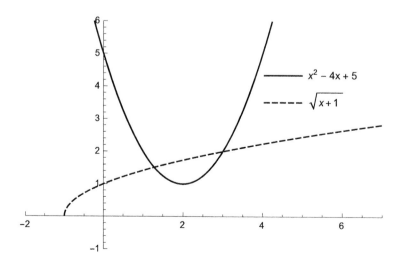

Figure 16.3 Objective functions in example problem.

subject to:

$$-1 \le x \le 10 \qquad (16.4e)$$

Let us note that this problem has a unique solution for each objective function, $x_1 = 2$ and $x_2 = 10$ for $f_1(x) = x^2 - 4x + 5$ and $f_2(x) = -\sqrt{x+1}$, respectively. Furthermore, it is also interesting to note that there are an infinite number of Pareto-optimal solutions, for different degrees of compromise between the two objective functions.

As already mentioned, this problem will be used to exemplify some of the solution strategies presented in this section. Let us start by presenting two general solution approaches to multiobjective optimization, the Weighted Sum Method and the ε-Constraint Method.

16.3.1 The Weighted Sum Method

The Weighted Sum Method, also known as the *scalarization method*, defines the following optimization problem:

$$\min \sum_{i=1}^{k} w_i f_i(x) \qquad (16.5)$$

subject to $x \in \mathcal{X}$, where $w_i \ge 0$ for all $i = 1, 2, \ldots, K$ and $\sum_{i=1}^{K} w_i = 1$.

The solution of the optimization problem in Equation (16.5) will produce a weakly Pareto-optimal solution, and if $w_i > 0$ for all $i = 1, 2, \ldots, K$ a Pareto-optimal solution.

The weighting method can be used as an *a posteriori* method so that different weights are used to generate a Pareto frontier and then the most satisfactory solution can be chosen. Alternatively, specific weights can be determined beforehand given the preferences of a DM, in which case the method is used as an *a priori* method.

Weighted Sum Method

This example can be reformulated by the Weighted Sum Method as follows:

$$\min \quad \alpha \left(x^2 - 4x + 5 \right) - (1 - \alpha) \sqrt{x+1} \qquad (16.6)$$

subject to:

$$-1 \le x \le 10 \qquad (16.7)$$

such that from varying values of $0 \le \alpha \le 1$ it is possible to obtain different Pareto-solutions. By scaling the objectives by $1 / \left(f_i^{\mathrm{nad}} - f_i^{\mathrm{ideal}} \right)$, the Pareto frontier in Figure 16.4 is obtained.

General Remarks on the Weighted Sum Method

It is advisable to normalize the objectives with some scaling so that all functions are of similar magnitudes. Systematic ways of perturbing the weights to obtain different Pareto-optimal solutions can be reviewed in [6]. Furthermore, a commonly used scalarization method is to divide each objective function by their corresponding range

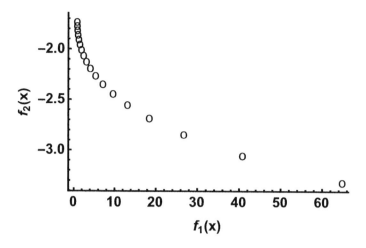

Figure 16.4 Pareto frontier by Weighted Sum Method.

of $f_i(x)$ in the Pareto-optimal set, characterized by the *nadir* and *utopian* objective values, that is, $f_i^{\text{nad}} - f_i^{\text{uto}}$ [1].

Although this method is easy to implement, it has some serious drawbacks. It can be proven that any Pareto-optimal solution can be found by altering the weights only if the problem is convex. If the problem is nonconvex, however, some Pareto-optimal solutions may not be found no matter how the weights are selected. The requirements for which the whole Pareto optimal set can be generated by the weighting method are presented in detail in [7].

Another issue with the use of the Weighted Sum Method is that if some of the objective functions correlate with each other, then different weight configurations may not produce expected solutions; this is further explored in [8] and [9].

Generally speaking, if the problem is convex, the Weighted Sum Method is a good alternative to generate a Pareto-frontier and provide information to the DM if an *a priori* approach is desired. It is also a preferable method if the DM has clear preferences for different objectives that can be translated into weights. If none of the previous are true, other alternatives might be more appealing to solve multiobjective optimization problems.

16.3.2 ε-Constraint Method

The ε-Constraint Method reformulates the optimization problem by selecting a single objective function, and converting the others into inequalities:

$$\min \quad f_l(x) \tag{16.8}$$

subject to:

$$f_j(x) \le \varepsilon_j \quad j = 1, .., k, \ j \ne l \tag{16.9}$$

$$x \in \mathcal{X} \tag{16.10}$$

where l is the selected objective function, and ε_j are upper bounds on objectives.

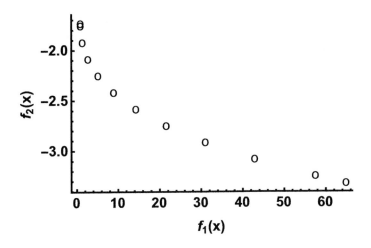

Figure 16.5 Pareto frontier by ε-Constraint Method.

The solution of the problem in Equations (16.8–16.10) will always be a weakly Pareto-optimal solution. Furthermore, in the nonconvex case, if one wants to guarantee a Pareto-optimal solution x^*, the problem must be solved for every $l = 1, 2, \ldots, K$, where $\varepsilon_j = f_j(x^*)$ for every $j = 1, 2, \ldots, K$, $j \neq l$. As a result, to ensure Pareto optimality K different problems must be solved, which incurs a large computational cost. Note that ε_j has no sign restriction, and can be positive or negative.

ε-Constraint Method

For this example we will arbitrarily choose $f_1(x) = x^2 - 4x + 5$ to be minimized, while changing the upper bound ε_2 for function $f_2(x) = -\sqrt{x+1}$.

$$\min \quad x^2 - 4x + 5 \tag{16.11}$$

subject to:

$$-\sqrt{x+1} \leq \varepsilon_2 \tag{16.12}$$

$$-1 \leq x \leq 10 \tag{16.13}$$

where $-3.31 \leq \varepsilon_2 \leq 0$ is chosen between the *ideal* and *nadir* values. The Pareto-frontier computed by this approach can be observed in Figure 16.5.

Let us highlight that the choice of function to minimize is arbitrary, and the reader is encouraged to solve the problem by minimizing the other objective function.

General Remarks on the ε-Constraint Method

If the number of objective functions is large, it may be difficult to determine upper bounds for each objective, as one might end up with an infeasible problem. Possible ways of obtaining a number of Pareto-optimal solutions by perturbing the upper bounds can be found in [6].

Similarly to the Weighted Sum Method, this method can be used in both the *a posteriori* and *a priori* approaches. In an *a posteriori* framework the information gathered from the inequality constraints can be of use to the DM. In an *a priori* approach, the approximate values of the upper bound on objective functions can be specified, and then these can be perturbed around these values.

Comparing against the Weighted Sum Method, specifying upper bounds can be more intuitive for the DM than specifying weights, because objective function values have physical significance in engineering applications. However, the drawback here is that if there is a promising solution really close to the bound but on the infeasible side, it will never be found. In other words, the bounds are a very stiff way of specifying preference information, while weights present more flexibility. However, while the Weighted Sum Method is not a good alternative for nonconvex problems, if the number of objective functions is small, the ε-Constraint Method would be a suitable method to solve such problem.

16.3.3 Neutral Preference Methods

Neutral preference methods solve multicriterion optimization problems and they try to find the best solution which represents a compromise between all objective functions.

Method of Global Criterion

The method of global criterion [10, 11] defines some reference point (generally the ideal or utopia point) and chooses the Pareto-point that is closest (with respect to some norm) to this reference point.

If the L_2 norm is used, then the Method of Global Criterion would define the following optimization problem:

$$\min_{x \in \mathcal{X}} \sum_{i=1}^{K} \left(f_i(x) - f^* \right)^2 \tag{16.14}$$

where f^* is the reference point. Solving problem (16.14) will result in a Pareto-optimal solution. This approach can be tailored to be an *a posteriori* approach where different norms are used to compute different solutions, hence giving more information to the DM to choose the best candidate.

Method of Global Criterion

For this example we will use the *ideal* point as the reference point and the L_2 norm is chosen. The Method of Global Criterion (with the previously mentioned scaling) yields the following optimization problem:

$$\min \left(\frac{x^2 - 4x + 5 - f_1^*}{G_{c1}} \right)^2 + \left(\frac{-\sqrt{x+1} - f_2^*}{G_{c2}} \right)^2 \tag{16.15}$$

subject to:

$$-1 \leq x \leq 10 \tag{16.16}$$

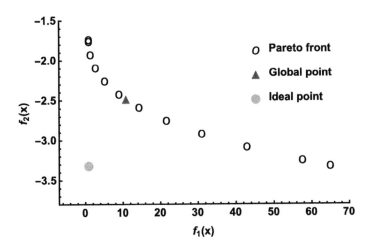

Figure 16.6 Pareto frontier and Method of Global Criterion point.

where $G_{c1} = f_1^{nad} - f_1^{ideal}$, $G_{c2} = f_2^{nad} - f_2^{ideal}$, $f_1^* = 1$ and $f_2^* = -3.31$. The *global point* computed by this approach, along with the Pareto frontier computed by the ε-Constraint Method and the ideal point are shown in Figure 16.6.

General Remarks on the Method of Global Criterion

- It must be noted that the norm used will affect the solution obtained.
- The method works properly only if the objective functions are scaled to a uniform magnitude.
- A preferred scaling method is to divide each term inside the norm by the corresponding range of $f_i(x)$ in the Pareto optimal set, characterized by the *nadir* and *utopian* objective values, that is, $f_i^{nad} - f_i^{uto}$ [1].
- It is possible to use this approach as an *a posteriori* method, in which case the DM defines different weight combinations to obtain different sets of Pareto fronts.

Neutral Compromise Solution Method

The idea behind this method is to project a point located in the middle of the ranges of objective values in the Pareto optimal set to become feasible. Components of such a point can be obtained as the average of the ideal (or utopian) and nadir values of each objective function. This approach defines the following optimization problem:

$$\min_{x \in \mathcal{X}} \max_{i=1,2,\dots,K} \left[\frac{f_i(x) - \frac{f_i^{uto} + f_i^{nad}}{2}}{f_i^{nad} - f_i^{uto}} \right] \tag{16.17}$$

In Equation (16.17) the utopian objective function value is used; however, the ideal point can be used instead. The solution of Equation (16.17) produces weakly Pareto-optimal solutions.

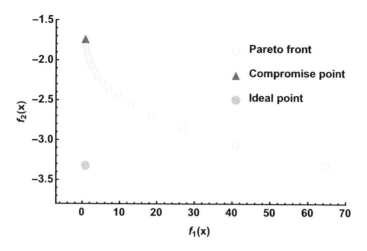

Figure 16.7 Pareto frontier and compromise point.

Neutral Compromise Solution Method

Let us now formulate the optimization problem by using the Neutral Compromise Solution Method:

$$\min \quad \max \left[\frac{x^2 - 4x + 5 - \frac{f_1^{uto} + f_1^{nad}}{2}}{f_1^{nad} - f_1^{uto}}, \quad \frac{-\sqrt{x+1} - \frac{f_2^{uto} + f_2^{nad}}{2}}{f_2^{nad} - f_2^{uto}} \right] \tag{16.18}$$

subject to:

$$-1 \leq x \leq 10 \tag{16.19}$$

The *compromise point* computed by the Neutral Compromise Solution Method is shown in Figure 16.7, along with the Pareto frontier computed by the ε-Constraint Method and the ideal point for comparison.

16.3.4 A Priori Methods

In the *a priori* approach, a DM has enough information (or knows the system well enough) to specify particular preferences before the solution process. Procedures that follow this approach are the following.

Value Function Method

In this approach, a *value function* is defined $v : \mathbb{R}^k \to \mathbb{R}$, such that it maps the preferences established by the DM. The *value function* is assumed to be non-increasing with the increase of objective values because we are considering that objective functions are to be minimized, while the value function is to be maximized. This means that the *value function* will increase (be better) if the value of an objective function decreases. The optimization problem can be defined as follows:

$$\max_{x \in \mathcal{X}} \; v\left(\mathbf{f}(x)\right) \tag{16.20}$$

where $\mathbf{f}(x) = \left[f_1(x), \; f_2(x), \dots, f_k(x) \right]$. This approach can be proven to yield Pareto-optimal solutions. Unfortunately, the difficulty in this approach is that it relies on the ability to construct the function v, which can be difficult, if not impossible. Furthermore, an appropriate *value function* can be difficult to optimize in itself because of its complex nature. Several of the methods presented here are of this nature (*e.g.* the Weighted Sum Method when $\alpha \neq 0$). For more discussion on the subject see [1].

Lexicographic Ordering Method

In this approach, all the objectives are arranged in order of importance, and the procedure follows a philosophy where a more important objective is infinitely more important than a less important objective. After the functions are ordered, the framework minimizes the most important objective. If this problem has a unique solution, it is the final one and the solution process stops. Otherwise, the second most important objective function is minimized, with a constraint to guarantee that the most important objective function preserves its optimal value. If this problem has a unique solution, the solution process stops. Otherwise, the process proceeds as above. The solution of the Lexicographic Ordering Method produces Pareto-optimal solutions. However, this method is not often used in practice due to several disadvantages, among which are:

1. It is very likely that the process stops before less important objective functions become relevant. This means that many of the objectives are not even taken into account, although they were initially thought of as important (hence their role as objectives).
2. It might happen that a small increase in an important objective function results in a very high decrease in a not-so-important objective, which would in general be appealing to obtain a better solution. This, however, is not taken into account by this method.

Lexicographic Ordering Method
Given that the example in Equation (16.4) has two objective functions, there are two possible solutions when Lexicographic Ordering is used. Both of these solutions along with the Pareto frontier are shown in Figure 16.8.

Goal Programming

In Goal Programming, the DM must specify aspiration levels for the objective function values z_i for $i = 1, 2, \dots, K$. The aim of the solution process is to minimize the difference between the aspired value and the objective functions' values. As a first step, goals are established as follows:

$$f_i(x) \leq z_i \quad i = 1, 2, \dots, K \tag{16.21}$$

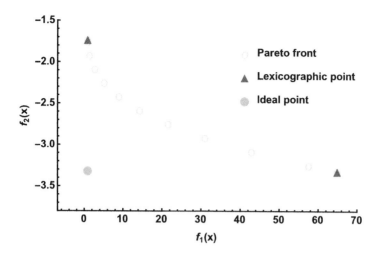

Figure 16.8 Pareto frontier and lexicographic points.

Aspiration levels in Equation (16.21) should be selected such that they cannot be satisfied simultaneously. Furthermore, deviations are defined as:

$$\delta_i = \max\left[0,\ f_i(x) - z_i\right] \tag{16.22}$$

which are to be minimized.

Then the following optimization problem is formulated:

$$\min \sum_{i=1}^{K} \delta_i \tag{16.23}$$

subject to:

$$f_i(x) - \delta_i \le z_i \quad i = 1, 2, \ldots, K \tag{16.24}$$

$$\delta_i \ge 0 \quad i = 1, 2, \ldots, K \tag{16.25}$$

$$x \in \mathcal{X}$$

The solution of Goal Programming is proven to yield Pareto-optimal solutions if either the aspiration levels form a Pareto-optimal reference point or all the variables δ_i have positive values at the optimum. This intuitively means that if all aspirations are met (and are hence feasible), the solution is equal to that reference point, which is very likely not Pareto-optimal, as it is probably the case that some objective can be improved. As a consequence, aspiration levels must be over-optimistic to ensure a Pareto-optimal solution. The interested reader is referred to [13], where further details are discussed.

This approach can be combined with previously mentioned approaches to yield variants, such as Weighted Goal Programming and Lexicographic Goal Programming.

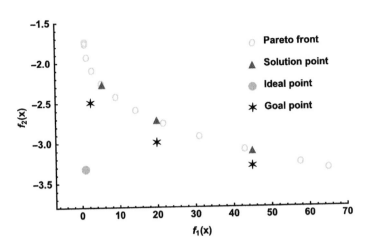

Figure 16.9 Pareto frontier and goal points.

Goal Programming

For this example, define three goals, with solution pairs:

$$[z_1, z_2] = [2.5, -2.5], \ [20, -3.0], \ [45, -3.3]$$

Each of these three goals will give alternative *goal points*. Using a scaling factor of $1/\left(f_i^{nad} - f_i^{ideal}\right)$ the solutions computed by this method are presented in Figure 16.9.

Let us note that the scaling factor may or may not be used, depending on the particular problem and on the intuition of the DM.

16.3.5 A Posteriori Methods

The aim of *a posteriori* methods is to generate a set of Pareto-optimal solutions, such that the DM can have an overview of different Pareto-optimal solutions, where if possible different procedures to generate these solutions can yield additional information and aid in the selection process. A serious limitation of this approach is that generating this set of Pareto-optimal solutions is generally computationally intensive. Furthermore, if the problem involves more than three objectives, representation and display of results can be challenging; this is discussed in detail in [14] and [1].

Method of Weighted Metrics

The Method of Weighted Metrics is a generalization of the Method of Global Criterion where different sets of weights are used to obtain a set of solutions. By using the L_2 norm, this method is defined by the following optimization problem:

$$\min_{x \in \mathcal{X}} \sum_{i=1}^{K} w_i \left(f_i(x) - f^{\dagger}\right)^2$$

where $f^\dagger$ is some reference point, which is not the ideal or utopian solution, but rather a designated point by the DM based on the system's information. This approach has the same properties as those mentioned in Section 16.3.3 as long as all weights are positive.

Achievement Scalarizing Function Approach

This approach uses scalarizing (achievement) functions, which are based on an arbitrary reference point $\bar{y} \in \mathbb{R}^k$, and the idea is to project the reference points onto the set of Pareto-optimal solutions. The difference with other approaches is that this reference point does not have to be the ideal or utopian point, but rather a reference the DM finds appealing. Different Pareto-optimal solutions can be then computed by different reference points, which provides information to the DM to choose a solution. There are many ways to produce achievement functions; one such formulation can be the following:

$$\min_{x \in \mathcal{X}} \max_{i=1,2,\dots,K} \; w_i \left(f_i(x) - \bar{y} \right) + \rho \sum_{i=1}^{k} f_i(x) - \bar{y} \tag{16.26}$$

where w_i are normalization weights, e.g. $w_i = 1/\left(f_i^{\text{nad}} - f_i^{\text{uto}}\right)$, and $\rho > 0$. This approach can be proven to produce proper Pareto-optimal solutions, and that any proper Pareto-optimal solution can be found. Achievement functions can be formulated in many ways and the interested reader can review them in [15, 16, 17, 18, 1], and different problems are formulated depending on the achievement function chosen.

16.3.6 Interactive Methods

Due to the rigid nature of *a priori* and *a posteriori* approaches, interactive methods emerge as a flexible alternative. Interactive methods can be viewed as the equivalent of a model predictive control in dynamic optimization, in the sense that it is a real-tine optimization strategy that can adapt as the problem develops.

In this approach, a DM plays a crucial part in the solution process, by searching for the most preferred solution by directly interacting with the solution algorithm(s). Hence, interactive methods iterate through both the DM, and a solution algorithm, such that they exchange information and gain insight into the problem at hand. The general framework of an interactive method is as follows:

1. Initialize: Generate a preliminary Pareto-optimal solution set (possibly by some neutral preference method) and calculate ideal, nadir, and utopia points.
2. Give computed information to the DM and ask for input *e.g.* aspiration levels and solution approach(es).
3. Generate new Pareto-optimal solution(s) given the specified preferences and information provided by the DM.
4. If one of these new solutions satisfies the DM then stop, otherwise return to step 2.

The greatest advantage of the interactive approach is that the DM can correct the preferences stated, and hence the DM does not need to imply some global preferences, but rather what is best suited as the problem unfolds. Furthermore, finding the final solution is not always the only task but it is also noteworthy that the DM gets to know the problem, and its possibilities and limitations [3]. Hence, the interactive method uses non-interactive methods as building blocks in an iterative algorithm.

Interactive methods are a broad class of methods, and an active area of research that is not reviewed in this book; for further study on this subject the reader is referred to the following excellent sources [1, 20].

16.4 PARETO SOLUTION SETS

In the previous sections we emphasized computing Pareto-optimal points; however, in many cases, we are interested in computing the *optimal Pareto set*. By optimal Pareto set we mean the set of solutions that conform the Pareto frontier. Let us note that computing the Pareto frontier is a difficult problem, largely of interest in nonconvex optimization, and hence although sometimes local optimization methods are used (*e.g.* gradient based methods), most of the emphasis for computing Pareto solution sets has been on evolutionary methods.

Similarly, as there are many algorithms that compute Pareto-optimal points, there are many solution procedures that compute Pareto solution sets. Before we talk about different approaches to obtain a Pareto solution set, we must define the quality of such a solution. In the case of continuous variables, a Pareto solution set is a hyper-area, and can be conformed of an infinity of points. This means that it is many times impossible to compute all the solutions in a Pareto solution set. Therefore, metrics such as the spread, clusters, and accuracy of the Pareto set are important. Finally, let us emphasize that in this section we will be talking about the general case of nonconvex optimization problems. This means that there is a difference between the *observed Pareto solution* set and the *true Pareto solution* set, where the former can be any Pareto set computed by an algorithm, while the latter must consist of only Pareto-optimal (efficient) solutions as described in Section 16.2.1. Notice that an observed Pareto solution set is an estimate (and most of the time a discrete representation) of a true Pareto solution set.

16.4.1 Quality Metrics of Observed Pareto Solution Sets

Several quality metrics for Pareto solution have been proposed, in this chapter we have opted to focus mostly on those reviewed and evaluated in [21]. Let us now define a few useful concepts to measure Pareto solutions.

Scaled objective space

If we are to combine different objectives to create a Pareto metric, we should make sure to give all objectives the same value, therefore comparisons between objectives must be done in the scaled objective space. We can scale each objective i by using the Ideal and Nadir points as follows:

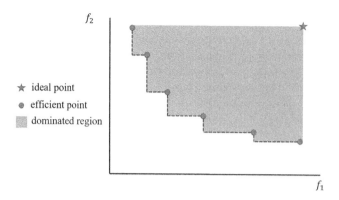

Figure 16.10 Representation of the dominated region of a Pareto solution set in the two dimensional case (two objectives)

$$\overline{f}_i(x) = \frac{f_i(x) - f_i^{ideal}}{f_i^{nad} - f_i^{ideal}}$$

Let us note that in the scaled objective space $f^{ideal} = (0, 0, ..., 0)$ and $f^{nad} = (1, 1, ..., 1)$

Dominant Region of a point

In the scaled objective space, we now define the dominant region: The dominant region of a solution $\overline{f}^j = \left(\overline{f}_1(x), \overline{f}_2(x), ..., \overline{f}_K(x)\right)$ is defined as the hyper-rectangle $\underline{HR}\left(\overline{f}^j\right)$ such that all points (solutions) inside this hyper-rectangle (i.e. $\overline{f}^i \in \underline{HR}\left(\overline{f}^j\right)$) must be $\overline{f}^i \prec \overline{f}^j$, where $\prec$ denotes entry-wise strict inequality. This means that any point $\overline{f}^i$ is better than $\overline{f}^j$.

Dominated Region of a point

The dominated region of a solution $\overline{f}^j = \left(\overline{f}_1(x), \overline{f}_2(x), ..., \overline{f}_K(x)\right)$ is defined as the hyper-rectangle $\overline{HR}\left(\overline{f}^j\right)$ such that all points (solutions) inside this hyper-rectangle (i.e. $\overline{f}^i \in \overline{HR}\left(\overline{f}^j\right)$) must be $\overline{f}^i \succ \overline{f}^j$, where $\succ$ denotes entry-wise strict inequality. This means that any point $\overline{f}^i$ is worse than $\overline{f}^j$.

Dominated Region of Pareto Solution Set

If a Pareto Solution Set is conformed by np number of efficient points $P = \left(\overline{f}^1, \overline{f}^2, ..., \overline{f}^{np}\right)$ the dominated region of the entire Pareto solution set, which we denote as $\overline{HR}(P)$, is defined as the union of the individual Pareto points' dominated region $\overline{HR}(P) = \bigcup_{j=1}^{np} \overline{HR}\left(\overline{f}^j\right)$. This is graphically represented in Figure 16.10.

Hyper-area Difference (HD)

Let us remind ourselves that when finding a Pareto solution set, we make a difference between the *observed* Pareto solution set and the *true* Pareto solution set. The hyper-area is used to measure the objective space dominated by an observed Pareto solution set. Therefore, if we have two algorithms, the difference between two hyper-areas can be used as a metric to asses which algorithm is better. If the true Pareto solution is know, then it is also possible to asses the difference in hyper-area between a observed Pareto solution set and the true Pareto solution set.

Pareto Spread

Another metric that is desirable for Pareto solution sets, is the spread of the *observed* Pareto points. An observed Pareto solution set that spreads over a wider range of the objective function values provides the decision maker (DM) with a broader range of options to select from. There are two metrics that quantify the spread of a Pareto solution [28], the overall Pareto spread, and the *kth* objective Pareto spread.

Overall Pareto spread (OS)

Defined as the ratio between the hyper-rectangle defined by the ideal and nadir point $HR_{I,N}$ and the hyper-rectangle defined by the observed Pareto points $\overline{HR}(P)$ as shown in Figure 16.10 (Dominated Region of Pareto Solution Set):

$$OS(P) = \frac{\overline{HR}(P)}{HR_{I,N}}$$

The larger the value of the ratio $OS(P)$, the better an observed Pareto set is considered.

kth objective Pareto spread (OS_k)

This metric provides insight on the range of the solution set with respect to each individual objective function. The kth objective Pareto spread is defined as follows:

$$OS_k(P) = \left| \max_{j=1}^{np} \overline{f}_k^j - \min_{j=1}^{np} \overline{f}_k^j \right|, k = 1, ..., K \tag{16.27}$$

Notice that Equation 16.27 gives information with respect to each k objective dimension. The higher its $OS_k(P)$ value, the better P is considered for this objective. The DM can then take preference if having more information in a specific (or a set of specific) objective dimensions is preferred.

16.4.2 Solution techniques for Pareto solution sets

In this section we give a brief overview of solution techniques that address computing Pareto solutions sets, and we give references for further study.

Evolutionary approaches

Evolutionary algorithms (EA) have been widely applied to solve these type of problems, their main advantage is that based on their population-like nature, they generate several elements of the Pareto optimal set in a single run. In addition, these approaches are effective when dealing with nonconvex, uncertain and noisy optimization problems, which are common in engineering practice. We give a general description of GAs in the case of single objective optimization.

Genetic algorithms (GA) are the discrete counterpart of EAs. GAs and EAs have become a popular method for solving large optimization problems with multiple local optima. The GA is named from the process of drawing an analogy between the components of a configuration vector x and the genetic structure of a chromosome.

The goal is to maximize a function $f(x)$ of the vector $x = (x_1, x_2, ..., x_N)$. The basic GA begins by randomly generating an even number n of x vectors, where each x_i entry is within the variable's bounds $x_i^{lb} \leq x_i \leq x_i^{ub}$. Each vector x would be defined as an individual in the initial population:

$$
\begin{matrix}
x_1^1 & x_2^1 & \cdots & x_N^1 \\
x_1^2 & x_2^2 & \cdots & x_N^2 \\
\vdots & \vdots & \ddots & \vdots \\
x_1^n & x_2^n & \cdots & x_N^n
\end{matrix}
$$

A fitness value g^j is calculated for each vector (individual) x^j:

$$g^j = f(x^j)$$

The vectors with the highest fitness values will be those most likely to be selected as the *parents* for the following iterations. The probability p^j of a vector x^j being selected as a parent is generally proportional to its fitness value g^j. The new parents are considered in pairs, and for each pair, a crossover operation is performed with a pre-selected probability p_c. If crossover occurs, an integer k is generated uniformly at random between 1 and $(N-1)$. The first k elements from a parent are chosen, and the last $N-k$ elements of the other parent are exchanged to create two new x vectors. For example, for the first pair of parents, suppose that crossover occurred; then two new vectors are:

$$
\begin{matrix}
x_1^1 & x_2^1 & \cdots & x_k^1 & x_{k+1}^2 & x_{k+2}^2 & \cdots & x_N^2 \\
x_1^2 & x_2^2 & \cdots & x_k^2 & x_{k+1}^1 & x_{k+2}^1 & \cdots & x_N^1
\end{matrix}
$$

If crossover does not occur, the parents are copied unchanged into the new generation. After the crossover operation, mutation is performed with a pre-selected probability p_m, which randomly changes values of entries from the new generation. The algorithm is allowed to continue for a specified number of generations. On termination, the x value with the highest $f(x)$ is returned, regardless of whether or not it is present in the last generation. EAs are generally referred to algorithm that adapt GAs for continuous optimization problems. Some popular EAs for multiobjective optimization are:

- A general multiobjective parallel evolutionary algorithm (GenMOP), developed at the Air Force Institute of Technology (AFIT) [32]. This algorithm employs numerous techniques to determine when different evolutionary operators should take place (e.g. mutation, mating), which makes it innovatory and efficient in many situations.
- Orthogonal Multi-Objective Evolutionary Algorithm (OMOEA) [33]. OMOEA is design to solver partiularly challenging constrained multi-objective problems. The main advantage is its strict definition of the constraints in the multiobjective problem and explicitly considering them when Pareto dominance is defined.
- Further sources for popular approaches are outlined in Section 16.5.1.

Metaheuristic approaches

Although EAs are the most widely used stochastic search techniques in multiobjective optimization, there are a few other algorithms that have been tailored for the same purpose. Many of them have shown to be quite effective in dealing with nonconvex difficult to solve problems. We briefly outline a few examples:

Simulated Annealing

Simulated annealing [34] is a stochastic local search technique that operates iteratively by choosing an element y from a neighbourhood $N(x)$ of the present configuration x; the candidate y is either accepted as the new configuration, or rejected. The acceptance probability is determined by the difference of the objective function between the current configuration $f(x)$ and the candidate configuration $f(y)$. A common choice for this acceptance probability $p(x, y, T)$ is the Metropolis acceptance probability. For a parameter $T > 0$, this is given by:

$$ p(x, y, T) = \left\{ \begin{array}{ll} 1 & \text{if } f(y) \geq f(x) \\ \exp\left[\frac{f(y) - f(x)}{T}\right] & \text{if } f(y) < f(x) \end{array} \right\} \tag{16.28} $$

From the form of Equation (16.28), it is evident that better moves are always accepted, but it is also possible to accept a move to a worse configuration than the present one. An in depth description of SA for multiobjective problems can be found in [35].

Particle Swarm Optimization

In Particle Swarm Optimisation (PSO) [36], a number of simple entities (particles) are placed in the search space of a problem or function, and each one evaluates the objective function at its current location. Each particle then determines its movement by combining its current position, the best position it has had and the position of one or more members of the swarm. The next iteration takes place after all particles have been moved. Eventually the swarm as a whole, like a flock of birds collectively foraging for food, is likely to move close to an optimum of the fitness function. The detailed description of PSO algorithms for multiobjective optimization can be reviewed in [37].

Other approaches

Other approaches such as Distributed Reinforcement Learning [38], Differential Evolution [39], Tabu Search [40], have also shown to be effective in computing the Pareto front. For the interested reader, these are reviewed in [30].

Gradient-based approaches

When multiobjective optimization problems are continuous, gradient-based algorithms to find the Pareto frontier can be efficient alternatives to their stochastic search counterparts. Gradient-based multiobjective optimization algorithms are of particular interest in these cases as they can be fast, precise and stable with respect to local convergence.

These methods generally rely on computing and using the gradient of the hypervolume indicators that represent approximation sets to the Pareto front. This has been a relatively new, but very active topic of research for computing Pareto fronts. The interested reader is directed to Section 16.5.1 for more references on the topic.

16.5 CONCLUSIONS AND FURTHER READING

16.5.1 Further reading

Multiobjective optimization is a very vast field in optimization, and there is still ongoing research to design new solution methods and more efficient algorithms. In this section we outline some resources that might be of interest to the reader who wishes to deepen into this topic.

- Theorems and further rigorous mathematical properties of different multiobjective methods can be found in the excellent book [1].
- In the case of nonconvex multiobjective optimization, finding optimal solutions for problems with multiple solution sets is an important subfield withing multiobjective optimization. It is furthermore particularly important in chemical processes. The interested reader in this regard is referred to [31]
- Given the multimodality and complexity of nonconvex multiobjective optimization problems, evolutionary algorithms and metaheuristics have flourished as solution procedures. The author is referred to the excellent book on these efficient solution procedures [30].
- Other commonly used *a posteriori* methods include Normal Boundary Intersection [3], Modified Normal Boundary Intersection [4], Normal Constraint [5, 6], Successive Pareto Optimization [7], Directed Search Domain [8], NSGA-II [9], or general evolutionary algorithms tailored for multiobjective optimization [2].
- A good reference for theoretical and algorithmic discussion on multiobjective optimization for general engineering applications can be found in [10].
- Other metrics for quality assessment of a multiobjective optimization algorithms such as *Accuracy of the Observed Pareto Frontier* (AC), Cluster (CL_μ), *Number of Distinct Choices* (NDC_μ) and scalar representation of solutions can be found in [28], and [29]
- The first approach that seems to have proposed using the gradient of the hypervolume indicator with respect to a set of decision vectors is [41]. Another efficient Gradient ascent algorithm in this same research direction can be found in [42].

16.6 PROBLEMS

Here we outline a few useful examples that will allow the reader to better understand the concepts presented (use of a computational software is advised).

1. Given the following multiobjective optimization problem:

$$\min_{x} \quad \left[x^2 - 4x + 5 + 4\sin(x), \ \frac{1}{10}(x-2)^3 + 4 \right] \tag{16.29}$$

subject to:

$$-2 \le x \le 6 \tag{16.30}$$

a) Find the Ideal and Nadir points
b) Use the the Weighted Sum Method to determine the Pareto front
c) Use the ε-Constraint Method to determine the Pareto front
d) Are the two Pareto fronts the same? Analyze the difference between both approaches
   hint: Both objective functions are nonconvex, since they are one-dimensional, you can plot them to better determine the global optimum of each

2. For the multiobjective optimization problem given above:

   a) Analyze the *Dominated Region* of the Pareto front the Weighted Sum Method and the ε-Constraint Method computed

   b) Compute the Hyper-area Difference (HD) and the Overall Pareto spread (OS) for each approach and compare

3. Implement an evolutionary algorithm (EA), with the following characteristics:

   a) A preliminary EA should take x as an input and evaluate $x^2 - 4x + 5 + 4\sin(x)$ and $\frac{1}{10}(x-2)^3 + 4$

   b) The select the fitness value of each individual based on Pareto dominance by the other individuals

   c) Crossover can be done by taking the mean between the individuals

   d) Use 20 individuals initially, after several iterations, let us say 100, your population should resemble an optimal Pareto set.

   e) Compare this solution with the Weighted Sum Method and the ε-Constraint Method

16.7 REFERENCES

[1] Miettinen, K. Nonlinear Multiobjective Optimization. Springer. 1999.

[2] Gade Pandu Rangaiah (2009). MULTI-OBJECTIVE OPTIMIZATION: Techniques and Applications in Chemical Engineering.

[3] Ehrgott, M. Multicriteria Optimization. Birkhäuser. 2012.

[4] Statnikov, R. B. Multicriteria Design: Optimization and Identification. Kluwer Academic Publishers. 1999.

[5] Rosenthal, R. E. "Principles of multiobjective optimization." Decision Science. 1985; 16(2):p. 133.

[6] Chankong, V. and Haimes, Y. Y. Multiobjective Decision Making: Theory and Methodology. Elsevier Science Publishing. 1983.

[7] Censor, Y. "Pareto optimality in multiobjective problems." Applied Mathematics and Optimization. 1977;4(1).

[8] Steuer, R. E. Multiple Criteria Optimization: Theory, Computation, and Application. Wiley. 1986.

[9] Tanner, L. "Selecting atext-processing system as a qualitative multiple criteria problem." European Journal of Operational Research. 1991;50(2):p. 179.

[10] Yu, P. L. "A class of solutions for group decision problems." Management Science. 1973;19(8):p. 946.

[11] Cochrane, J. L. and Zeleny, M. Multiple Criteria Decision Making. University of South Carolina Press. 1973;p. 262–301.

[12] Fishburn, P. C. "Lexicographic orders, utilities and decision rules: A survey." Management Science. 1974;20(11):p. 1442–1471.

[13] Jones, D. F., Tamiz, M., and Mirrazavi S. K. "Intelligent solution and analysis of goal programmes: The GPSYS system." Decision Support Systems. 1998;23(4): p. 329–332.

[14] Tanino, T., Tanaka, T., and Inuiguchi M. Multi-Objective Programming and Goal Programming: Theory and Applications. Springer. 2003.

[15] Wierzbicki, A. P. "A mathematical basis for satisficing decision making." Mathematical Modelling. 1982;3:p. 391–405.

[16] Wierzbicki, A. P. "On the completeness and constructiveness of parametric characterizations to vector optimization problems." OR Spectrum. 1986;8(2):p. 73–87.

[17] Gal, T., Stewart T. J., and Hanne, T. Multicriteria Decision Making: Advances in MCDM Models, Algorithms, Theory, and Applications. Kluwer. 1999;p. 9–39.

[18] Wierzbicki, A. P., Makowski, M., and Wessels, J. Model-Based Decision Support Methodology with Environmental Applications. Kluwer Academic Publishers. 2000; p.71–89.

[19] Haimes, Y. Y., Tarvainen, K., Shima T., and Thadathil, J. Hierarchical Multiobjective Analysis of Large-Scale Systems. Hemisphere Publishing Corporation. 1990.

[20] Statnikov, R. B. Multicriteria Design: Optimization and Identification. Kluwer Academic Publishers. 1999.

[21] Wu, J. and Azarm S. "Metrics for quality assessment of a multiobjective design optimization solution set." Journal of Mechanical Design. 2001;123(1).

[22] M. Fleischer, The Measure of Pareto Optima Applications to Multi-objective Meta-heuristics, C.M. Fonseca et al. (Eds.): EMO 2003, LNCS 2632, pp. 519–533, 2003. c Springer-Verlag Berlin Heidelberg 2003.

[23] Van Veldhuizen, David A., Multiobjective Evolutionary Algorithms: Classifications, Analyses, and New Innovations, thesis, AIR FORCE INST OF TECH WRIGHT-PATTERSONAFB OH SCHOOL OF ENGINEERING, 1999.

[24] S. Zeng, L. Ding, Y. Chen, and L. Kang. A New Multiobjective Evolutionary Algorithm: OMOEA. In Proceedings of the 2003 Congress on Evolutionary Computation (CEC'2003), volume 2, pages 898–905, Canberra, Australia, December 2003. IEEE Press.

[25] S. Kirkpatrick, C. D. Gelatt, Jr., M. P. Vecchi. Optimization by Simulated Annealing, Science, 13 May 1983, Volume 220, Number 4598.

[26] P. Serafini. Simulated Annealing for Multiple Objective Optimization Problems. In G. Tzeng, H. Wang, U. Wen, and P. Yu, editors, Proceedings of the Tenth International Conference on Multiple Criteria Decision Making: Expand and Enrich the Domains of Thinking and Application, volume 1, pages 283–292, Berlin, 1994. Springer-Verlag.

[27] J. Kennedy and R. Eberhart. Particle swarm optimization. In Proceedings of ICNN'95 - International Conference on Neural Networks, volume 4, pages 1942 1948. IEEE. ISBN 0-7803-2768-3. doi: 10.1109/ICNN.1995.488968.

[28] T. Ray and K. Liew. A Swarm Metaphor for Multiobjective Design Optimization. Engineering Optimization, 34(2):141–153, March 2002.

[29] C. E. Mariano Romero. Aprendizaje por Refuerzo en Optimización Multiobjetivo. PhD thesis, Departamento de Ciencias Computacionales, Instituto Tecnológico y de Estudios Superiores de Monterrey, Cuernavaca, Morelos, México, Marzo 2001. (In Spanish).

[30] C. Chang, D. Xu, and H. Quek. Pareto-optimal set based multiobjective tuning of fuzzy automatic train operation for mass transit system. IEE Proceedings on Electric Power Applications, 146(5):577–583, September 1999.

[31] X. Gandibleux, N. Mezdaoui, and A. Fréville. A Tabu Search Procedure to Solve Combinatorial Optimisation Problems. In R. Caballero, F. Ruiz, and References 667 R. E. Steuer, editors, Advances in Multiple Objective and Goal Programming, volume 455 of Lecture Notes in Economics and Mathematical Systems, pages 291–300. Springer-Verlag, 1997.

[32] Emmerich, M., Deutz, A.: Time complexity and zeros of the hypervolume indicator gradient field. In: Schütze, O., Coello, C.A.C., Tantar, A.-A., Tantar, E., Bouvry, P., Moral, P.D., Legrand, P. (eds.) EVOLVE - A Bridge between Probability, Set Oriented Numerics, and Evolutionary Computation III. SCI, vol. 500, pp. 169–193. Springer (2014).

[33] Hao Wang, Andre Deutz, Thomas Back, and Michael Emmerich, Hypervolume Indicator Gradient Ascent Multi-objective Optimization, EMO 2017, LNCS 10173, pp. 654–669, 2017.

[34] S. Kirkpatrick, C. D. Gelatt, Jr., M. P. Vecchi. Optimization by Simulated Annealing, Science, 13 May 1983, Volume 220, Number 4598.

[35] P. Serafini. Simulated Annealing for Multiple Objective Optimization Problems. In G. Tzeng, H. Wang, U. Wen, and P. Yu, editors, Proceedings of the Tenth International Conference on Multiple Criteria Decision Making: Expand and Enrich the Domains of Thinking and Application, volume 1, pages 283–292, Berlin, 1994. Springer-Verlag.

[36] J. Kennedy and R. Eberhart. Particle swarm optimization. In Proceedings of ICNN'95 - International Conference on Neural Networks, volume 4, pages 1942 1948. IEEE. ISBN 0-7803-2768-3. doi: 10.1109/ICNN.1995.488968.

[37] T. Ray and K. Liew. A Swarm Metaphor for Multiobjective Design Optimization. Engineering Optimization, 34(2):141–153, March 2002.

[38] C. E. Mariano Romero. Aprendizaje por Refuerzo en Optimizacion Multiobjetivo. PhD thesis, Departamento de Ciencias Computacionales, Instituto Tecnologico y de Estudios Superiores de Monterrey, Cuernavaca, Morelos, Mexico, Marzo 2001. (In Spanish).

[39] C. Chang, D. Xu, and H. Quek. Pareto-optimal set based multiobjective tuning of fuzzy automatic train operation for mass transit system. IEE Proceedings on Electric Power Applications, 146(5):577–583, September 1999.

[40] X. Gandibleux, N. Mezdaoui, and A. Freville. A Tabu Search Procedure to Solve Combinatorial Optimisation Problems. In R. Caballero, F. Ruiz, and References 667 R. E. Steuer, editors, Advances in Multiple Objective and Goal Programming, volume 455 of Lecture Notes in Economics and Mathematical Systems, pages 291–300. Springer-Verlag, 1997.

[41] Emmerich, M., Deutz, A.: Time complexity and zeros of the hypervolume indicator gradient field. In: Schuetze, O., Coello, C.A.C., Tantar, A.-A., Tantar, E., Bouvry, P., Moral, P.D., Legrand, P. (eds.) EVOLVE - A Bridge between Probability, Set Oriented Numerics, and Evolutionary Computation III. SCI, vol. 500, pp. 169–193. Springer (2014).

[42] Hao Wang, Andre Deutz, Thomas Back, and Michael Emmerich, Hypervolume Indicator Gradient Ascent Multi-objective Optimization, EMO 2017, LNCS 10173, pp. 654–669, 2017.

16.8   FURTHER READING

Multiobjective optimization is a very vast field in optimization, and there is still ongoing research to design new solution methods and more efficient algorithms.

Theorems and Further Rigorous Mathematical Properties of Different Multiobjective Methods

Miettinen, K. (1999). Nonlinear Multiobjective Optimization. Springer.

Nonconvex Multiobjective Optimization

Tarafder, A., Rangaiah, G. P., and Ray, A. K. (2007). "A study of finding many desirable solutions in multi-objective optimization of chemical processes." Computers and Chemical Engineering. 31, p. 1257–1271.

Evolutionary Algorithms and Metaheuristics

Coello Coello, C., Lamont, G. B., and van Veldhuisen, D. A. (2007). Evolutionary Algorithms for Solving Multi-Objective Problems. Springer.

Normal Boundary Intersection

Das, I. and Dennis, J. E. (1998). "Normal-boundary intersection: A new method for generating the pareto surface in nonlinear multicriteria optimization problems." SIAM Journal on Optimization. 8(3):p. 631.

Modified Normal Boundary Intersection

Motta, R. S., and Afonso, S. M. B., and Lyra, P. R. M. (2012). "A modified NBI and NC method for the solution of N-multiobjective optimization problems." Structural and Multidisciplinary Optimization. 46(2).

Normal Constraint

Messac, A., Ismail-Yahaya, A., and Mattson, C. A. (2003). "The normalized normal constraint method for generating the Pareto frontier." Structural and Multidisciplinary Optimization. 25(2).

Messac, A. and Mattson, C. A. (2004). "Normal constraint method with guarantee of even representation of complete Pareto frontier." AIAA Journal. 42(10):p. 2101.

Successive Pareto Optimization

Mueller-Gritschneder, D., Graeb, H., and Schlichtmann, U. (2009). "A successive approach to compute the bounded pareto front of practical multiobjective optimization problems." SIAM Journal on Optimization. 20(2):p. 915–934.

Directed Search Domain

Erfani, T. and Utyuzhnikov, S. V. (2011). "Directed search domain: A method for even generation of pareto frontier in multiobjective optimization." Journal of Engineering Optimization. 43(5):p. 1–18.

NSGA-II

Deb, K., Pratap, A., Agarwal, S., and Meyarivan, T. (2002). "A fast and elitist multiobjective genetic algorithm: NSGA-II." IEEE Transactions on Evolutionary Computation. 6(2): p. 182.

General evolutionary algorithms for multiobjective optimization

Rangaiah, G. P. and Bonilla-Petriciolet, A. (2013). Multi-Objetive Optimization in Chemical Engineering. Wiley.

17

Optimization under Uncertainty

17.1 INTRODUCTION

During the course of this book, it has been assumed that the models we wish to optimize have parameters known *a priori* with full certainty; however, this is rarely the case in real problems. This means that the parameters given in our models can vary within a range of values. Taking this into account, and making some assumptions about these variations, optimization under such uncertainty can be carried out. Optimization problems involving stochastic models occur in many applications in chemical engineering, and in reality there are very few situations in which there is no uncertainty. Therefore, the theory and application of optimization under uncertainty will be presented in this chapter.

First, it is necessary to understand the sources of uncertainty.

Uncertainty arises due to many reasons; for example, future uncertain events such as product demand in the following years (or even day!), weather conditions, stock prices, *etc.* One source of uncertainty can be attributed to difficulties to model systems accurately, giving rise to uncertain system parameter values as well as uncertain relations within the model. An example of this in a chemical engineering context is the kinetics inside reactors, for which reaction parameters and kinetic equations are unknown. Yet another source of uncertainty is human error, or machine precision, such as when measuring chemical of physical properties of a system.

Second, it is required to know how uncertainty can be modeled.

As engineers we know how to model reactors, distillation columns, heat exchangers, among many other operating units and physical structures without taking into account uncertainty. One of the most common approaches to represent uncertainty, and a quite successful one, is to make certain parameters in our models uncertain. In this way, we can construct a model, assigning deterministic values to those parameters we know behave without uncertainty (or whose uncertainty is insignificant), and assign randomness to parameters that we know (or suspect) are uncertain. These uncertain parameters in our model are often represented as a random variable with a known probability distribution. This probability distribution can be determined by statistical data, knowledge about the system, or by experience. It is also possible that we do not have enough information to assign a probability distribution; in this case, most of the time an interval within a uniform probability distribution is assumed.

Two of the main approaches to address optimization under uncertainty will be presented next, namely, *Stochastic programming* and *Robust optimization*.

We first define a deterministic optimization problem as follows:

$$\min \quad f(x, \xi)$$

subject to:

$$g_i(x, \xi) \leq 0, \ i = 1, \ldots, m_g$$

where $x \in \mathbb{R}^n$ are the decision variables and $\xi \in \mathbb{R}^k$ are the problem parameters. In this case, the deterministic optimization case, ξ are fixed uncertainty-free parameters.

17.1.1 Stochastic Programming

If one wishes to model the ξ parameters such that they capture uncertainty from the problem, one approach can be to use *stochastic optimization* (or *stochastic programming*). A stochastic optimization problem can be defined as follows:

$$\min \ \mathbb{E}_{\xi \in \Xi} \left[f(x, \xi) \right]$$

subject to:

$$g_i(x, \xi) \leq 0, \ \xi \in \Xi - \text{a.s.} \ \forall i = 1, \ldots, m_g$$

In this case ξ is a random vector with some probability distribution defined by the set Ξ. For stochastic optimization problems, some performance index taking into account this distribution is optimized; in this case, the expected value of the objective function with respect to the probability distribution is minimized. The constraints are required to be satisfied almost surely. This entails that they will be satisfied with probability 1. Loosely speaking, this means that there can be exceptions; however, they happen with probability zero.

Some Remarks on Stochastic Programming

- If the parameter distribution $\mathbb{P}$ is well known, a rule of thumb is that stochastic programs are a good solution approach.
- Unfortunately, in general, stochastic programming can only handle small and medium-sized problems.
- More details on this modeling approach are presented in this chapter; however, the interested reader is also directed to the excellent book [1] for a more extensive coverage on this subject.

17.1.2 Robust Optimization

Robust optimization is an alternative to model problems under uncertainty. In this case we also assume the parameters ξ are unknown; however, the difference with stochastic programming is that now we assume the worst possible parameter vector in response to the decision variables x.

$$\min \ \max_{\xi \in \Xi} f(x, \xi)$$

subject to:

$$g_i(x, \xi) \leq 0, \ \forall \xi \in \Xi \ \forall i = 1, \ldots, m_g$$

Here, Ξ is the uncertainty set. This can be interpreted as playing against the devil's advocate, where no matter what decision variables x are chosen, ξ will be such that the worst outcome for that decision happens. Therefore we wish to choose the actions that have the least worst possible outcome.

Some Remarks on Robust Optimization

- If no information on the parameter distributions is known (or can be assumed with confidence), then robust optimization is the suggested approach.
- Recent advances in robust optimization have made relatively large problems tractable.

Following this brief introduction, let us now see some examples on how to address optimization problems with underlying uncertainty.

17.2 DIFFERENT APPROACHES TO ADDRESS OPTIMIZATION UNDER UNCERTAINTY

Parametric uncertainty is a way to represent problems where some of the parameters in the models are assumed not to be known exactly. Depending on how much we know about this uncertainty, or to what level of confidence we wish to satisfy constraints, different approaches might be employed. The following will describe different frameworks to address an optimization problem with uncertainty.

Let us assume that we are running a manufacturing firm and we must produce at least h_1 units of product p_1 and h_2 units of product p_2. To manufacture these two products we use raw materials x_1 and x_2, which we have to order in advance from a third party. We know that the proportions of our raw materials to products is given by:

$$6x_1 + 11x_2 = h_1 \tag{17.1}$$

$$4x_1 + 3x_2 = h_2 \tag{17.2}$$

further to the above, we know x_1 is $5/6$ the cost of x_2 and we want to minimize the cost or purchasing the required raw materials. Henceforth, we formulate the following (highly simplified) optimization problem:

$$\min \quad 5x_1 + 6x_2 \tag{17.3}$$

subject to:

$$6x_1 + 11x_2 \geq h_1 \tag{17.4}$$

$$4x_1 + 3x_2 \geq h_2 \tag{17.5}$$

$$x_1, \ x_2 \geq 0 \tag{17.6}$$

Notice that we have inequalities as we do not care if we surpass the specific h_1 and h_2 values, as long as we are at the minimum cost. Furthermore, let us now assign values such that $h_1 = 200$ and $h_2 = 100$.

This would result in our standard linear program (LP) discussed in Chapter 10. However, assume now that we are not completely sure about our demand h_1 and h_2. In fact, we know that $h_1 = 200 + \varepsilon_1$ and $h_2 = 100 + \varepsilon_2$, where $\varepsilon_1 \sim \mathcal{N}(0, 10)$ and $\varepsilon_2 \sim \mathcal{N}(0, 7)$. This implies that we know the mean value of our demand is h_1 and h_2; however, since they are normally distributed we are unsure of the exact value they will have. Let us further assume that the proportion between raw material x_1 and h_1

is also uncertain, similarly the proportion of raw material x_2 in h_2, and that they have the following relation:

$$(6 + \alpha_1)x_1 + 11x_2 = h_1 \tag{17.7}$$

$$4x_1 + (3 + \alpha_2)x_2 = h_2 \tag{17.8}$$

such that $\alpha_1 \sim U(-0.8, 0.8)$ and $\alpha_2 \sim U(-0.6, 0.6)$. We can now reformulate our original LP as:

$$\min \quad f(x) = 5x_1 + 6x_2 \tag{17.9}$$

subject to:

$$(6 + \alpha_1)x_1 + 11x_2 \geq 200 + \varepsilon_1 \tag{17.10}$$

$$4x_1 + (3 + \alpha_2)x_2 \geq 100 + \varepsilon_2 \tag{17.11}$$

$$x_1, \ x_2 \geq 0 \tag{17.12}$$

Given that we will order x_1 and x_2 in advance we must find a way to cope with this uncertainty.

17.2.1 Average Value Approach

The first possibility is to simply assume the expected value for all uncertain variables and solve our initial LP in Equations (17.3–17.6). We can see a graphical representation of this problem in Figure 17.1.

The solution found in Figure 17.1 is: $f(x) = 142.308$, $x_1 = 19.23$, $x_2 = 7.69$.

As this is, however, a very crude approach, and uncertainty is not taken into account; other alternatives are considered next.

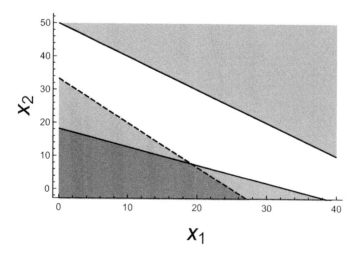

Figure 17.1 Graphic representation of LP in Equations (17.3–17.6).

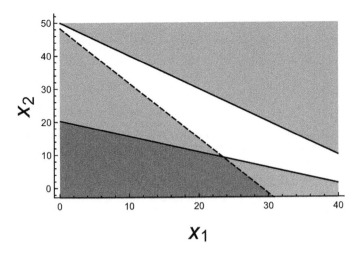

Figure 17.2 Graphic representation of worst-case scenario LP.

17.2.2 Worst-Case Scenario

Given that we have several parameters for our optimization problem, and ideally, we would like to satisfy our demand h_1 and h_2, let us consider the worst-case approach. We hence consider $\alpha_1 = -0.8$ and $\alpha_1 = -0.8$ and since ε_1 and ε_2 are unbounded we can consider a 99% confidence, as such $\varepsilon_1 = 23.3$ and $\varepsilon_2 = 16.0$. We are then left once again with an LP that can readily be solved. This worst-case scenario LP is visualized in Figure 17.2.

It is possible to observe that the feasible space has shrunk significantly from the previous case, and now the following solution is obtained: $f(x) = 172.602$, $x_1 = 23.47$, $x_2 = 9.20$.

This can be considered a robust optimization approach, as we are taking the worst possible scenario, and not using the information of probability distributions, but rather an interval within which our parameter values are contained.

With this result we satisfy all demands to at least a 99% confidence; however, we might have been overly pessimistic and hence obtain a higher cost of 172. Different methods that allow for handling uncertainty differently are considered next.

17.2.3 Probabilistic Constraints (Chance Constraints)

Using probabilistic constraints enables solving an optimization problem, and assigning the degree of certainty to which we would like to satisfy our constraints.

Returning to our example problem, assume we want to assign the following probability to satisfy the constraints:

$$p_1(x) = Pr(6x_1 + 11x_2 \geq h_1) \tag{17.13}$$

$$p_2(x) = Pr(4x_1 + 3x_2 \geq h_2) \tag{17.14}$$

This means that we want to satisfy $6x + 11x_2 \geq h_1$ with a probability $p_1(x)$, as well as satisfy $4x_1 + 3x_2 \geq h_2$ with a probability $p_2(x)$. If we set $p_{1,2}(x) = 90\%$, then we have the following problem:

$$\min \quad f(x) = 5x_1 + 6x_2 \tag{17.15}$$

subject to:

$$p_1(x) \geq 0.9 \qquad (17.16)$$

$$p_2(x) \geq 0.9 \qquad (17.17)$$

$$x_1, \, x_2 \geq 0 \qquad (17.18)$$

It is now important to find an analytical expression for the constraints in Equations (17.16) and (17.17). We can define $F_h(Tx) = Pr(Tx \geq h)$ to be the *cumulative distribution function* (CDF). Where T is the coefficient matrix that multiplies our vector $[x_1, x_2]^T$. Remember that the CDF of a real-valued random variable τ (in our case $\tau = Tx$), is the probability that τ will take a value less than or equal to v (in our case $v = h$). Hence, if $F_h(Tx) \geq 0.9$ we will have satisfied our problem in Equations (17.13–17.18). Finding an explicit form of the constraint $F_h(\tau) \geq v$ is not always possible (and in fact it can rarely be done). However, given scalar normally distributed variables (which is our case), we can compute this constraint explicitly as follows:

$$F_h(Tx) \geq 0.9$$

we have

$$Tx \geq F_h^-(0.9)$$

where F_h^- is the generalized inverse CDF, which can be defined as:

$$F_h^-(\eta) = \inf[Tx : F_h \geq \eta]$$

where for scalar $h \sim \mathcal{N}(\hat{h}, \sigma^2)$ we have

$$F_h^-(\eta) = \hat{h} + \phi^{-1}(\eta) \, \sigma$$

where ϕ^{-1} is the inverse CFD of $\mathcal{N}(0, 1)$, a function that produces a scalar with a probability as an argument.

Therefore, reformulating our problem, we now have

$$\min \quad f(x) = 5x_1 + 6x_2 \qquad (17.19)$$

subject to:

$$6x_1 + 11x_2 \geq 200 + \phi^{-1}(0.9) \; 10 \qquad (17.20)$$

$$4x_1 + 3x_2 \geq 100 + \phi^{-1}(0.9) \; 7 \qquad (17.21)$$

$$x_1, \, x_2 \geq 0 \qquad (17.22)$$

The above optimization problem is now a standard LP, which can be easily solved, and the solution is $f(x) = 153.299$, $x_1 = 21.54$, $x_2 = 7.59$.

It is important to note that we can handle this problem explicitly because we have normal distributions, scalar constraints, and linear equations; however, this is one of the few cases that can be analytically computed. In the case of dependent (vector) constraints, a covariance matrix would be computed to handle their interactions. When we have more difficult problems, other approaches must be considered. One such approach is presented in Section 17.4.

17.2.4 Scenario Approach

Let us think now of a slightly different scenario to handle uncertainty in our example problem. As in our previous problem formulation we must pre-order the raw material quantities x_1 and x_2. However, if we do not meet the demand for h_1 and/or h_2 we are allowed to order more raw material, which we will denote as y_1 and y_2. These extra materials will incur higher costs than the pre-ordered quantities. The problem can be formulated as follows:

$$\min \quad 5x_1 + 6x_2 + 15y_1 + 18y_2 \tag{17.23}$$

subject to:

$$(6 + \alpha_1)x_1 + 11x_2 + 6y_1 + 11y_2 \geq h_1 \tag{17.24}$$

$$4x_1 + (3 + \alpha_2)x_2 + 4y_1 + 3y_2 \geq h_2 \tag{17.25}$$

$$x_1, \ x_2, \ y_1, \ y_2 \geq 0 \tag{17.26}$$

In this way, the optimization can be conducted in two steps. In the first optimization step the variables x_1 and x_2 are determined, and once the uncertain parameters α_1, α_2, h_1, and h_2 occur, the variables y_1 and y_2 can adjust for any violated constraints in a second optimization. However, when we initially optimize for the x variables, we must assign a value to the risk of using y_1 and y_2. A natural choice for this is the expected value. Then the objective function in Equation (17.23) can be replaced by $5x_1 + 6x_2 + \mathbb{E}\left[15y_1 + 18y_2\right]$. However, now we have the problem that computing an expected value for a general distribution of α_1, α_2, h_1, and h_2 can be very expensive computationally; furthermore, we would have an infinite number of constraints given that our distributions are continuous, hence, in this form, our problem is intractable. A general approach to solve this type of problems is by scenario sampling. This means that we randomly (taking our distribution into account) sample N values of $\alpha_{1,2}, h_{1,2}$ and formulate the following problem:

$$\min \quad 5x_1 + 6x_2 + \sum_{i=1}^{N} p_i \left(15y_{1i} + 18y_{2i}\right) \tag{17.27}$$

subject to:

$$(6 + \alpha_{1i})x_1 + 11x_2 + 6y_1 + 11y_2 \geq h_{1i} \tag{17.28}$$

$$4x_1 + (3 + \alpha_{2i})x_2 + 4y_1 + 3y_2 \geq h_{2i} \tag{17.29}$$

$$x_1, \ x_2, \ y_{1i}, \ y_{2i} \geq 0 \tag{17.30}$$

$$\text{for } i = 1, \ldots, N$$

where we have N scenarios, and p_i is the probability of scenario i. We can therefore see that the problem in Equations (17.27–17.30) is an LP and can be solved with readily available methods. However, the constraints (and terms in the objective function) will be multiplied by the number of scenarios we consider. Therefore, we have a trade-off: the larger the number of scenarios, the more confident we are that we will satisfy the constraints, but this will make the LP larger and more difficult to solve.

When we solve the problem in the previous equations for 100 scenarios the following solution is obtained: $f(x) = 163.806$, $x_1 = 20.90$, $x_2 = 8.90$.

There are many aspects about this problem that have not been mentioned, such as complexity, quality of the solution, and possible issues. This is further discussed in Section 17.4. Furthermore, this problem is known as a *two-stage stochastic linear program*, and can be formulated more generally; this is reviewed in the following subsection.

17.2.5 General Two-Stage Stochastic Linear Programming Formulation

In the two-stage stochastic programming approach, decision variables are split into two groups. The first group consists of variables whose value must be decided before uncertain events, hence these variables are not uncertain (also sometimes referred to as the "here and now variables"). When the values of these "deterministic" variables are decided, one must make some assumptions about how the uncertain variables will behave, so as to optimize the whole problem. Once the value for the first group of variables has been decided, uncertain events take place and then the second group of variables is optimized (also called the "wait and see variables"). It should be noted that the decisions made for the first group of variables will affect the possible outcomes of the second group of decision variables.

The formulation of two-stage stochastic linear programs is as follows:

$$\min_{x} \quad c^{T}x + \mathbb{E}_{\mathbb{P}}[Q(x,\xi)] \tag{17.31}$$

subject to:

$$x \in X$$

where

$$Q(x,\xi) \quad = \quad \min_{y} \quad q(\xi)^{T}y$$
$$\text{subject to:} \quad W(\xi)y \geq H(\xi) - T(\xi)x \tag{17.32}$$
$$y \geq 0$$

where $X \subseteq \mathbb{R}^{n_x}$ and $Y \subseteq \mathbb{R}^{n_y}$ are polyhedral sets, and $c \in \mathbb{R}^{n_x}$ and ξ are random variables (or parameters). Hence, the first stage decision variables are x, which will be decided before the uncertain parameter values ξ are known. The second stage occurs when variables y are chosen after the realization of ξ. Note that to make a decision on x the expected value of $Q(x,\xi)$ is computed. W, H, and T are matrices of appropriate dimensions, the same as those that would appear in a standard linear program, the difference being that they are also affected by the random parameters ξ.

There are two main solution strategies to solve the problem given in Equation (17.31).

In the first approach an assumption of discrete distributions of the uncertain parameters is made. Following this approach a large-scale linear program is formulated, where each uncertainty realization is a linear program for a different discrete value of the uncertain parameters. This is the approach that was discussed in Section 17.2.4, and is further discussed throughout the rest of this chapter.

In the second common approach continuous parameter distributions are considered, and in this case sampling-based decomposition and approximation schemes have been developed. These solution strategies make ample use of convex analysis and measure theory, which are out of the scope of this chapter; the interested reader can refer in more detail to these formulations in the following references [2, 3, 4, 5, 6], as well as the material outlined in Section 17.10.

17.3   ROBUST OPTIMIZATION

The general form of a robust optimization problem can be as follows:

$$\min \quad \max_{\xi \in \Xi} f(x, \xi) \tag{17.33}$$

subject to:

$$g_i(x, \xi) \leq 0, \ \forall \xi \in \Xi, \ \forall i = 1, \ldots, m_g$$

where $x \in \mathbb{R}^n$ are the decision variables, $\xi \in \mathbb{R}^k$ are the uncertain problem parameters and Ξ is the uncertainty set. Problem (17.33) in its general form can be non-linear, nonconvex. Depending on the application, the maximization problem over the parameters ξ might require global optimization, which further complicates the problem and hence, even the solution of small problems might be intractable. In this book we will focus on introducing linear robust problems. In the Further Reading section (17.10) of this chapter we outline additional material on this topic.

17.3.1 Linear Robust Optimization

In this section we describe the general solution procedure for (single-stage) linear robust optimization problems; let us define this as follows:

$$\min \quad \max_{a_0, b_0 \in \Xi_0} a_0^T x + b_0 \tag{17.34}$$

subject to:

$$a_i^T x + b_i \leq 0 \qquad \forall (a_i, b_i) \in \Xi_i, \ \forall i = 1, \ldots, m$$

$$x \in \mathcal{X}$$

where $\mathcal{X} \subseteq \mathbb{R}^n$ and the uncertainty sets $\Xi_i \subseteq \mathbb{R}^n \times \mathbb{R}$ are polyhedral, $i = 1, \ldots, m$. For simplicity we will assume $\Xi_0 = \Xi_1 = \cdots = \Xi_m$. Here, the Ξ_i are the uncertainty sets. As mentioned earlier, this problem can be interpreted as a game between two players. Player A chooses the worst possible parameters $a_j, b_j \ j = 0, \ldots, m$, and in response, player B must make the best selection for x in response to player's B move.

For a better illustration of this problem, let us augment the decision vector x with an additional component x_{n+1} to replace the affine functions with a linear formulation:

$$\min \quad \max_{\xi_0 \in \Xi_0} \xi_0^T x$$

subject to:

$$\xi_i^T x \leq 0 \qquad \forall \xi_i \in \Xi_i, \ \forall i = 1, \ldots, m$$

$$x \in \mathcal{X} \times \mathbb{R}, \qquad x_{n+1} = 1$$

Now, we re-formulate the problem to its epigraph from:

$$\min \quad \tau$$

subject to:

$$\tau \geq \xi_0^T x \qquad \forall \xi_0 \in \Xi_0$$

$$\xi_i^T x \leq 0 \qquad \forall \xi_i \in \Xi_i, \ \forall i = 1, \ldots, m$$

$$x \in \mathcal{X} \times \mathbb{R}, \qquad x_{n+1} = 1$$

$$\tau \in \mathbb{R}$$

Here we note that by changing notation of the problem we can obtain the following:

$$\min \quad c^T x \tag{17.35}$$

subject to:

$$\xi_i^T x \leq 0 \qquad \forall \xi_i \in \Xi_i, \ \forall i = 1, \ldots, m$$

$$x \in \mathcal{X}$$

where $c \in \mathbb{R}^n$, $\mathcal{X} \subseteq \mathbb{R}^n$ and the uncertainty sets satisfy $\Xi_i = \left\{ \xi_i \in \mathbb{R}^{k_i} : F_i \xi_i \leq g_i \right\}$ for $F_i \in \mathbb{R}^{l_i \times k_i}$ and $g_i \in \mathbb{R}^{l_i}$, $i = 1, \ldots, m$.

Let us note that Problem (17.35) is not the standard LP problem we learned in Chapter 10, where we presented Linear Programs. In this case, each constraint $\xi_i^T x \leq 0$ must hold for an infinite number of constraint realizations $\forall \xi_i \in \Xi_i$, this kind of optimization problem is termed *semi-innite programming*. To solve Problem (17.35) we do as follows:

First we note that a constraint $\xi_i^T x \leq 0 \quad \forall \xi_i \in \Xi_i$ is satisfied if and only if

$$\max_{\xi_i \in \Xi_i} \quad \xi_i^T x \leq 0 \tag{17.36}$$

From LP theory, presented in Chapter 10, we know that for a fixed x the maximum is attained at an extreme point of Ξ_i. We can therefore replace constraints $\xi_i^T x \leq 0 \quad \forall \xi_i \in \Xi_i$ by

$$\max_{\xi_i \in \mathrm{ext} \Xi_i} \quad \xi_i^T x \leq 0$$

where $\mathrm{ext} \Xi_i$ denotes the extreme points of $\mathrm{ext} \Xi_i$. This problem seems easier to solve than its semi-infinite dimensional equivalent; however, this is far from ideal because the number of extreme points of the polyhedron is typically exponential in the (half-space) description of Ξ_i. The benefit is that we can make use of duality theory from linear programming. We remember that:

$$\max \left\{ \xi_i^T x : F_i \xi_i \leq g_i, \ \xi_i \in \mathbb{R}^{k_i} \right\} \leq 0$$
$$\Leftrightarrow \quad \min \left\{ g_i^T \lambda_i : F_i^T \lambda_i = x, \ \lambda_i \in \mathbb{R}_+^{l_i} \right\} \leq 0$$
$$\Leftrightarrow \quad \exists \lambda_i \in \mathbb{R}_+^{l_i} : g_i^T \lambda_i \leq 0, \ F_i^T \lambda_i^T = x$$

We can now replace this constraint in problem (17.35), as follows:

$$\min \quad c^T x \tag{17.37}$$

subject to:

$$g_i^T \lambda_i \leq 0, \ \forall i = 1, \ldots, m$$

$$F_i^T \lambda_i = x, \ \forall i = 1, \ldots, m$$

$$x \in \mathcal{X}$$

$$\lambda_i \in \mathbb{R}_+^{l_i} \ \forall i = 1, \ldots, m$$

We can now tractably solve the robust linear program (17.34). The price to pay is that the size of problem (17.37) grows polynomially with respect to the original deterministic problem and the description of the uncertainty sets.

17.3.2 Distributionally Robust Optimization

Distributionally Robust Optimization is a relatively new paradigm for decision making under uncertainty. In this type of problems, the uncertain problem data are governed by a probability distribution that is itself subject to uncertainty. The distribution is then assumed to belong to an ambiguity set comprising all distributions that are compatible with the decision maker's prior information. We can define this problem as follows:

$$\min_{} \ \sup_{\mathbb{P} \in \mathcal{D}} \mathbb{E}_{\mathbb{P}} \left[f(x, \xi) \right] \tag{17.38}$$

subject to:

$$g_i(x, \xi) \leq 0 \ \mathbb{P} - \text{a.s.} \ \forall \mathbb{P} \in \mathcal{D}, \ \forall i = 1, \ldots, m$$

$$x \in \mathcal{X}$$

where $\mathcal{D}$ is the ambiguity set. If $\mathcal{D}$ were to have only one element (*i.e.* $\mathcal{D} = \{\mathbb{P}\}$), then the problem reduces to a stochastic program because we would be minimizing over the expected value of ξ. Therefore, distributionally robust optimization can be thought of as a robust formulation for stochastic programming. If, however, $\mathcal{D} = \{\delta_\xi : \xi \in \Xi\}$ problem (17.38) reduces to a robust optimization problem, where δ_ξ is the Dirac distribution that places unit mass onto the parameter realization ξ. Loosely speaking, in robust optimization we are interested in the worst possible realization of ξ, whereas in distributionally robust optimization we are interested in the worst possible distribution $\mathbb{P}$ from which ξ is drawn. Therefore, we have a sup rather than a max in the problem description, because we are maximuzing with respect to a set, and therefore yield a least upper bound.

For a broader description on Distributionally Robust Optimization and its solution techniques, the reader is referred to [7] as well as the Further Reading section of this chapter.

17.4 SAMPLE AVERAGE APPROXIMATION METHOD

Optimization under uncertainty is a broad topic, and a variety of efficient algorithms have been developed to address specific problem types by exploiting their structure. In this chapter, we will focus on the *Sample Average Approximation (SAA)* method, which is one of the most efficient algorithms to address general optimization problems under uncertainty. In the further reading section of this chapter, various references are provided for the reader interested in more specific solution approaches.

The SAA principle is very general, and can be applied to unconstrained or constrained problems with uncertainty. It finds application in stochastic programming, chance constraints, stochastic-dominance constraints, complementarity constraints, robust optimization, and others.

The main idea behind SAA is to take samples from random variables using *Monte Carlo*, and compute the expected objective and constraint values by averaging the samples. The resulting approximate problem can then be solved by conventional deterministic optimization algorithms. Furthermore, the process can be repeated with different samples to obtain candidate solutions along with statistical estimates of their optimality gaps.

If we consider exactly solving the two-stage stochastic linear program in Equations (17.31–17.32) there are two main difficulties:

1) To compute exactly $\mathbb{E}_{\xi \in \Xi}[Q(x, \xi)]$, even in the case of finite scenarios $\{\xi_1, \xi_2, \ldots, \xi_{|\Xi|}\}$ with probabilities $\{p_1, p_2, \ldots, p_{|\Xi|}\}$, we would have to calculate

$$\mathbb{E}_{\xi \in \Xi}[Q(x, \xi)] = \sum_{i=1}^{|\Xi|} p_i Q(x, \xi_i)$$

which grows combinatorially with the size of the data. Hence, it becomes intractable even for medium-sized problems with finite scenarios (notice this worsens when we have continuous distributions describing the random parameters).

2) $\mathbb{E}_{\xi \in \Xi}[Q(x, \xi)]$ is a nonlinear function, which further increases the difficulty of solving the problem, particularly when the number of variables and constraints increases.

The SAA proposes an efficient alternative that approximates $\mathbb{E}_{\xi \in \Xi}[Q(x, \xi)]$ by taking N samples $\{\xi^1, \xi^2, \ldots, \xi^N\}$ where we follow the convention from [8] to use superscript for sample values. We can compute our approximation of the objective function as:

$$\frac{1}{N} \sum_{j=1}^{N} Q\left(x, \xi^j\right)$$

and we denote z_N the optimal value for our approximate minimization problem:

$$z_N = \min_{x \in X} \quad c^T x + \frac{1}{N} \sum_{j=1}^{N} Q\left(x, \xi^j\right) \tag{17.39}$$

Inequality constraints can be reformulated in an equivalent fashion. The resulting problem can then be solved by standard methods for deterministic optimization.

[9] and [8] present a procedure to provide lower bounds or "goodness of solution" for this problem. For this procedure, (17.39) should be solved multiple times, and generate M independent problems, and their solutions, each with N scenarios. We obtain

M solution values $\{z_N^1, z_N^2, \ldots, z_N^M\}$ and M candidate solutions $\{x_N^1, x_N^2, \ldots, x_N^M\}$ where we can obtain the average of these solution values as:

$$\bar{z}_N = \frac{1}{M} \sum_{m=1}^{M} z_N^m$$

We know that statistically speaking $\mathbb{E}[\bar{z}_N] \leq z^*$(further explained in Section 17.5.1), where by z^* we denote the "true" solution, as if we were solving the problem exactly. Hence, $\mathbb{E}[\bar{z}_N]$ is a lower bound for our optimal solutions value. We also know that for any feasible point $\hat{x} \in X$ the value of $c^T \hat{x} + \mathbb{E}_{\xi \in \Xi}[Q(\hat{x}, \xi)]$ is an upper bound on z^*, hence we estimate this upper bound by

$$\hat{z}_{N'}(\hat{x}) = c^T \hat{x} + \frac{1}{N'} \sum_{j=1}^{N'} Q(\hat{x}, \xi^j) \tag{17.40}$$

where $\{\xi^1, \xi^2, \ldots, \xi^{N'}\}$ are the N' scenarios and should be chosen such that $N' > N$. With this, we have that $\hat{z}_{N'}(\hat{x})$ is an unbiased estimator of $c^T \hat{x} + \mathbb{E}_{\xi \in \Xi}[Q(\hat{x}, \xi)]$ and hence we have that $\mathbb{E}[\hat{z}_{N'}(\hat{x})]$ is a lower bound on z^*. Furthermore, we can estimate the variances, and calculate confidence intervals, for $\bar{z}_N$ and $\hat{z}_{N'}(\hat{x})$ as follows:

$$\hat{\sigma}_{\bar{z}_N}^2 = \frac{1}{(M-1)M} \sum_{m=1}^{M} (z_N^m - \bar{z}_N)^2$$

and

$$\hat{\sigma}_{\hat{z}_{N'}(\hat{x})}^2 = \frac{1}{(N'-1)N'} \sum_{n=1}^{N'} \left(c^T \hat{x} + Q(\hat{x}, \xi^n) - \hat{z}_{N'}(\hat{x})\right)^2$$

Given all of this, we can now determine the optimal solution to our problem, and this is chosen to be the smallest among the candidate solutions $\{x_N^1, x_N^2, \ldots, x_N^M\}$, hence we define our optimal objective function value as

$$x_N^* \in \operatorname{argmin}\left\{\hat{z}_{N'}(\hat{x}) \mid \hat{x} \in \{x_N^1, x_N^2, \ldots, x_N^M\}\right\}$$

With the solution x_N^* we can compute the optimality gap by

$$\hat{z}_{N'}(x_N^*) - \bar{z}_N$$

where we take $\hat{z}_{N'}(x_N^*)$ to be recomputed with an independent sample to obtain an unbiased estimate. The variance of this optimality gap can be estimated by

$$\hat{\sigma}_{\hat{z}_{N'}(x_N^*) - \bar{z}_N}^2 = \hat{\sigma}_{\hat{z}_{N'}(x_N^*)}^2 + \hat{\sigma}_{\bar{z}_N}^2 \tag{17.41}$$

This terminates the SAA method, which enables us to estimate optimal solutions to problems under uncertainty, and approximate the possible duality gap.

Notice that computing a duality gap is most useful for convex problems, however it can also be done for nonconvex problems. Furthermore, no statements about sample sizes and statistical guarantees were made, as well as whether constrained problems are present. This will be addressed in the subsequent sections.

17.5 SCENARIO GENERATION AND SAMPLING METHODS

In Section 17.4 we presented the SAA method to solve optimization problems under uncertainty. The procedure presented relied on computed expected values by sampling scenarios from our optimization problem. In this section we will give a more detailed explanation behind the sampling rationale, as well as the guarantees and properties the SAA solution procedure has. Before we look into our optimization problem, let us delve into the problem of computing an approximate expected value from samples. In other words, in this section we will only consider the approximation of a function by a sampling scheme, without taking into account any optimization.

We denote one Monte Carlo sample of a random vector as $x^j = \left(x_1^j, x_2^j, \ldots, x_m^j\right)$. Let us now draw N independent and identically distributed samples $x^1, \ldots, x^N$ to compute

$$Q_N = \frac{1}{N} \sum_{j=1}^{N} Q(x^j)$$

In probability theory, there is a theorem called the strong law of large numbers (sLLN) that describes the result of sampling the same experiment a large number of times. This theorem states that by averaging samples as $N \to \infty$, Q_N will converge *almost surely* to the expected value $\mathbb{E}[Q(x)]$, as stated here:

$$\lim_{N \to \infty} Q_N = \mathbb{E}[Q(x)]$$

Therefore, by relying on the strong law of large numbers we can use Q_N as a reasonable approximation of $Q^* = \mathbb{E}[Q(x)]$.

This means that if we take $N = \infty$, then the SAA method would be accurately computing the expected value. It should be kept in mind that having a sample of infinite size is not practical. Therefore, the following can be done when N is of finite size.

We know that:

$$\mathbb{E}[Q_N] = \mathbb{E}\left[\frac{1}{N} \sum_{j=1}^{N} Q(x^j)\right] = \frac{1}{N} \sum_{j=1}^{N} \mathbb{E}\left[Q(x^j)\right] = Q^*$$

and

$$\sigma_{Q_N}^2 = \text{Var}\left(\frac{1}{N} \sum_{j=1}^{N} Q(x^j)\right) = \frac{1}{N^2} \sum_{j=1}^{N} \text{Var}\left(Q(x^j)\right) \approx \frac{1}{N} \text{Var}(Q(x))$$

where we denote the variance of Q_N by $\sigma_{Q_N}^2$. In probability theory, there is a theorem called the *central limit theorem* (CLT), which establishes that, in almost all situations, when independent random variables are added, their normalized sum tends to a normal distribution $\mathcal{N}(0, 1)$. Therefore, by the central limit theorem, with N sufficiently large

$$\frac{\sqrt{N}(Q_N - Q^*)}{\sigma} \approx \text{Normal}(0, 1)$$

where $\sigma^2 = \mathrm{Var}(Q(x))$. Where with N sufficiently large we can approximate σ^2 as

$$\sigma^2 \approx \sigma_{Q_N}^2 = \frac{\sum_{j=1}^N \left(Q(x^j) - Q_N\right)^2}{N - 1}$$

Using the above, we have

$$P\left(Q_N - 1.96\frac{\sigma}{\sqrt{N}} \le Q^* \le Q_N + 1.96\frac{\sigma}{\sqrt{N}}\right) = 0.95$$

that is out of 100 experiments, on average in 95 of those the interval given by

$$\left[Q_N - 1.96\frac{\sigma}{\sqrt{N}}, \; Q_N + 1.96\frac{\sigma}{\sqrt{N}}\right]$$

will contain the true value Q^*. Notice that the constant 1.96 is a value dependent on the desired confidence level for in this case normal distributions.

This presents an understanding behind using sampling techniques by the SAA method to approximate functions. In the following section we will look at these approximations now applied in an optimization context.

17.5.1 Scenario Generation Applied to Optimization

In the previous section we were mainly concerned with understanding how generating scenarios can lead to estimating an expected value and its deviations; however, nothing was mentioned about how this will affect an optimization problem. In this section we present the implications this has in the context of optimization.

As mentioned previously, we want to generate Monte Carlo samples $\{\xi^1, \xi^2, \dots \xi^N\}$ and solve the optimization problem by the SAA approach.

We denote SP to be the stochastic optimization problem with optimal solution x^*, set of optimal solutions s^* and z^* to be the optimal value, assuming the problem was solved exactly. We denote SP_N to be the approximation to the stochastic optimization problem with optimal solution x_N, set of optimal solutions s_N and z_N to be the optimal value. We now want to know if $x_N \to x^*$, $s_N \to s^*$ and $z_N \to z^*$ as we sample more scenarios.

Before we go any further let us define *uniform convergence*, which will be useful as a measure for our approximate problem leading to a solution similar to that of the true problem.

Uniform convergence is a type of convergence of functions such that a sequence of functions f_n converges uniformly to a limiting function f on a set E if, given any arbitrarily small positive number ε, a number N can be found such that each of the functions $f_N, f_{N+1}, \dots$ differ from f by no more than ε at every point x in E. In other words, *uniform convergence* implies that a sequence f_n converges to a desired function f at a uniform pace regardless of the point in the domain.

Now that the measure for convergence is defined, let us examine if our approximated problem converges to our original one if an infinite sample size is present, this clearly being an optimistic case.

We draw a sequence of samples called the *sample path* $\{\xi^1, \xi^2, \dots\}$; then, for each N, we construct an approximation

$$Q_N(\cdot) = \frac{1}{N}\sum_{j=1}^N Q(\cdot, \xi^j)$$

using the first N terms of the sample path we solve

$$\min_{x \in X} \ Q_N(x) \tag{17.42}$$

If $N \to \infty$ and our functions Q are convex, then we have uniform convergence. This means that $z_N \to z^*$, and under mild additional assumptions $s_N \to s^*$ (which itself implies $x_N \to x^*$). However, we highlight that all this applies for $N \to \infty$, and will not be useful in most practical cases. Furthermore, this has some consequences that are useful in practical instances. Under additional assumptions of compactness of X, Lipschitz continuity of $Q(\cdot, \xi)$, and uniqueness of the solution x^* the following holds:

$$\sqrt{N}\left(z_N - z^*\right) \overset{d}{\to} \mathcal{N}\left(0, \sigma\left(x^*\right)\right) \tag{17.43}$$

where $\sigma^2(x) = \mathrm{Var}(Q(\cdot, \xi))$ and by $\overset{d}{\to}$ we denote "converges in distribution to."

This means that as we sample more scenarios from our original function, the distribution of our solutions is normal, and although $\sigma(x^*)$ is unknown, approximations can be made. Furthermore, from Equation (17.43) we know convergence is of rate $1/\sqrt{N}$, and hence we know the cost of improving our approximate solution.

Another important insight we get is that there is a bias such that the minimum value is underestimated; in other words:

Given that:

$$\mathbb{E}\left[\min_{x \in X} \ Q(x, \xi)\right] \leq \min_{x \in X} \ \mathbb{E}\left[Q(x, \xi)\right]$$

then we have:

$$\mathbb{E}\left[z_N\right] \leq z^*$$

This we will call *optimization bias*.

As we mentioned, having $N \to \infty$ is impractical, so we now outline some of the convergence properties for a finite N.

The upside of getting more samples is that the difference between $\mathbb{E}[z_N]$ and z^* decreases as N grows.

In addition, under the additional assumption of strong convexity on the function $Q(x, \xi)$, given any $\varepsilon > 0$ we know constants $\beta(\varepsilon) > 0$ and C exist such that

$$\mathrm{P}\left(\left\|x_N - x^*\right\| > \varepsilon\right) \leq Ce^{-N\beta(\varepsilon)}$$

In loose terms, given all the previous assumptions, this implies that our solution x_N approaches the true solution x^* exponentially fast with respect to our sample size N.

With the this can understand the rationale of using the SAA method based on sampling to approximate the original problem.

As a summary of this section, when estimating the solution of the true problem by Monte Carlo sampling, we obtain an estimated value z_N of our real value z^*. We know that we will have an *optimization bias* $\mathbb{E}[z_N] \leq z^*$, and that given certain properties z_N will converge to z^*. Furthermore, we know that given our number of samples N we can derive statistical estimators $\sqrt{N}(z_N - z^*) \overset{d}{\to} \mathcal{N}(0, \sigma(x^*))$ and $\mathrm{P}(\|x_N - x^*\| > \varepsilon) \leq Ce^{-N\beta(\varepsilon)}$ of the closeness between our approximated solution and the true solution.

17.5.2 Scenario Generation Applied to Constrained Optimization Problems

In this section, let us deal with problems that include constraints that are a function of the random variable ξ. These can be formulated as follows

$$\min_{x \in X} \mathbb{E}[Q(x,\xi)] \tag{17.44}$$

subject to:

$$g_i(x,\xi) \leq 0 \quad i = 1,\ldots,m \tag{17.45}$$

Given this problem, an intuitive solution could be to simply replace Equation (17.45) by its expected value as follows:

$$\mathbb{E}[g_i(x,\xi)] \leq 0 \quad i = 1,\ldots,m \tag{17.46}$$

This, under the sampling approach, would translate into:

$$\frac{1}{N}\sum_{j=1}^{N} g(\cdot,\xi^j) \leq 0$$

Unfortunately, this means that we want our constraint to be satisfied on average. For example, if $g(\cdot,\xi^j)$ was randomly distributed, this would mean that Equation (17.45) would be satisfied on average only 50% of the time. To ameliorate this problem, a practical alternative is to add a constant on the right-hand side of the inequality and replace Equation (17.46) with:

$$\mathbb{E}[g_i(x,\xi)] \leq \varepsilon \quad i = 1,\ldots,m \tag{17.47}$$

where ε is a slack value that enlarges the probability that solutions from this approach are within the feasible region.

Another convenient approach could be modeling it as a chance constraint:

$$P(g_i(x,\xi) > 0) \leq \alpha \quad i = 1,\ldots,m \tag{17.48}$$

One of the most effective approaches for difficult optimization problems is to treat these constraints by the *scenario approach*; as discussed in Section 17.2.4 we can reformulate the probabilistic constraints as:

$$g\left(x,\xi^j\right) \leq 0 \quad i = 1,\ldots,N \tag{17.49}$$

Hence this establishes that we want the constraint to hold for all scenarios that are sampled. If the objective function of the optimization problem and the functions $g_i(x,\xi)$ are convex, then we can derive the following equivalence between Equations (17.48) and (17.49):

Given a confidence level $\beta \in (0,1)$ and a dimension d_x of x. If the following is true

$$N \geq \frac{2}{\alpha}\left(\ln\left(\frac{1}{\beta}\right) + d_x\right)$$

then with probability at least $1 - \beta$ we have that x_N satisfies all constraints in the original problem by at most a fraction α. This result is of great use as it is regardless of the distribution of ξ. Stated as a chance constraint we have:

$$P(g(x_N,\xi) > 0) \leq \alpha$$

This assumes we do not want to violate any constraints, at least in our optimization problem. Furthermore, we may want to allow some explicit violation of our constraints. For this, we could state a mixed-integer program as follows:

$$\min_{x \in X} \frac{1}{N} \sum_{j=1}^{N} Q(x, \xi^j) \tag{17.50}$$

subject to:

$$g\left(x, \xi^j\right) - M z_i \leq 0 \quad i = 1, \ldots, N \tag{17.51}$$

$$\frac{1}{N} \sum_{i=1}^{N} z_i \leq \gamma \tag{17.52}$$

$$z \in \{1, 0\}^N \tag{17.53}$$

In this problem, M is given a large value such that z_i will be given the value 0 if the constraint for scenario i is not violated, and $z_i = 1$ if for that scenario the constraint is violated. Equation (17.52) will limit the amount of constraints that are allowed to be violated. For simplicity it was assumed that only one constraint is present, which can be directly extended to multiple constraints. Hence, in this approach we allow there to be a number γ of violations out of all N scenarios sampled. Note that big M formulations such as that in this problem are sometimes difficult to solve, and reformulations could be done to ameliorate convergence difficulties. For solution approaches to mixed-integer problems we refer the reader to Chapter 18.

17.6 SAMPLING METHODS FOR SCENARIO GENERATION

In the previous sections we have focused on the properties of generating scenarios with Monte Carlo sampling. In this section we will see that there are some other methods that may yield better results with regards to the sampling procedure.

When dealing with an optimization problem under uncertainty, in general, objective function values z_N are supplied along with some confidence region or interval, using the variance $\sigma_{z_N}^2$ of this optimal solution. For example, a solution could be in an interval of plus and minus one standard deviation:

$$\left[z_N - \sigma_{z_N}, \ z_N + \sigma_{z_N}\right]$$

where σ_{z_N} is the square root of the variance (*i.e.* the standard deviation). As we have mentioned before, we can approximate the variance as:

$$\sigma_{z_N}^2 = \frac{\sum_{i=1}^{N} \left(Q_N(x, \xi^i) - z_N\right)^2}{N - 1} \tag{17.54}$$

When stating an optimal result as an interval, it is always natural to want to make this interval as small as possible, and this is done by reducing the variance. Unfortunately, the variance decreases linearly with the number of samples in the Monte Carlo sampling case (as seen in Equation (17.54)), which means it decreases as a square root with respect to the standard deviation. This can become very expensive for large sample sizes, and therefore other more efficient sampling techniques must be used to ensure a lower variance.

Before we start with sampling techniques, let us mention the two important points that can be addressed when reducing variance:

- Exploit correlations: as we see later on, exploiting correlations can have a positive effect on reducing the variance
- Sampling the space "uniformly": This means that there are methods (that although random) perform certain procedures to sample the space with certain uniformity, such that there are no large spaces left without samples, while others are cluttered with them, and hence get more relevant information about the problem overall.

A third approach is to sample "important" regions more: If there are scenarios that are more prone to happen that are not included in the distribution of samples, or scenarios that involve greater risk and therefore should be mitigated, it is many times worth sampling these regions more extensively than the rest of the space. Although this approach does not reduce variance, it has practical uses for obvious reasons, and a sampling method that follows this philosophy will also be introduced in the following sections.

17.6.1 Antithetic Variates

Assume that we want to estimate $f = \mathbb{E}\left(Q(x, \xi)\right) = \mathbb{E}\left(y\right)$. To estimate this expected value, we draw two samples y_1 and y_2. To estimate the f value, we compute the mean of our samples:

$$f_N = \frac{f_1 + f_2}{2}$$

to estimate the variance of our prediction we compute:

$$\text{Var}\left(f_N\right) = \frac{\text{Var}\left(f_1\right) + \text{Var}\left(f_2\right) + 2\text{Cov}\left(f_1, f_2\right)}{4}$$

with this, it is obvious that we can reduce the variance of $\text{Var}\left(f_N\right)$ by making the value of $\text{Cov}\left(f_1, f_2\right)$ negative.

Translating this into a case where we have a set of samples $\left\{\xi^1, \xi^2, \ldots, \xi^N\right\}$ of length N, it is possible to correlate negatively every two variables, to lower the variance of the overall sample path.

This property can be exploited, and a framework that uses this fact is called *Antithetic Variates* (AV). This sampling scheme works as follows:

1. Sample observations $\left\{U^1, U^2, \ldots, U^{N/2}\right\}$, notice we are assuming an even number of samples.
2. Obtain the remaining $N/2$ samples by calculating the antithetic pair of each sample $\left\{U^{N/2+1}, U^{N/2+2}, \ldots, U^N\right\} = \left\{1 - U^1, 1 - U^2, \ldots, 1 - U^{N/2}\right\}$.
3. Apply the inverse cumulative distribution function (assuming it is invertible) to obtain N observations from the problem distribution, and obtain $\left\{\xi^1, \xi^2, \ldots, \xi^N\right\}$.

With this procedure we can obtain negatively correlated random variables for almost any distribution.

213

Following this approach, now we estimate $\mathbb{E}[Q(x, \xi)]$ as:

$$Q_{N,AV}(x) = \frac{2}{N} \sum_{j=1}^{N/2} \frac{Q(x, \xi^j) + Q(x, \xi^{j'})}{2}$$

where $\xi^{j'}$ is the antithetic of ξ^j. Furthermore, its variance can be calculated by:

$$\text{Var}\left(Q_{N,AV}(x)\right) = \frac{\text{Var}\left(Q_{N,AV}(x, \xi)\right)}{N} + \frac{1}{N}\text{Cov}\left(Q_{N,AV}\left(x, \xi^j\right), Q_{N,AV}\left(x, \xi^{j'}\right)\right)$$

The first term corresponds to the variance of the standard MC method; if the second term is negative, then the Antithetic Variates method reduces overall covariance. Because we want this second term to be negative, we induce negative correlation when drawing up our uniform random sample pairs $\left(U^i, 1 - U^i\right)$. However, the fact that we have negative correlation in these terms does not always imply that we will have negative correlation in the end. This is because, in many cases, some "negative correlation" is lost when transforming from $U^i, 1 - U^i$ to ξ^j and $\xi^{j'}$. Furthermore, after applying $Q(x, \xi^j)$ negative correlation can be further lost.

For the function $Q(x, \xi)$ we will preserve negative correlation if it is bounded and monotone in each component of ξ; an example of this is with two-stage stochastic linear programs with fixed recourse matrix when the random variables ξ are independent. Let us note that these properties are very important to apply AVs successfully. If one applies this method when monotonicity is lost, the method can actually increase the variance, which is the complete opposite of what we want.

A nice side effect of Antithetic Variates is that it can sometimes decrease optimization bias (*i.e.* the computed expected value is closer to the "true" expected value).

17.6.2 Latin Hypercube Sampling

Latin Hypercube Sampling (LHS) addresses the idea of sampling the space "uniformly" to obtain a better approximation of the stochastic function. Let us illustrate the method in a one-dimensional space. Suppose we want to obtain N random samples $\{\xi^1, \xi^2, \ldots, \xi^N\}$ by LHS, by the cumulative distribution presented in Figure 17.3.

To obtain a relatively "uniform" sampling, we partition our distribution into $N = 5$ segments as shown in Figure 17.4.

Then in each segment, a uniformly random number is chosen, and to these numbers the inverse cumulative distribution function is applied to obtain the random sampling shown in Figure 17.5.

In this way we obtain five random samples from the desired distribution, making sure that they are "spread out" along the domain we wish to sample.

For two-dimensional LHS, one can generate two one-dimensional independently drawn LHS samples, as shown in Figure 17.6.

Notice that dots are randomly generated inside the squares. Afterward, the squares are randomly permuted to create the two-dimensional LHS represented in Figure 17.7.

Notice how each column and each row have only one sample; this is where the "Latin Hypercube" name came from, and this can be extended to higher dimensions, with one sample on each hyperplane.

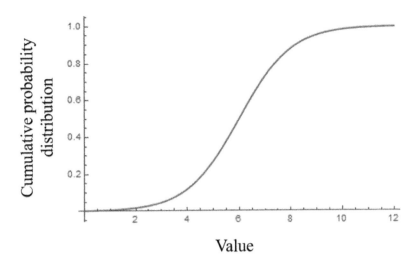

Figure 17.3 Cumulative distribution of random variable ξ.

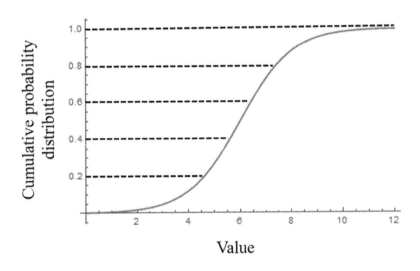

Figure 17.4 Cumulative distribution of random variable ξ divided into five segments.

In summary, LHS works as follows:

1. For each component of ξ

   1. Sample uniformly $u^i \sim U\left(\frac{i-1}{N}, \frac{i}{N}\right)$ for $i = 1, \ldots, N$
   2. Randomly permute the observations

2. Apply the inverse cumulative distribution to u
3. Obtain N LH samples $\{\xi^1, \xi^2, \ldots, \xi^N\}$

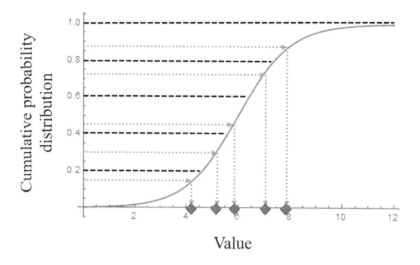

Figure 17.5 LHC sampling in one dimension.

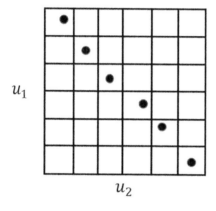

Figure 17.6 LHC sampling in two dimensions.

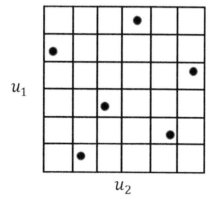

Figure 17.7 LHC sampling in two dimensions.

Furthermore, LHS is guaranteed to have a lower variance than MC sampling, described by the following relation:

$$\text{Var}\left(Q_{N,LHS}\left(x\right)\right) \le \frac{N}{N-1}\text{Var}\left(Q_{N,MC}\left(x\right)\right)$$

A nice side-effect of LHS is that, in practice, it helps reduce optimization bias.

17.6.3 Quasi-Monte Carlo

Quasi-Monte Carlo methods use what is known as low discrepancy sequences. Roughly speaking, low discrepancy sequences are sequences that cover the space uniformly and hence are used in many cases as sampling techniques. Contrary to MC or LHS these sequences do not have to be random (although some of them are).

Some ways to generate sequences are Halton sequence, Sobol sequence, Lattice rule, among others. These methods have similar advantages to that of LHS in the sense that they can reduce variance because they cover the space more "uniformly." It is also possible to induce randomness in the sequences used to compute Quasi-Monte Carlo points, giving rise to Randomized Quasi-Monte Carlo methods. The derivation of these sequences is complex and beyond the scope of this chapter; however, the interested reader is referred to adequate references at the end of this chapter.

17.6.4 Importance Sampling

The main goal of an Importance Sampling (IS) approach is to collect more information on, and hence model better, areas that have low probability of happening, but are important. For example, when modeling risk, we are concerned with the scenarios that have high inherent risk. These scenarios might be rare, but since the whole point of a "risk model" is to model risk, then we should concentrate sampling in the areas with high risk (even though these areas are unlikely).

However, if we simply get many samples from the risky part of the distribution and then average their values with other samples we obtain, this would not get a correct value, as we are saying that the risky values are in fact more probable than they actually are. Therefore, we need one distribution to sample out scenarios, and another distribution to weight them depending on how probable they actually are. This is done as follows:

Let us assume that we have a problem where random events ξ have a density function $f(\xi)$. We then compute the expected value, as follows:

$$\mathbb{E}\left[Q\left(x,\xi\right)\right] = \int_{\Xi} Q\left(x,\xi\right)\ f(\xi)\,d\xi$$

However, we do not want to sample from the original distribution $f(\xi)$, but rather from a distribution $q(\xi)$, which gives higher probability of sampling unlikely but risky scenarios. Therefore we use the likelihood ratio $\mathcal{L}(\xi) = \frac{f(\xi)}{q(\xi)}$ to map out sampling $q(\xi)$ to the real scenario distribution $f(\xi)$. Hence we have

$$\mathbb{E}\left[Q\left(x,\xi\right)\right] = \int_{\Xi} Q\left(x,\xi\right)\ \mathcal{L}(\xi)q(\xi)\,d\xi$$

therefore, instead of having N samples by a MC approach sampled from density $f(\xi)$, we would get N samples $\{\xi_q^1, \xi_q^2, \ldots, \xi_q^3\}$ sampled from density $q(\xi)$. We compute the expected value by:

$$Q_{N,IS}(x) = \frac{1}{N} \sum_{j=1}^{N} Q(x, \xi_q^j) \mathcal{L}(\xi_q^j)$$

In this way, we are able to sample risky scenarios more intensively, without compromising the accuracy of our approximation.

17.7 SOLUTIONS ON AVERAGE APPROXIMATION ALGORITHM

By bringing all of this together, we can now construct the SAA algorithm.

Algorithm 17.1 Sample Average Approximation algorithm
Initialize: Determine the sample size N, the sample size N' to determine the upper bound in Equation (17.40), initial number M of SAA replications and tolerance ε. Additionally, it might be a good idea to determine a scheme to increase or decrease N, N' and M, to adjust for some level of confidence. Determine the sampling scheme; this will be problem dependent to best exploit the properties of the chosen sampling framework.

1. For $m = 1, \ldots, M$

    1. *Generate a sample of size N and solve the SAA in Equation (17.42) to obtain the optimal objective function value z_N and the optimal values x_N*
    2. *Estimate the optimality gap $z_N - \hat{z}_{N'}(\hat{x})$, and the variance for the gap estimation in Equation (17.41)*
    3. *If the optimality gap and the optimality gap variance are sufficiently small go to Step 3. Otherwise, continue to Step 2.*

2. *If the optimality gap and/or the optimality gap variance are too large, increase N and/or N', and return to Step 1.*
3. *Choose the best solution x_N^* among all candidates produced and report optimality gap, its confidence interval, and confidence intervals for the optimal solution. Stop*

Algorithm 17.1 presents a general framework to solve optimization problems involving uncertainty through SAA. Different implementations can be derived and extended to solution domains with integer or mixed-integer domains, and further reading material on this topic can be found in Section 17.10.

17.8 FLEXIBILITY ANALYSIS OF CHEMICAL PROCESSES

Flexibility analysis is a topic in optimization under uncertainty that has been developed and addressed particularly by the Chemical Engineering Society. The problem that flexibility analysis addresses is to give a measure of the "operability of a plant," that is, the ability of the plant to operate over a range of conditions while still satisfying performance specifications. As mentioned in the original paper [10]: "A good process design must not only exhibit an optimal balance between capital and operating costs, it must also exhibit operability characteristics which will allow the economic performance to be realizable in a practical operating environment."

More concisely, flexibility analysis provides a measure of the feasible operation space, where process adjustments can be made to accommodate different parameter realizations. The more flexible a plant design is, the greater its ability to adjust to variations in conditions that may happen during operation. These sources of variations can range from feed quality, product requirements, new environmental or safety measurements, heat exchanger fouling, catalyst deactivation, and many others.

17.8.1 Flexibility Index

Swaney and Grossmann proposed to define the flexibility index F as the maximum range over which the parameters may vary independently of each other while still remaining inside the feasible region R. For a two-dimensional parameter space, this is schematically represented in Figure 17.8.

Unfortunately, as with general uncertainty in optimization, obtaining the flexibility index is no easy task. The procedure outlined by Swaney and Grossmann is as follows.

Let us assume that our chemical process or plant is described by the following equations:

$$h(d, u, x, \theta) = 0$$

$$g(d, u, x, \theta) \leq 0$$

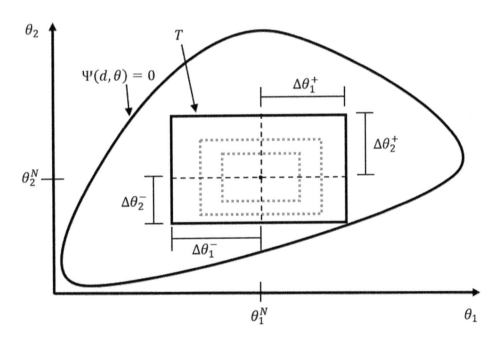

Figure 17.8 Maximum hyperrectangle T inscribed within a feasible region $\Psi(d, \theta)$ where d represents the design variables selected (after [10]).

where h are typically equations that describe the system (mass and energy balances), while g is a vector of inequalities such as safety, environmental, or process constraints, d are the design variables, u are the control variables, x are the state variables, and θ are the parameters. We must first get rid of the equality constraints and incorporate them into the inequalities. This is done by solving for the state variables x:

$$h(d, u, x, \theta) = 0 \rightarrow x = x(d, u, \theta)$$

and then inserting these equations into the inequality set:

$$g(d, u, x(d, u, \theta), \theta) = f(d, u, \theta) \leq 0$$

Let us now note that the control variables u represent the degrees of freedom that may be adjusted during operation to suit prevailing conditions; feasibility for a given d and θ requires only that some u satisfies $f(d, u, \theta) \leq 0$.

To obtain the flexibility index F we can then formulate the following optimization problem:

$$F = \max \delta$$

subject to:

$$\forall \theta \in T(\delta)\, \{\exists u \mid f(d, u, \theta) \leq 0\}$$

$$T(\delta) = \left\{\theta \mid \left(\theta^N - \delta \Delta \theta^-\right) \leq \theta \leq \left(\theta^N + \delta \Delta \theta^+\right)\right\}$$

where we wish to find the maximum δ such that for all parameters values inside the hyperrectangle $T(\delta)$ satisfy the feasible set for some control value. Notice that δ grows equally for both sides, as seen in Figure 17.8, even if there is room for growth on the left-hand side of the figure; since the right-hand side of the hyperrectangle touches the feasible region, δ cannot be increased further.

It was shown by Halemane and Grossmann in [11] that the feasibility condition can be replaced by:

$$\max_{\theta \in T(\delta)} \min_u \max_{i \in I} \quad f_i(d, u, \theta) \leq 0$$

replacing this expression in our optimization problem yields:

$$F = \max \delta$$

subject to:

$$\max_{\theta \in T(\delta)} \min_u \max_{i \in I} \quad f_i(d, u, \theta) \leq 0 \tag{17.55}$$

$$T(\delta) = \left\{\theta \mid \left(\theta^N - \delta \Delta \theta^-\right) \leq \theta \leq \left(\theta^N + \delta \Delta \theta^+\right)\right\}$$

As one can probably guess, it is very challenging to solve this problem, which many times leads to a nondifferentiable global optimization problem. Therefore, a general procedure to find the exact solution for this problem is still a topic of present day research. Fortunately, there are some assumptions that make a solution attainable, or at least an approximation of it.

One assumption that can be made is that the solution lies on a vertex of the hyperrectangle. For this to be certainly true, the following must hold.

We first define the following problem, which can be thought of as providing a measure of feasibility for a design d at the parameter realization θ:

$$\psi (d, \theta) = \min_{u,z} z$$

$$\text{s.t. } f_i (d, u, \theta) \leq z, \ i \in I$$

Then, using this definition, we know that if $\psi (d, \theta)$ is continuous and one-dimensional quasi-convex (1-DQC) in θ, then the solution θ^* of the problem in Equation (17.55) must lie at a vertex of the hyperrectangle $T (\delta)$. Where a function $\psi (d, \theta)$ is 1-DQC in θ if and only if for $\theta^1, \theta^2 \in R$, where $\theta^2 = \theta^1 + \beta e^i$, β is a nonzero scalar, and e^i is a coordinate direction given by the jth column of the identity matrix $I(\dim p \times p)$, the following condition holds: $\max \{ \psi (d, \theta^1), \psi (d, \theta^2) \} \geq \psi (d, a\theta^1 + (1 - a) \theta^2) \forall a \in [0, 1]$ [12].

Following from this, how can we relate the constraint function $f_i (d, u, \theta)$ present in our profile to the function $\psi (d, \theta)$? This is done as follows.

If the constraint functions $f_i (d, u, \theta), i \in I$ are jointly quasi-convex in u and one-dimensional quasi-convex in θ, then the function $\psi (d, \theta)$ is one-dimensional quasi-convex in θ.

The intuitive interpretation of this is that the constraint functions $f_i (d, u, \theta)$ must define a convex region in the space of the controls u and each individual parameter taken one at a time. If this is the case, then it can be guaranteed that the critical point of the hyperrectangle will lie at a vertex. If this is not the case, then the solution may not lie on a vertex, although it is possible that it does. It should be noticed that in practice, it might be very difficult to determine whether $\psi (d, \theta)$ is continuous and one-dimensional quasi-convex (1-DQC).

17.8.2 Solution Procedures

Even when the conditions for the solution of the problem in Equation (17.55) are not satisfied, there is still a strong tendency in engineering problems for the solution of Equation (1) to lie at a vertex. Therefore, the solution procedures outlined here will assume the solution lies on a vertex, and will find the most efficient way to determine the optimal vertex without necessarily enumerating all 2^p of them.

Let us first define the problem in Equation (17.55) in an equivalent way that will allow us to explain better the solution procedures:

$$F = \max_{\tilde{\theta} \in \tilde{T}} \delta^* (\tilde{\theta}) \tag{17.56}$$

subject to:

$$\tilde{T} = \left\{ \tilde{\theta} \mid -\Delta \theta^- \leq \tilde{\theta} \leq \Delta \theta^+ \right\}$$

where $\delta^* (\tilde{\theta})$ is a function given by the nonlinear program:

$$\delta^* (\tilde{\theta}) = \max_{\delta, u} \delta \tag{17.57}$$

subject to:

$$f(d, u, \theta) \leq 0 \tag{17.58}$$

$$\theta = \theta^N + \delta\tilde{\theta} \tag{17.59}$$

Under the assumption that the optimal solution lies on a vertex, we replace our continuous search space in $\tilde{\theta}$ by the search on discrete vertices as $\tilde{\theta} \in V$ where $V = \{\tilde{\theta}^K, k \in K_v\}$ and $K_v = \{1, \ldots, 2^p\}$. Then it is possible to model these discrete choices for $\tilde{\theta}$ by introducing binary variables $y_j, \ j = 1, \ldots, p$. In this way, when $y_j = 0$ component θ_j will imply a negative deviation from parameter θ, while if $y_j = 0$ this will imply a positive deviation from the original parameter value. Hence we now have the following nonlinear integer program:

$$F = \max_y \delta^* \left(\tilde{\theta}(y) \right)$$

$$y \in \{0, 1\}$$

where y is a p–dimensional vector.

Enumeration

In a first approach, we could simply enumerate all 2^p vertices. Note that to evaluate each vertex we need to solve the resulting NLP; therefore we would have to solve 2^p NLPs and choose the best one among them. That is, solving the following problem 2^p times for a different value of y:

$$F = \max_{k \in K_v} \delta^* \left(\tilde{\theta}^k \right)$$

This is not such a bad idea if there are six or seven parameters; however, when you get to just 20 parameters, you already have to solve over one million NLPs. Therefore, this approach is by no means general, and alternative algorithms must be used when the number of parameters is not small. If the problem is monotonic in θ, a branch and bound approach as introduced in Chapter 18 can be conducted.

A Heuristic Approach

To avoid full enumeration, and when a branch and bound approach is not adequate, Swaney and Grossmann proposed a heuristic framework to determine the optimal vertex, as follows:

Given a candidate solution vertex $\delta^* (\tilde{\theta}^k)$, we can linearize the function $\delta^* (\tilde{\theta})$ at our current point k, and we obtain the linearization:

$$\delta^* \left(\tilde{\theta}^l \right) \approx \delta^* \left(\tilde{\theta}^k \right) \left(1 + v^T \left(\tilde{\theta}^l - \tilde{\theta}^k \right) \right)$$

where v^T is the gradient of $f(d, u, \theta)$ with respect to θ multiplied by its multiplier, that is:

$$v = -\lambda^T \nabla_\theta f^T (d, u, \theta)$$

According to this linearization $v^T \tilde{\theta}^l < 0$, this means that v_j must have opposite sign to $\tilde{\theta}^l_j$ for $j = 1, \ldots, p$, this is:

$$\text{sign}\left(\tilde{\theta}^l_j\right) = -\text{sign}\left(v_j\right), \quad j = 1, \ldots, p \tag{17.60}$$

Therefore, if we assume all $\tilde{\theta}^l$ values are positive, we find a vertex $\tilde{\theta}'$ in the direction $-v$ such that (given the linearization) $\delta^*(\tilde{\theta}') \leq \delta^*(\tilde{\theta}^l)$ and vertex $\delta^*(\tilde{\theta}^l)$ becomes a candidate for examination. However, it might happen that $\tilde{\theta}^k = \tilde{\theta}^l$, which means that $\tilde{\theta}^k$ is a local minimizer of the problem in Equation (17.56). If this is the case, we can use another approach to look for a search direction. The previous strategy assumes that the active set of constraints will not vary as we move toward $\tilde{\theta}'$; however, if our active set were to change, it is possible that another direction exists in which we find $\delta^*(\tilde{\theta}') \leq \delta^*(\tilde{\theta}^l)$. We hence search for constraints that might become inactive by linearizing each active constraint as:

$$\triangle f_i \approx \bar{\delta} \frac{\partial f_i}{\partial \theta^T} \left(\tilde{\theta}^l - \tilde{\theta}^k\right)$$

where $\bar{\delta}$ is the value for the current upper-bound estimate of solution $\delta^*(\tilde{\theta}^*)$ (i.e. best known feasible solution so far). Using the sign of the gradient term for each parameter and constraint $\frac{\partial f_i}{\partial \theta_j}$ we can determine for each constraint f_i the vertex $\tilde{\theta}^{*,i}$ which will cause the greatest increase in $\triangle f_i$, this is:

$$\tilde{\theta}^{*,i}_j = \begin{cases} \triangle \theta^+_j & \frac{\partial f_i}{\partial \theta_j} \geq 0 \\ -\triangle \theta^-_j & \frac{\partial f_i}{\partial \theta_j} < 0 \end{cases} \quad \text{for } j = 1, \ldots, p$$

Given this, we can define the nonnegative variable

$$w_i = \frac{\partial f_i}{\partial \theta^T} \left(\tilde{\theta}^{*,i} - \tilde{\theta}^k\right)$$

and estimate the increase for each constraint by:

$$\triangle f_i \approx \bar{\delta} w_i$$

the following predictor can be computed:

$$\rho_i = \frac{\triangle f_i}{-f_i} = \frac{\bar{\delta} w_i}{-f_i}$$

where we can think of $-f_i$ as the slack for constraint i and of ρ_i as the ratio of the possible increase for constraint i divided by the slack that constraint has. Therefore, if the value of ρ_i is much greater than one, there is a large possibility of this constraint being violated at the vertex $\tilde{\theta}^{*,i}$, which means there will be a change on the active set and maybe the decrease of $\delta^*(\tilde{\theta})$; we are hence looking for high values of ρ_i. Swaney and Grossmann proposed introducing a parameter ρ^{MAX} such that all vertices $\tilde{\theta}^{*,i}$ with $\rho_i > \rho^{MAX}$ became candidates for evaluation.

The algorithm to determine the flexibility of a candidate design is then the following

223

Algorithm 17.2 *Flexibility Analysis Algorithm*

    Initialization: Select a vertex $\tilde{\theta}^k$, initialize set of examined vertices $V' = \emptyset$, set upper bound $\bar{\delta} = \infty$

1. *Step 1:*

    1. *Examine candidate vertex $\tilde{\theta}^k$ by solving (17.57–17.59)*
    2. *Set $V' = V' \cup \left\{\tilde{\theta}^k\right\}$*
    3. *If $\delta^*\left(\tilde{\theta}^k\right) < \bar{\delta}$, set best known solution $\tilde{\theta}_{bks} = \tilde{\theta}^k$, and $\bar{\delta} = \delta^*\left(\tilde{\theta}^k\right)$, proceed to Step 2.*
    4. *If $\delta^*\left(\tilde{\theta}^k\right) \geq \bar{\delta}$, proceed to Step 3.*

2. *Step 2:*

    1. *Use Equation (17.60) to project the θ-gradient to determine $\tilde{\theta}^l$.*
    2. *If $\tilde{\theta}^l = \tilde{\theta}^k$, go to Step 3, otherwise set $\tilde{\theta}^k = \tilde{\theta}^l$ and return to Step 1*

3. *Step 3:*

    1. *At the solution of Equations (17.57–17.59) for the current estimate $\tilde{\theta}_{bks}$ compute $\rho_i = \frac{\bar{\delta} w_i}{-f_i}$ for each inactive constraint $f_i < 0$.*
    2. *Select the largest ρ_i such that $\tilde{\theta}^{*,i}$ has not been examined, where our chosen index $m \mid \rho_m = \max\left\{\rho_i \mid i \in I, \tilde{\theta}^{*,i} \notin V'\right\}$ if this $\rho_i < \rho^{MAX}$ or no unexamined $\tilde{\theta}^{*,i}$ remain, go to Step 4.*
    3. *Otherwise, set $\tilde{\theta}^k = \tilde{\theta}^{*,m}$, go to Step 1.*

4. *Step 4: Stop*

    1. *Set $\tilde{\theta}^* = \tilde{\theta}_{bks}$, $F = \bar{\delta} = \delta^*\left(\tilde{\theta}^*\right)$*

This finishes the flexibility analysis as presented by Swaney and Grossmann. Although we present the original idea and rationale behind flexibility analysis, there have been new developments, and determining the flexibility of a process is still an area of active research.

17.9 REFERENCES

[1] Shapiro, A., Dentcheva, D., and Ruszczynski, A. Lectures on Stochastic Programming: Modeling and Theory. SIAM. 2009.

[2] Oliveira, W., Sagastizábal, C., and Scheimberg S. "Inexact bundle methods for two-stage stochastic programming." SIAM Journal of Optimisation. 2011;21(2):p. 517–544.

[3] Elmaghraby, S. "Allocation under uncertainty when the demand has continuous D.F." Management Science. 1960;10:p. 270–294.

[4] Ziemba, W. "Computational algorithms for convex stochastic Programs with simple recourse." Operations Research. 1970;18:p. 414–431.

[5] Ziemba, W. "Solving nonlinear programming problems with stochastic objective functions." The Journal of Financial and Quantitative Analysis. 1972;7(3):p. 1809–1827.

[6] Rockafellar, R., and Wets, R. B. "A Lagrangian finite generation technique for solving linear-quadratic problems in stochastic programming." Mathematical Programming Study. 1986;28:p. 63–93.

[7] Wiesemann, W., Kuhn, D., and Sim, M. "Distributionally robust convex optimization." Operations Research. 2014;62(6):p. 1358–1376.

[8] Verweij, B., Ahmed, S., and Kleywegt, A. J., Nemhauser, G., and Shapiro, A. "The sample average approximation method applied to stochastic routing problems: A computational study." Computational Optimization and Applications. 2003;24(2):p. 289–333.

[9] Mak, W. K., Morton, D. P., and Wood, R. K. "Monte Carlo bounding techniques for determining solution quality in stochastic programs." Operations Research Letters. 1999;24:p. 47–56.

[10] E. Grossman. "An index for operational flexibility in chemical process design, R. E. Swaney, 1." AlChE Journal 1985;31(4).

[11] Halemane, K. P. and Grossmann, I. E. "Optimal process design under uncertainty." American Institute of Chemical Engineers Journal. 1983;29:p. 425.

[12] Swaney, R. E. and Grossmann, I. E. "An index for operational flexibility in chemical process design part I: Formulation and theory"; American Institute of Chemical Engineers Journal. 1985;31(4):p. 621–630.

17.10 FURTHER READING RECOMMENDATIONS

Optimization under uncertainty is a very broad field in optimization. In this chapter we decided to present Sample Average Approximation, a general framework that can be used to solve many applications where problems with uncertain parameters must be addressed. However, there are many methods, algorithms, and applications that we did not mention. Therefore, in this section we outline material that could help an interested reader in learning more about this area.

Survey on Monte Carlo Sampling-Based Methods

Homem-de-Melloa, T. and Bayraksan, G. (2014). "Monte Carlo sampling-based methods for stochastic optimization." Surveys in Operations Research and Management Science. 19:p. 56–85.

Full Presentation on the Topic of Stochastic Programming

Shapiro, A. and Dentcheva, D. (2009). Lectures on Stochastic Programming: Modeling and Theory. SIAM.

Survey on Robust Optimization

Bertsimas, D., Brown, D. B., and Caramanis, C. (2011). "Theory and applications of robust optimization." SIAM Review. 53(3):p. 464–501.

Ben-Tal, A., El Ghaoui, L., and Nemirovski, A. (2009). Robust Optimization. Princeton Series in Applied Mathematics.

Introduction to Distributionally Robust Optimization

Wiesemann, W. & Kuhn, D. & Sim, M. (2014). "Distributionally Robust Convex Optimization." Operations Research. 62(6):pp. 1358–1376.

Delage, E. & Ye, Y. (2010). "Distributionally robust optimization under moment uncertainty with application to data-driven problems." Operations Research. 58(3).

Stochastic Integer Programming

Luedtke, J. (2007). "Integer Programming Approaches for Some Non-convex and Stochastic Optimization Problems." PhD Dissertation.

Ahmed, S., Tawarmalani, M., and Sahinidis, N. V. (2003). "A finite branch-and-bound algorithm for two-stage stochastic integer programs." Mathematical Programming.

Higle, J. L. and Sen, S. (2000). "The C3 theorem and a D2 algorithm for large scale stochastic integer programming: Set convexification." Technical report, University of Arizona.

Sherali, H.D. and Fraticelli, B. M. P. (2002). "A modified Benders partitioning approach for discrete subproblems: An approach for stochastic programs with integer recourse." Journal of Global Optimization. 22:p. 319–342.

An Efficient Algorithm to Solve Large Stochastic Nonlinear Programming

Shastri, Y. and Diwekar, U. (2006). "An efficient algorithm for large scale stochastic nonlinear programming problems." Computers & Chemical Engineering. 30(5):p. 864–977.

Fuzzy Set Theory with Particular Emphasis on Fuzzy Mathematical Programming as an Application

Zimmermann, H. J. (1991). Fuzzy Set Theory and Its Application. Kluwer Academic Publishers.

Stochastic Dynamic Programming

Sutton, R. S. and Barto, A. G. (2018). Reinforcement Learning: An Introduction, Second Edition. The MIT Press.

Multiparametric Programming under Uncertainty

Dua, V., Bozinis, N. A., and Pistikopoulos, E. N. (2002). "A multiparametric programming approach for mixed-integer quadratic engineering problems." Computers & Chemical Engineering. 26:p. 715–733.

Pistikopoulos, E. N. and Dua, V. (1998). "Planning under uncertainty: A parametric optimization approach." Proceedings of third international conference on foundations of computer-aided process operations. p. 164–169.

17.11 EXERCISES

1. Given the following LP:

$$\max \quad 6x_1 + 5x_2$$

subject to:

$$8x_1 + 11x_2 + \varepsilon_1 \leq 28$$

$$4x_1 + 3x_2 + \varepsilon_2 \leq 8$$

$$x_1, x_2 \geq 0$$

(i) Where $\varepsilon_1 = 0, \varepsilon_2 = 0$, solve the above Linear Program.

(ii) Where $\varepsilon_1 \in [-3, 3], \varepsilon_2 \in [-2, 2]$, solve the above problem using Linear Robust Optimization.

(iii) Where $\varepsilon_1 \sim \mathcal{N}(1, 1), \varepsilon_2 \sim \mathcal{N}(0, 4)$, solve the above problem using to a 95% confidence level using SAA with random sampling.

2. Given the following nonlinear optimization problem:

$$\min \quad \frac{(x_1 - 2.8x_2)^2}{x_1 + x_2 + 10} - 2x_2 - x_1$$

subject to:

$$5x_1 + 3x_2 + \varepsilon_1 \leq 14$$

$$2x_1 + x_2 + \varepsilon_2 \leq 7$$

$$x_1, x_2 \geq 0$$

where $\varepsilon_1 \sim \mathcal{N}(0, 2), \varepsilon_2 \sim \mathcal{N}(0, 9)$. Solve the above optimization problem using SAA with the following sampling schemes:

(i) Random sampling.
(ii) Latin hypercube sampling.
(iii) Antithetic variates.
(iv) Compare the solution and confidence for the three approaches.

3. For Problem 2, let us now assume that ε_1 takes value $\varepsilon_1 = 0$ with probability 0.999 and $\varepsilon_1 = 12$ with probability 0.001. Furthermore, constraint $5x_1 + 3x_2 + \varepsilon_1 \leq 14$ is regarding the safety of a process, and must hence be satisfied to a high degree of certainty 0.99. Constraint $2x_1 + x_2 + \varepsilon_2 \leq 7$ requires only 0.9 degree of certainty. Solve this problem using SAA with the following sampling schemes:

(i) Random sampling.
(ii) Latin hypercube sampling.
(iii) Importance sampling.
(iv) Compare the solutions and whether the certainty criterion is met.

18

Mixed-Integer Programming Problems

Optimization problems presented in this book have so far involved mostly continuous variables. However, chemical engineering is full of decision making problems that comprise both integer and continuous variables. For example, typically, process synthesis can be modeled as an optimization problem with binary variables $\{0, 1\}$ to represent the presence or absence of a unit in the process flowsheet. Some other examples are heat exchanger networks, distillation systems, reactor networks, waste-water treatment systems, among many others. Furthermore, even the simplest case of mixed-integer programming that is a mixed-integer linear programming (MILP) problem belongs to the class of NP-hard combinatorial problems [1], which means they are much harder to solve than their continuous counterparts, and their difficulty is further increased when nonlinearities are present in the system (*i.e.* mixed-integer nonlinear programming (MINLP) problems).

Let us define a general MINLP problem:

$$\min_{x,y} \quad f(x, y) \tag{18.1}$$

subject to:

$$h(x, y) = 0 \tag{18.2}$$

$$g(x, y) \leq 0 \tag{18.3}$$

$$x \in X \subseteq \mathbb{R}^n \tag{18.4}$$

$$y \in \mathbb{Z} \tag{18.5}$$

where x represents continuous variables and y represents integer variables. It can be seen that this optimization problem follows a similar structure to those that have been previously presented throughout this book, and that the only difference is the inclusion of the y vector of integer variables and its feasible set $y \in \mathbb{Z}$. An important subclass of mixed-integer programming in chemical engineering is when systems have integer variables with value either 0 or 1. This is particularly useful because processes or decisions can be either turned on (one) or off (zero).

Furthermore, general mixed-integer programs can be reformulated to have only binary variables (*i.e.* integer variables only being either 1 or 0) by the following transformation:

$$y = y^L + z_1 + 2z_2 + 4z_3 + \cdots + 2^{N-1}z_N \tag{18.6}$$

where the variable y^L is the lower bound of y, and y can now be replaced by variables $z_1, \ldots, z_N$. This means that solution methods may focus solely on binary problems and still be able to solve general mixed-integer programming problems. However, the number N will depend on the lower bound–upper bound range that the y variables covers, and is given by:

$$N = 1 + \text{int} \left[\frac{\log \left(y^U - y^L \right)}{\log 2} \right] \tag{18.7}$$

which means that if $y^U - y^L$ is large for a particular problem this might not be an appropriate approach.

The solution techniques to mixed-integer programming are different than those for continuous problems, hence, before we present the solution strategies to this type of problems some background material is necessary. These key building blocks are explained in the following section.

18.1 PRELIMINARIES TO SOLVING MIXED-INTEGER PROGRAMMING PROBLEMS

Given that enumerating all integer solutions for mixed-integer problems is computationally intractable even for medium-sized problems, many approaches to solve MILP and MINLP rely on computing lower and upper bounds to estimate the quality of integer solutions. For any integer configuration (at intermediate points of an algorithm), it is possible to calculate some lower and upper bound. This gives insight into whether or not it is worth exploring this solution (or related solutions) further. While upper bounds can simply be computed from any feasible point, strategies to compute lower bounds are more complex, and specific to each algorithm. To compute these lower bounds, *relaxation* is a crucial technique, and although the full details on lower bound construction are presented in Sections 18.2 and 18.3, here we give an overview of relaxation strategies.

18.1.1 Relaxations

Relaxations are computed to produce lower bounds on possible solutions. This is done by enlarging the feasible set of a given preliminary feasible point (possible solution); therefore, the optimal solution of this relaxed feasible point can only be lower (better) or equal to the true solution of the feasible point. We emphasize that for relaxation to be useful in this context it must satisfy three key points:

1. The relaxations must be easier to solve than the original problem.
2. Any feasible solution of the original problem must also be feasible in the relaxation.
3. The relaxation must always underestimate (or be equal to) the solution of the original problem.

More formally:

Definition An optimization problem $\min \{ \Psi_R(x) : x \in X_R \}$ is a relaxation of the problem $\min \{ \Psi(x) : x \in X \}$ if $X_R \supseteq X$ and $\Psi_R \leq \Psi(x)$ for all $x \in X$.

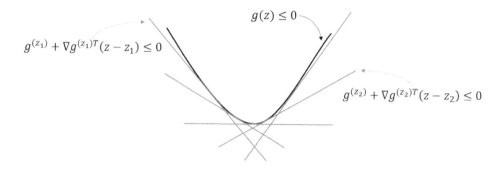

Figure 18.1 Relaxation of convex constraints.

Therefore, the role of these relaxations is to provide an easier problem to solve, and to obtain an optimal solution that can be used as a lower bound for the original problem. An example could be relaxing an MILP into an LP, or an MINLP into an MILP or an NLP. The aim is that the relaxed problem is easier to solve and underestimates the solution of the original problem. Let us review different relaxation approaches.

Integrality Relaxation

Integrality constraints (*i.e.* $y \in \mathbb{Z}$) can be relaxed to $y \in \mathbb{R}$. This procedure casts an MILP into an LP and an MINLP into an NLP.

Relaxation of Convex Constraints

Convex functions $g(z) \leq 0$, as per Equation (18.3), can be relaxed into a set of supporting hyperplanes derived from a linearization, possibly using a first-order Taylor series expansion, which would be as follows:

$$g^{(k)} + \nabla g^{(k)T} \left(z - z^{(k)} \right) \leq 0 \tag{18.8}$$

for a set of points $z^{(k)}$, $k = 1, 2, \ldots, K$. This linearization forms a collection of hyperplanes that constitute a *polyhedral relaxation* of these original constraints $g(z) \leq 0$. This can be visualized as in Figure 18.1.

Relaxing Nonconvex Constraints

If constraints $g(z) \leq 0$ are nonconvex, it is slightly harder to relax them. A popular approach is to derive some convex underestimator $\breve{g}(z)$ such that constraint $g(z) \leq 0$ can be relaxed by these convex functions, noting that it is further possible to relax these functions $\breve{g}(z)$ to a set of linear constraints such as the approach shown in Section 18.1.1. Convex underestimators for nonconvex functions are further reviewed in Chapter 19, as in this chapter we will be focusing on the convex case.

All these relaxation approaches enlarge the feasible set, and hence create problems that are easier to solve. Different relaxation approaches are schematically presented in Figure 18.2.

In Figure 18.2 the image on the left represents a convex domain, where the dots indicate the integer solutions. From this image, three different relaxations are shown.

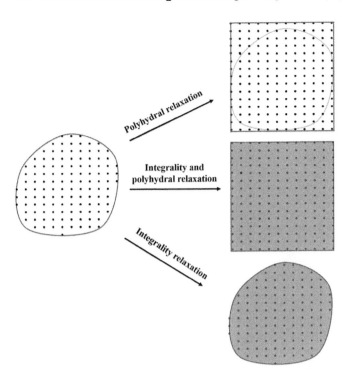

Figure 18.2 The figure on the left denotes a nonlinear feasible space where only integral solutions are allowed. The figures on the right show: (a) Polyhedral relaxations where the feasible space is no longer nonlinear; however, only integral solutions are allowed. (b) Integrability and polyhedral relaxations where the feasible space is linear and continuous. (c) Integrality relaxation where the solution space is still nonlinear but continuous.

The bottom relaxation is an integrality relaxation, where the domain has changed from discrete to continuous, although it is still convex (as opposed to linear). The top relaxation shows the feasible domain being changed to linear, although continuous solutions are still not permitted. The center image has a linear and continuous feasible domain.

Constraint Enforcement

Constraint enforcement is a strategy that relies on relaxations, and it is used for mixed-integer programming to exclude infeasible solutions. If there is a point $\hat{x}$, which is feasible with respect to the relaxed problem, but is not feasible to the original one, constraint enforcement strategies ensure the exclusion of these solutions. In this way, the optimization algorithm can find an optimal solution that is feasible to the original problem. There are two broad classes of constraint enforcement strategies: relaxation refinement and branching.

Relaxation Refinement

Relaxation refinement is in general implemented once a new point $\hat{x}$ is found, which satisfies constraints of the relaxed problem; however, it does not satisfy the constraints

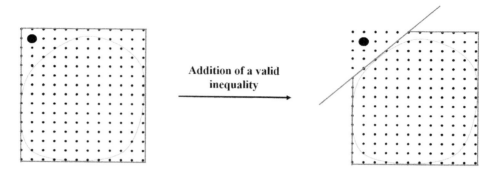

Figure 18.3 The figure on the left shows a solution point that is feasible to the relaxed problem (square), but infeasible to the original problem (convex figure inside the square). The figure on the right shows the addition of a new constraint that excludes this infeasible point from the relaxed problem's feasible space.

of the non-relaxed problem (the original problem). In this case, this new infeasible point gives information that highlights what domain the relaxed problem sees as feasible while the original problem does not.

Methods generally proceed by adding a new constraint to the relaxed problem. This new constraint makes point $\hat{x}$ infeasible, but must also not exclude any feasible solutions, such that all feasible points from the mixed-integer programming problem must satisfy this inequality. This type of inequality is referred to as a *valid inequality* [2]. Valid inequalities are generally linear constraints; however, this is not a requirement. Figure 18.3 shows how this approach may exclude infeasible solutions while still maintaining all the solutions from the original MINLP problem as feasible.

18.1.2 Branching

Branching partitions the relaxed feasible region into subproblems (SPs), such that all feasible solutions from the original MINLP problem (P) remain in at least one of these subproblems, while excluding part of the infeasible points from all subproblems. A simple example could be portrayed with an integrality relaxation approach. With integrality relaxation, a binary variable can be relaxed to be $0 \le \hat{x} \le 1$. Furthermore we can assign to this variable the value of $\hat{x} = 0$ in a subproblem (SP$_1$), and $\hat{x} = 1$ in another subproblem (SP$_2$). In this way, all feasible solutions from P ($\hat{x} \in [0, 1]$) can be found in either SP$_1$ or SP$_2$, while non-feasible solutions ($0 < \hat{x} < 1$) are excluded from them.

In some cases, both branching and relaxation refinement can be used in conjunction, such that the relaxed domain is split, while at the same time enforcing new constraints to reduce the infeasible regions of the relaxed domain [2].

18.1.3 Fathoming

When a branching approach is implemented, many SPs can be generated, and in many cases it can be assured that some of these SPs do not contain the optimal solution. In this case, there is no further need to explore them and efforts can be focused on the remaining SPs. This is called *fathoming*. Concisely, a candidate subproblem SP$_i$ will be considered fathomed if one of the following two conditions is true:

1. It can be assured that SP$_i$ does not contain a better solution than the best solution found so far.
2. An optimal solution to the problem has been found.

In this way, branching creates new subproblems while fathoming discards those that are not needed anymore, until the termination criterion is met.

18.2 SOLUTION TECHNIQUES FOR MIXED-INTEGER LINEAR PROGRAMMING PROBLEMS (MILP)

MILPs are employed as subproblems in the optimization of MINLPs, which makes their efficient solution important. Furthermore, MILPs themselves have found a variety of applications in chemical engineering, particularly in heat exchanger synthesis [9], distillation column synthesis [10, 11, 12], scheduling [13, 14, 15, 16, 17, 18], among others [3].

A general MILP can be stated as follows:

$$\min_{x,y} \quad c^T x + d^T y \tag{18.9}$$

subject to:

$$Ax + By = b \tag{18.10}$$

$$x \geq 0, \quad x \in \mathbb{R}^n_+ \tag{18.11}$$

$$y \in \mathbb{Z}_+ \tag{18.12}$$

where inequalities can be generalized as equalities by adding slack variables. The optimization problem in Equations (18.10)–(18.12) comprises a linear objective and linear constraints with respect to continuous and integer variables. If the elements in c and A were all zero, it would constitute an integer programming problem (IP). If all elements in d and B were zero the resulting problem would be a linear program (LP).

Several approaches have been developed to address these types of problems, with emphasis given to branch and bound and Cutting plane approaches in this chapter.

18.2.1 Branch and Bound

Branch and bound methods generate a decision tree to represent the different integer combinations that are feasible solutions of the problem. This feasible region is later partitioned into subproblems (by branching) and upper and lower bounds are systematically computed for the different solutions. If an upper bound (meaning the worst value) of a solution is computed, any solutions that have a higher lower bound (meaning their best solution) can be excluded from the decision tree, as clearly the optimum will not be in that decision branch (fathoming). This is explained in more detail later.

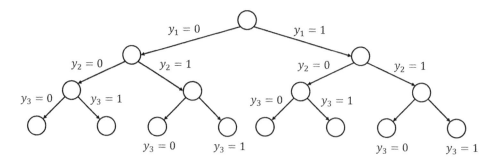

Figure 18.4 Binary tree representation on MILP.

Assume we have the following problem:

$$\min_{y} \quad y_1 + y_2 + y_3 \tag{18.13}$$

$$y \in [0, 1] \tag{18.14}$$

The binary decision tree representation of this problem is shown in Figure 18.4.

Figure 18.4 represents a graph or the possible solutions of the MILP, more specifically it is an acyclic connected graph, also known as a *tree graph*. The very top node is referred to as the *root* of the tree, while the bottom nodes are known as the *leaves*. Nodes that derive from an earlier node are known as *child nodes* (of the node they derived from), while the earlier node is known as the *parent node* (of those nodes derived from it).

As it can be observed from Figure 18.4 that the number of solutions for an MILP grows exponentially with the number of variables. Given that this problem presents only binary variables (*i.e.* $y \in [0, 1]$), the problem grows as 2^n, where n is the number of variables. It can be observed that in Figure 18.4 there are three variables, and hence $2^3 = 8$ nodes are at the bottom (leaves) of the tree. Hence, computing all solutions for a MILP would become intractable even with a small number of variables.

The general idea of the branch and bound approach is to explore the feasible space in search of the optimal solution without having to compute all possible solutions through the use of branching and fathoming. A branch and bound strategy will explore as few solutions as it needs to, before deciding that it has reached an optimal solution or can assure that no feasible solution exists.

The outline of the branch and bound approach is the following:

The first step is to branch the initial problem into one or more subproblems, and create a list to keep track of these candidate subproblems. We select one of these subproblems, we find a solution, and, if this solution is integer (and feasible) we take note of it, and go back to the list of candidate subproblems and select a new subproblem (in the hope of finding a better solution). If the solution is not integer we once again branch this subproblem to a second level, add the subproblems to the list of candidates and proceed by once again selecting a new candidate (note that the newly created subproblems will also be in this list). We continue with this strategy until we either find an optimal solution or determine that none exists. With the use of fathoming we avoid enumerating all solutions, hence, the more solutions we are able to exclude by fathoming, the more efficient the algorithm will be. This makes creating good upper and lower bounds important so that the tree can be pruned as fast as possible. However, if computing the bounds is too expensive, the algorithm will still be slow. Hence, there is a trade-off between creating good upper and/or lower bounds and evaluating a higher number of solutions.

Algorithm 18.1 *Branch and Bound*

1. *Initialization: Set the best value so far $f^* = \infty$, initialize the subproblem candidate list by appending the original MILP.*
2. *If the subproblem candidate list is empty return the best solution found so far f^*, if this solution is $f^* = \infty$ declare the problem as infeasible. Else:*

   1. *Select one of the subproblems in the candidate list (SP_i)*
   2. *Relax SP_i to create RSP_i, denote this solution f_{RSP_i}*
   3. *Apply fathoming*

      1. *If RSP_i is infeasible, then SP_i is also infeasible, hence go back to step 2.*
      2. *If $f_{RSP_i} \geq f^*$ then SP_i has no feasible solution which is better than the best solution found so far, hence go back to step 2.*
      3. *If the optimal solution of RSP_i is feasible for SP_i this is a feasible solution for the original problem. If $f_{RSP_i} \leq f^*$ this is the new best solution found (i.e. $f^* = f_{RSP_i}$). Go to step 2.*
      4. *If none of the above hold, proceed to (d).*

   4. *Branch the current subproblem SP_i and add the subproblems to the candidate subproblem list. Go to step 2.*

In Algorithm 18.1 no specification was given concerning the order in which the subproblems were chosen. This is because there are several alternatives:

1. **Depth-first search**: (also known as Last-In-First-Out) this approach chooses the last nodes added to the list to be explored first. It is known as depth-first, as it intends to get to the bottom of the tree (or as deep as it can) before exploring more shallow solutions (higher up in the tree). The advantages are that of easy book-keeping, and that most likely optimal solutions appear deeper into the tree.
2. **Breadth-first search**: this approach considers all nodes in a given level, before moving on to the next generation of child nodes.
3. **Best bound search**: the selected candidate is the one with the least lower bound (most promising candidate). This approach has the potential of resulting in fewer candidate evaluations.

There is no general approach that dominates the others, so generally heuristic rules are used to choose an approach depending on the information known of the problem at hand. The most common way of relaxing an MILP is by converting its integer variables into continuous ones, hence the relaxed problem is an LP.

Example

$$\min_{x,y} \quad 4x_1 - 1.1y_1 - 3.2y_2 - 2.5y_3 \tag{18.15}$$

subject to:

$$5x_1 + y_1 + y_2 + 2y_3 \leq 2.5 \tag{18.16}$$

$$x_1 \geq 0 \tag{18.17}$$

$$y_1, y_2, y_3 \in [0, 1] \tag{18.18}$$

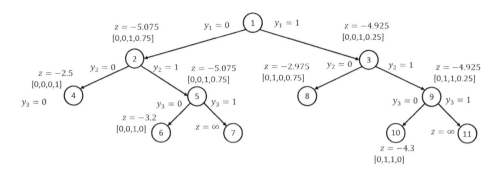

Figure 18.5 Best bound search branch and bound.

A best bound search procedure is shown in Figure 18.5.

The number inside the nodes of Figure 18.5 indicates in which order the solutions were explored. Let us follow the algorithm as it solved the problem:

Initially, the MILP was relaxed to an LP and yielded the solution $z = -5.075$, $[0, 0, 1, 0.75]$. It is always worthwhile to check whether a simple relaxation could solve the problem. Given that this is not the case, the branch and bound begins. All solutions z are the results of solving the relaxed LP. The algorithm commences by solving for nodes 2 and 3 and hence adding its children nodes to the candidate list. Given that this is a best bound search, the algorithm explores the children from node 2 first, as they present the best lower bound (node 2 has a lower bound of -5.075, while node 3 has a lower bound of -4.925). In this case, we have opted for always computing the $y = 0$ choice (rather than $y = 1$), which is an arbitrary choice, and hence node 4 is computed. With node 4 we find our first feasible solution, and note it down as the best answer found so far. Let us note that from now on, any node with a lower bound above -2.5 will be immediately discarded.

We see that node 4 has the worst lower bound ($z = -2.5$) and continue with node 5. Node 5 becomes the new most promising candidate and hence we explore its children. Note that when node 7 is explored, we realize that this solution is actually infeasible (even in the case that node 7 had any children, these would be discarded, as they would be also infeasible). At this point node 3 has become the most promising candidate. The algorithm ensues until the optimal solution is found $z = -4.3$, $[0, 0, 1, 0]$. Let us note that neither node 8 nor node 4 are further explored, as they could only result in worst objective function values than the already computed solution.

To add to this example, Figure 18.6 shows a breadth-first search approach to solving this problem.

In this case, breadth-first search was a less efficient approach than best bound search, but we must highlight that this will not always be the case.

General Remarks on Branch and Bound Algorithms

- Setting variables to 0 and 1 is not the only way to generate subproblems. Other frameworks create cuts within constraints. A detailed review on different ways to generate subproblems can be found in [3].

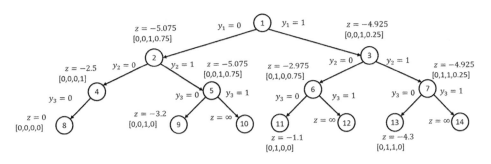

Figure 18.6 Breadth-first search branch and bound.

- In this section we discussed the LP relaxation of the MILP; however, other relaxations exist, such as removing one or several constraints and setting one or more free coefficients of binary variables in the objective function to 0. Furthermore, LP relaxation is the most frequently used.
- In general, there is a tradeoff between ease of solution of the relaxed MILP and the gap between the relaxation and the solution of problems. That is, in many cases, the easier the relaxed problem is to solve, the lower the quality of the lower bound created, and hence more iterations are needed to obtain a good lower bound. On the other hand, if the relaxed problem is similar to the original problem, it might be harder to solve, but will provide good quality lower bounds and hence need fewer iterations.

18.2.2 Cutting Planes

Cutting Planes is another method to find optimal solutions to MILPs. The general idea of these methods is to transform the MILP into a relaxed LP. Then, by introducing inequality constraints (which are termed as *cuts*), the feasible region of the LP is progressively narrowed down to exclude infeasible and suboptimal regions that are not in the desired integer domain. This is better illustrated in Figure 18.7, where points indicate integer solutions, lines are inequality constraints, and the shaded area is the infeasible region.

Figure 18.7(a) shows an LP with its solution at the intersection of two constraints, the point being $y_1 = \frac{84}{31}$, and $y_2 = \frac{46}{31}$, which is not an integer solution. However, by the addition of a cut (or a set of cuts), the LP shown in Figure 18.7(b) is constructed. This new LP has a fully integer solution of $y_1 = 3$, $y_2 = 1$, which would result in successfully solving an Integer Programming problem.

Different Cutting Plane methods have been developed; however, we will be focusing on what most research agrees to be the first method of this type, developed by Ralph Gomory in the 1950s. Although this method is not the most effective, it provides intuition, and modern Cutting Plane algorithms are based on the principle of the Gomory Cuts method [6].

Gomory Cuts

Let us first derive the Gomory Cut for Integer Programs (IPs). Assume we have an IP that we relax into an LP. By solving the LP we obtain an optimal solution that is

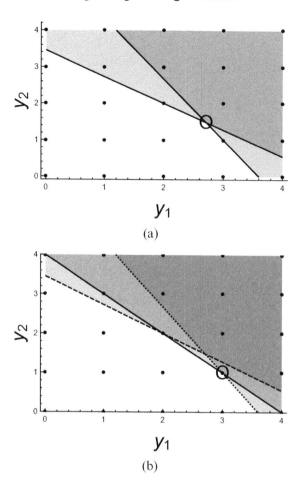

(a)

(b)

Figure 18.7 Addition of cuts to obtain an optimal feasible solution from a relaxed LP.

not an integer solution. Hence we wish to introduce an inequality (cut) that excludes a part of the infeasible region from the LP, and that this removed domain contains the infeasible solution found by the LP; however, this inequality must not exclude any feasible solutions from the original IP. This would allow us to re-solve the relaxed LP (with the new added constraint). If the solution is once again not integer, we repeat the process until we find a fully integer optimal solution by the relaxed LP, which would mean we have found an optimal solution for the original IP. The question now is, how to derive said cuts. Ralph Gomory proposed the following approach:

We solve the relaxed LP by the simplex method and we obtain the simplex tableau configuration at the solution (the reader is encouraged to revise Chapter 11 for this derivation). We then select a row where one of the integer variables (which is not integer at the solution) is basic, which would yield the equation:

$$x_i + \sum a_{i,j} x_j = b_i \qquad (18.19)$$

where x_i is the basic variable and the x_j are the nonbasic variables. Let us denote $\lfloor a_{i,j} \rfloor$ as the integer part of $a_{i,j}$, and $\lfloor b_i \rfloor$ as the integer part of b_i, then rewrite

Equation (18.19) and have all fractional parts in the right-hand side (RHS), while keeping all integer parts on the left-hand side (LHS):

$$x_i + \sum \lfloor a_{i,j} \rfloor x_j - \lfloor b_i \rfloor = b_i - \lfloor b_i \rfloor - \sum a_{i,j} x_j + \sum \lfloor a_{i,j} \rfloor x_j \qquad (18.20)$$

For any feasible integer point, the RHS must be less than 1, while the LHS must be an integer. To satisfy these two conditions we obtain:

$$b_i - \lfloor b_i \rfloor - \sum a_{i,j} x_j + \sum \lfloor a_{i,j} \rfloor x_j \leq 0 \qquad (18.21)$$

We have just produced the Gomory Cut, which is shown in Equation (18.21). This discussion explains why this inequality will not exclude any feasible region from the IP; however, how does this guarantee that we exclude the undesired non-integer solution from the relaxed LP?

We must note that $b_i - \lfloor b_i \rfloor > 0$; furthermore, for the optimal solution of the LP we have all basic variables $x_j = 0$, since the term $\sum \lfloor a_{i,j} \rfloor x_j \geq 0$ Equation (18.21) cannot hold for the original LP. It is then that we ensure that Equation (18.21) is an inequality that does not exclude any feasible solutions from the IP while cutting the infeasible solution obtained by the relaxed LP.

Let us now present the Cutting Planes algorithm by Gomory Cuts accompanied by an example to illustrate how these cuts are obtained in practice.

Algorithm 18.2 *Cutting Planes by Gomory Cuts*

1. *Relax the MILP into an LP and solve it using the Simplex tableau.*

    1. *If relaxed the LP is infeasible then declare the MILP is also infeasible*
    2. *If relaxed the LP is feasible, and has a fully integer solution for the integer variables, return the optimal solution. This is the optimal solution of the MILP.*
    3. *If relaxed the LP is feasible, but integer variables have fractional values go to Step 2.*

2. *Select a row where one of the integer variables has a fractional value.*
3. *Perform the Gomory Cut procedure to obtain a valid inequality.*
4. *Add this inequality to the MILP and return to Step 1.*

Let us now go through the following example and illustrate how this procedure is conducted.

Example   Solve the following Integer Programming problem:

$$\min \quad -6y_1 - 5y_2 \qquad (18.22)$$

subject to:

$$8y_1 + 11y_2 \leq 38 \qquad (18.23)$$

$$5y_1 + 3y_2 \leq 18 \qquad (18.24)$$

$$y \in \mathbb{Z}_+$$

Gomory's method is as follows:
Step 1.
Relax the above IP into an LP. Figure 18.8 illustrates the relaxed problem, where the infeasible regions are shaded and integer points are shown.

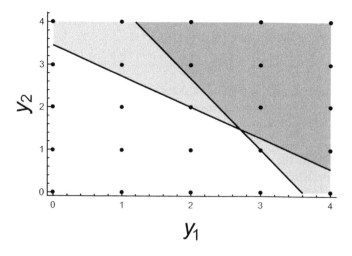

Figure 18.8 Linear Program resulting from the relaxed Integer Program.

Step 2.
Solve the relaxed LP through the Simplex tableau. The first step for this process is to add slack variables to the original problem, which yields the following:

$$\min \quad -6y_1 - 5y_2 \tag{18.25}$$

subject to:

$$8y_1 + 11y_2 + s_1 = 38 \tag{18.26}$$

$$5y_1 + 3y_2 + s_2 = 18 \tag{18.27}$$

$$y_1, y_2, s_1, s_2 \geq 0 \tag{18.28}$$

For this problem, the initial configuration of the tableau is the following:

	y_1	y_2	s_1	s_2	b
s_1	8	11	1	0	38
s_2	5	3	0	1	18
reduced cost	-6	-5	0	0	0

The solution of this LP yields the following Simplex tableau:

	y_1	y_2	s_1	s_2	b
y_1	1	0	$-\frac{3}{31}$	$\frac{11}{31}$	$\frac{84}{31}$
y_2	0	1	$\frac{5}{31}$	$-\frac{8}{31}$	$\frac{46}{31}$
reduced cost	0	0	$\frac{7}{31}$	$\frac{26}{31}$	23.677

We can observe that this solution yields $y_1 = \frac{84}{31}$, and $y_2 = \frac{46}{31}$, which is not an integer solution.

Step 3. Form a cut from the solution point of the relaxed LP.

We use the Simplex tableau to obtain a new constraint (cut) which we can introduce into the relaxed LP and exclude solutions that are infeasible for the IP. Let us choose the second row of the Tableau (both rows will yield the same cut):

$$0y_1 + 1y_2 + \frac{5}{31}s_1 - \frac{8}{31}s_2 = \frac{46}{31} \tag{18.29}$$

Now we separate all improper fractions into their integer and fractional part as follows:

$$0y_1 + 1y_2 + \frac{5}{31}s_1 + \left(-1 + \frac{23}{31}\right)s_2 = 1 + \frac{15}{31} \tag{18.30}$$

Notice how the negative number is dealt with. We next eliminate all the integers from the equation (notice we count zero as an integer), switching the equality for an inequality:

$$\frac{5}{31}s_1 + \frac{23}{31}s_2 \geq \frac{15}{31} \tag{18.31}$$

Substituting s_1 and s_2

$$\frac{5}{31}(38 - 8y_1 - 11y_2) + \frac{23}{31}(18 - 5y_1 - 3y_2) \geq \frac{15}{31} \tag{18.32}$$

We obtain the following cut

$$5y_1 + 4y_2 \leq 19 \tag{18.33}$$

Step 4. Add the cut to the previous LP formulation.

Our new relaxed LP is the following:

$$\min \quad -6y_1 - 5y_2 \tag{18.34}$$

subject to:

$$8y_1 + 11y_2 + s_1 = 38 \tag{18.35}$$

$$5y_1 + 3y_2 + s_2 = 18 \tag{18.36}$$

$$5y_1 + 4y_2 + s_3 = 19 \tag{18.37}$$

$$y_1, y_2, s_1, s_2, s_3 \geq 0 \tag{18.38}$$

The addition of this new constraint is shown in Figure 18.9.

Figure 18.9(a) shows the original relaxed LP, where the optimal point is shown in the intersection between the two constraints, while Figure 18.9(b) portrays the relaxed LP with the addition of the constraint in Equation (18.33). This constraint excludes the past optimal LP solution, which was infeasible to the IP.

On the downside it can be seen that the infeasible area excluded from this constraint is quite small. The process is repeated for a number of iterations to eventually yield the LP in Figure 18.10, which has an optimal solution of $y_1 = 3$, $y_2 = 1$, which is the optimal solution to the original IP.

We will now look at Gomory's derivation for the case of an MILP [7].

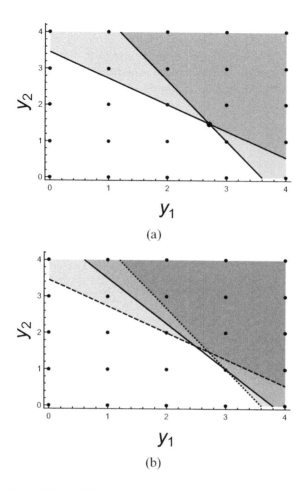

(a)

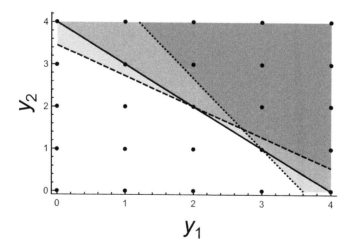

(b)

Figure 18.9 Addition of a Gomory cut.

Figure 18.10 Relaxed LP with optimal solution for IP.

Let us assume we have an MILP, and its feasible domain is denoted by

$$S = \left\{(y, x) \in \mathbb{Z}_+^n \times \mathbb{R}_+^p : Ax + Gy \leq b\right\} \tag{18.39}$$

where variables y can only take non-negative integer values. We then relax this MILP into an LP, and solve the LP by using the simplex tableau. Then, using a row from one of the integer variables (which did not yield an integer solution) from the simplex tableau at the solution, we get the following equation:

$$\sum_{j=1}^{n} a_j y_j + \sum_{j=1}^{p} g_j x_j = b, \quad y \in \mathbb{Z}_+, \; x \in \mathbb{R}_+ \tag{18.40}$$

Let b be a non-integer formed of its integer part $\lfloor b \rfloor$ and its non-integer part $0 < f_0 < 1$, such that $b = \lfloor b \rfloor + f_0$. We define the same for a_j such that $a_j = \lfloor a_j \rfloor + f_j$, where $0 \leq f_j < 1$. Then we can rewrite Equation (18.40) as:

$$\sum_{f_j \leq f_0} f_j y_j + \sum_{f_j > f_0} (f_j - 1) y_j + \sum_{j=1}^{p} g_j x_j = k + f_0 \tag{18.41}$$

where k is an integer number that includes the integer parts of the a_js and b. Then, the following disjunction is satisfied (depending on whether k is zero, positive, or negative):

$$\sum_{f_j \leq f_0} \frac{f_j}{f_0} y_j - \sum_{f_j > f_0} \frac{1 - f_j}{f_0} y_j + \sum_{j=1}^{p} \frac{g_j}{f_0} x_j \geq 1 \tag{18.42}$$

or

$$-\sum_{f_j \leq f_0} \frac{f_j}{1 - f_0} y_j + \sum_{f_j > f_0} \frac{1 - f_j}{1 - f_0} y_j - \sum_{j=1}^{p} \frac{g_j}{1 - f_0} x_j \geq 1 \tag{18.43}$$

The disjunction in Equations (18.42) and (18.43) is of the form:

$$a_1 z \geq 1 \tag{18.44}$$

or

$$a_2 z \geq 1 \tag{18.45}$$

which implies $\sum_j \max(a_{1j}, a_{2j}) z_j \geq 1$ for any $z \geq 0$. Looking at Equations (18.42) and (18.43) we must find the maximum coefficient for each variable between these two equations. Fortunately, in every case one is negative and one is positive, and hence with the positive terms we obtain the following:

$$\sum_{f_j \leq f_0} \frac{f_j}{f_0} y_j + \sum_{f_j > f_0} \frac{1 - f_j}{1 - f_0} y_j + \sum_{j=1}^{p} \frac{g_j}{f_0} x_j - \sum_{j=1}^{p} \frac{g_j}{1 - f_0} x_j \geq 1 \tag{18.46}$$

We have just proved that inequality (18.46) is valid for our original MILP domain S, and this equation is the Gomory mixed-integer inequality (GMI inequality) [7]. Let us note that in the IP case inequality (18.46) reduces to:

$$\sum_{f_j \leq f_0} \frac{f_j}{f_0} y_j + \sum_{f_j > f_0} \frac{1 - f_j}{1 - f_0} y_j \geq 1 \tag{18.47}$$

and given that fact that $\frac{1-f_j}{1-f_0} < \frac{f_j}{f_0}$ for $f_j > f_0$ the GMI reduces to:

$$\sum_{f_j \leq f_0} f_j y_j \geq f_0 \tag{18.48}$$

also known as *fractional cut*.

General Remarks on Cutting Plane Methods

- Note that the Gomory Cuts method is among the most inefficient of the Cutting Plane methods, and many improvements have been developed since, including Disjunctive Cutting Plane methods, Mixed Integer Rounding (MIR) cuts, Implied Bound Cuts, and GUB cover cuts.
- Improvements on using Gomory Cuts can be obtained by performing several rounds of cutting. This means solving the relaxed LP, generating a cut for each fractional basic variable, and updating the linear program by adding these cuts. Further improvements can be achieved by a lift-and-project approach [7].
- Branch-and-cut algorithms, which combine a branching approach with Gomory Cuts, have been found to be very efficient in practice [19].

18.2.3 General Remarks on MILP Algorithms

- In practice, branch and bound methods are deemed more efficient than cutting plane methods, even though a cutting plane method was the first method that proved that an MILP solution could be found in a finite number of steps.
- Cutting plane strategies have been combined with branch and bound strategies to yield what are called *branch-and-cut* algorithms, which are highly efficient and used in many commercial MILP solvers.

18.3   SOLUTION TECHNIQUES FOR MINLP PROBLEMS

Mixed-integer nonlinear programming (MINLP) problems have presented a challenge for the field of mathematical optimization, as they combine the combinatorial difficulty of optimizing over discrete variable sets with the challenge of handling nonlinear functions. Furthermore, this approach has a wide range of applications in engineering, and particularly in chemical engineering. In MINLP, integer variables can represent things like equipment size, availability or not of a stream/unit, events, among many others, while continuous variables model flow rates, reactions, concentrations, temperature, pressure, *etc.*

A general MINLP is of the following form:

$$\min_{x,y} \quad f(x,y) \tag{18.49}$$

subject to:

$$h(x,y) = 0 \tag{18.50}$$

$$g(x,y) \leq 0 \tag{18.51}$$

$$x \in X \subseteq \mathbb{R}^n \tag{18.52}$$

$$y \in \mathbb{Z} \tag{18.53}$$

The most widely used approaches to solving convex MINLPs will be presented in the remainder of this chapter, while solution methods for nonconvex MINLPs are addressed in Chapter 19.

18.3.1 Branch and Bound on MINLP Problems

Branch and bound was introduced as a solution procedure to address MILPs. Furthermore, the same building blocks can be applied to MINLPs and guarantee global optimality as long as the problem is convex with respect to the continuous variables, a constraint qualification (*e.g.* Slater's) holds at the solution of every NLP, and integer variables do not appear as bilinear terms in the problem's equations.

Branch and bound on MINLPs starts by relaxing the MINLP at the root node. If this relaxation is infeasible, then the MINLP is also infeasible. If the solution of the relaxation is feasible, then depending on the relaxation and the solution obtained, the MINLP may be declared solved (*e.g.* fully integer solution for an integrality relaxation). Otherwise, branch and bound searches a tree whose nodes correspond to the relaxed subproblems (*e.g.* NLPs for an integrality relaxation), and whose edges correspond to branching decisions. Optimality and feasibility of subproblems is used to prune nodes in the tree until an optimal solution is found, or the problem is declared infeasible.

The solution procedure is then equivalent to the one presented in Section 18.2.1, with the difference mainly in the suproblems that the relaxations yield (*e.g.* MINLPs yield NLPs by an integrality relaxation, instead of LPs).

The algorithms is outlined here.

Algorithm 18.3 *Branch and bound for MINLP*

1. *Initialization: Set the best value so far $f^* = \infty$, initialize the subproblem candidate list by appending the original MINLP.*
2. *If the subproblem candidate list is empty return the best solution found so far f^*, if this solution is $f^* = \infty$ declare the problem as infeasible. Else:*

    1. *Select one of the subproblems in the candidate list (SP_i)*
    2. *Relax SP_i to create RSP_i, denote this solution f_{RSP_i}*
    3. *Apply fathoming*

        1. *If RSP_i is infeasible, then SP_i is also infeasible, hence go back to step 2.*
        2. *If $f_{RSP_i} \geq f^*$ the then SP_i has no feasible solution which is better then the best solution found so far, hence go back to step 2.*
        3. *If the optimal solution of RSP_i is feasible for SP_i this is a feasible solution for the original problem. If $f_{RSP_i} \leq f^*$ this is the new best found solution (i.e. $f^* = f_{RSP_i}$). Go to step 2.*
        4. *If none of the above hold, proceed to (d).*

    4. *Branch the current subproblem SP_i and add the smaller subproblems to the candidate subproblem list. Go to step 2.*

Algorithm 18.3 can only guarantee optimality under the aforementioned convexity conditions; however, heuristic measures can be implemented so that it is able to solve nonconvex MINLPs. Although a global solution cannot be guaranteed, these heuristics have yielded good results for practical applications [8].

Branch and Bound for Nonconvex MINLP Problems

The following heuristics have been implemented in the state-of-the-art MINLP solver Bonmin [8].

- Given that a problem is not convex, if the integer variables in the MINLP are relaxed to yield an NLP, this NLP might find different local solutions depending on the starting points given. A "low quality" local solution for a node that has a much better relaxed solution can cause the optimal solution to be cut off. For this reason, it is convenient to evaluate each node a number of times, and select the best solution among them.
- Since the solution of the NLP might not produce the true lower bound of that node (given that a better local solution might exist), it is possible to allow some tolerance for fathoming a point, even if this lower bound is greater than the best known solution.

Rigorous approaches to solve nonconvex MINLPs to global optimality have also been developed (*e.g.* α-branch and bound), which are presented in Chapter 19.

General Remarks on Branch and Bound for MINLPs

- Nonlinear branch and bound presents a good generic choice to solve an MINLP problem.
- Good heuristic strategies to choose the next node to be explored combine depth-first and best-bound search. Two such strategies are the two-phase method which can be reviewed in [22] and the diving method proposed by [23].

18.3.2 General Benders Decomposition (GBD)

Generalized Benders Decomposition (GBD) starts with a problem such as the following:

$$\min_{x,y} \quad f(x, y) \tag{18.54}$$

subject to:

$$h(x, y) = 0 \tag{18.55}$$

$$g(x, y) \leq 0 \tag{18.56}$$

$$x \in X \subseteq \mathbb{R}^n \tag{18.57}$$

$$y \in Y = [0, 1]^q \tag{18.58}$$

where y is defined as *complicating variables*. When GBD is used to solve MINLPs these complicating variables turn out to be the binary variables $y \in Y = \{0, 1\}^q$. Hence, the use of GBD to solve MINLPs is a particular use of this method, although it can be used with any other class of problems with complicating variables.

The following example illustrates this method applied to a MINLP problem.

Example

$$\min \quad -y + 3x + \frac{1}{2}x^2 \tag{18.59}$$

subject to:

$$-2.1\,x - \ln\left(\frac{1}{5}x\right) + y \leq 0 \tag{18.60}$$

$$y \in [0, 1] \tag{18.61}$$

We can then decompose the objective and constraints by their dependence on x or y as follows:

$$f_x = 3x + \frac{1}{2}x^2 \tag{18.62}$$

$$f_y = -y \tag{18.63}$$

$$g_x = -2.1\,x - \ln\left(\frac{1}{5}x\right) \tag{18.64}$$

$$g_y = y \tag{18.65}$$

Let us note that this problem is linear in y and convex in x, and hence GBD is an appropriate technique to solve it.

We can now formulate the relaxed Master problem in the following form:

Relaxed Master Problem

$$\min_{y,\theta} \quad \theta \tag{18.66}$$

subject to:

$$\theta \geq -y + \mu^k y + L_1^k, \qquad k = 1, 2, \ldots, K \tag{18.67}$$

$$0 \geq \bar{\mu}^l y + \bar{L}_1^l, \quad l = 1, 2, \ldots, L \tag{18.68}$$

where

$$L_1^k = \min_x \quad 3x + \frac{1}{2}x^2 + \mu^k \left(-2.1\,x - \ln\left(\frac{1}{5}x\right)\right) \tag{18.69}$$

$$\bar{L}_1^l = \min_x \quad \mu^l \left(-2.1\,x - \ln\left(\frac{1}{5}x\right)\right) \tag{18.70}$$

where μ^k and μ^l could be thought of as Lagrange multiplier values of the constraint g_x, found at the $k$ith iteration of the NLP and a feasibility subproblem at iteration l, respectively.

Let us now proceed with the GBD algorithm.

Step 1. Let us select $y^1 = 0$

Solve the primal problem **GBD.P**

$$\min \quad f_x = 3x + \frac{1}{2}x^2 \tag{18.71}$$

subject to:

$$-2.1\,x - \ln\left(\frac{1}{5}x\right) \leq 0 \tag{18.72}$$

the solution of which is

$$x = 0.846 \tag{18.73}$$

$$\mu^1 = 3.282 \tag{18.74}$$

$$f_x = 2.89594 \tag{18.75}$$

Hence, we update the upper bound $UBD = 2.89594$
Step 2.

$$L_1^1 = \min_{x} \quad 3x + \frac{1}{2}x^2 + 3.282\left(-2.1\,x - \ln\left(\frac{1}{5}x\right)\right) \tag{18.76}$$

Given that strong duality holds we know that $L_1^1 = 2.89594$
Step 3.
The relaxed Master problem **GBD.M** is now of the form

$$\min_{y,\theta} \quad \theta \tag{18.77}$$

subject to:

$$\theta \geq -y + 3.282y + 2.89594$$

with its solution being

$$y = 0, \quad \theta = 2.89594 \tag{18.78}$$

We update our lower bound to be $LBD = 2.89594$. Given that our lower bound and our upper bound are the same, we can certify we have found the optimal solution to the problem.

General Remarks on Generalized Benders Decomposition Applied to MINLP

- Several variants of the GBD method exist, which provide better performance for particular types of problems; these variants are discussed in detail in [3].
- GBD has the advantage of only using the integer and objective variables to perform the cuts; however, these cuts are almost always dense.
- If a problem has the structure adequate for Outer Approximation (OA), presented in the following section, it is in general preferred to use OA instead of GBD, as the cuts performed by GBD are weaker than those by OA. This means that GBD might take more outer (master) iterations to converge.

18.3.3 Outer Approximation (OA)

Outer approximation is an alternative solution strategy to address MINLPs with the following structure:

$$\min_{x,y} \quad c^T y + f(x) \tag{18.79}$$

subject to:

$$g(x) + By \leq 0 \tag{18.80}$$

$$x \in X = \left\{x : x \in \mathbb{R}^n, \ A_1 x \leq a_1\right\} \subseteq \mathbb{R}^n \tag{18.81}$$

$$y \in Y = \left\{y : y \in \{0, 1\}^q, \ A_2 y \leq a_2\right\} \tag{18.82}$$

where $f(x)$ and $g(x)$ are once continuously differentiable convex functions with respect to x, and a constraint qualification (*e.g.* Slater's) holds at the solution of every NLP resulting from the problem in Equations (18.79–18.82) by fixing y.

OA follows the same basic principle as branch and bound and GBD, in that at every iteration it generates upper and lower bounds on the MINLP. It can be shown that this sequence of upper bounds is non-increasing and the sequence of lower bounds is non-decreasing, and hence that the sequences converge within a finite number of iterations to a tolerance ϵ [3].

The upper bounds are generated by fixing the y variables (*e.g.* $y = y^k$) of the original problem. The lower bounds are the result of solving the Master problem. In this case, the Master problem is derived using primal information at the current solution point x^k of the primal problem. The Master problem consists of an outer approximation (linearization) of the nonlinear objective and constraints around the primal solution x^k.

The difference between the OA and the GBD approaches lies in the way the Master problem is formulated. In this case, the OA constructs the Master problem based on primal information and an outer approximation (hence the name).

18.3.4 Outer Approximation Approach Development

Primal Problem

The primal problem corresponds to the original problem while fixing the y variables to 0–1 values, which are denoted as y^k for iteration k. The formulation is as follows:

$$\min_{x} \quad c^T y^k + f(x) \tag{18.83}$$

subject to:

$$g(x) + B y^k \leq 0 \tag{18.84}$$

$$x \in X = \left\{ x : x \in \mathbb{R}^n, \ A_1 x \leq a_1 \right\} \subseteq \mathbb{R}^n \tag{18.85}$$

Depending on the problem and the fixed points y^k the primal problem might result to be feasible or infeasible, which leads to two different cases.

Feasible Primal
If we have a feasible primal solution at iteration k, then this solution provides an upper bound for our problem, with $UBD = c^T y^k + f(x^k)$.

Furthermore, given that functions $f(x)$ and $g(x)$ are convex, we can linearize these functions and obtain the following:

$$f(x) \geq f(x^k) + \nabla f(x^k)(x - x^k), \quad \forall x^k \in X \tag{18.86}$$

$$g(x) \geq g(x^k) + \nabla g(x^k)(x - x^k), \quad \forall x^k \in X \tag{18.87}$$

which will be used in subsequent steps of the algorithm.

Infeasible Primal
If the optimization solver has identified an infeasible primal, then we, unfortunately, do not have a new upper bound. Furthermore, we need to identify a feasible point, and this is done by analyzing the constraint set, as follows:

$$g(x) + B y^k \leq 0 \tag{18.88}$$

Hence, we formulate a feasibility problem, similarly to the GBD case, where we minimize the violation of the constraints and hence identify a feasible solution. As in the GBD case, the norm used in the feasibility problem can be selected by the user; here we illustrate an l_1-minimization:

$$\min_{x \in X} \sum_{j=1}^{p} a_j \tag{18.89}$$

subject to:

$$g_j(x) + B y^k \le a_j, \quad j = 1, 2, \ldots, p \tag{18.90}$$

$$a_j \ge 0 \tag{18.91}$$

The solution of the problem in Equations (18.89–18.91) will provide a feasible point x^f, on which we can linearize the constraints (due to the fact that $g(x)$ is convex):

$$g(x) \ge g(x^f) + \nabla g(x^f)(x - x^f), \quad \forall x^l \in X \tag{18.92}$$

and hence obtain a linear constraint, which can be added to the primal formulation.

The Master Problem

The Master problem is formulated by relying on two key ideas:
1) The projection of the original problem (in Equations (18.79–18.82)) onto the y-space
2) An outer approximation (linearization) of the objective function and the constraints

Projection of Original Problem onto the y-space
The problem in Equations (18.79–18.82) can also be rewritten as:

$$\min_{y} \ \inf_{x} \ c^T y + f(x) \tag{18.93}$$

subject to:

$$g(x) + By \le 0 \tag{18.94}$$

$$x \in X \tag{18.95}$$

$$y \in Y \tag{18.96}$$

The minimum is replaced by the infimum with respect to x to cover the case where the solution could be unbounded for a fixed value of y.

Given that $c^T y$ does not depend on x, it can be taken outside the infimum and hence we can define $v(y)$:

$$v(y) = c^T y + \inf_{x} \ f(x) \tag{18.97}$$

subject to:

$$g(x) + By \le 0 \tag{18.98}$$

$$x \in X \tag{18.99}$$

Hence, $v(y)$ corresponds to the optimal value of the original problem for fixed y (*i.e.* the primal problem $P(y^k)$).

Now, it is useful to define the set of y variables for which a feasible solution in the x variables exists:

$$V = \{y : g(x) + By \leq 0, \quad \forall x \in X\}$$

We then rewrite the problem in Equations (18.79–18.82) as:

$$\min_{y} \quad v(y) \tag{18.100}$$

subject to:

$$y \in Y \cap V \tag{18.101}$$

The problem in Equations (18.100–18.101) is now the projection of the original problem (the problem in Equations (18.79–18.82)) onto the y-space. It is, however, difficult to solve this problem, as both V and $v(y)$ are known implicitly. Because of this [4] considered outer linearization of $v(y)$ and a specific representation of V, which is presented next.

Outer Approximation of $v(y)$

The outer approximation (linearization) of $v(y)$ is performed by means of an infinite set of supporting functions. Let us remember that $f(x)$ and $g(x)$ are convex and can hence be approximated around a point x^k by a linearized form. We then assume that an infinite number of linearizations of $f(x)$ can approximate the function itself. This can be better understood by observing Figure 18.2. Hence, the supporting functions for $v(y)$ correspond to linearizations of $f(x)$ and $g(x)$ at all $x^k \in X$:

$$f(x) \geq f(x^k) + \nabla f(x^k)(x - x^k), \quad \forall x^k \in X \tag{18.102}$$

$$g(x) \geq g(x^k) + \nabla g(x^k)(x - x^k), \quad \forall x^k \in X \tag{18.103}$$

This approximation of $v(y)$ results in an MILP due to the linearization of nonlinear functions $f(x)$ and $g(x)$. The constraint qualification assumption, which holds at the solution of every primal problem for fixed $y \in Y \cap V$, coupled with the convexity of $f(x)$ and $g(x)$, imply the following property [3]:

$$v(y) = \begin{bmatrix} \min_x & c^T y^k + f(x) \\ s.t. & g(x) + By \leq 0 \\ & x \in X \end{bmatrix} \tag{18.104}$$

$$= \begin{bmatrix} \min_x & c^T y + f(x^k) + \nabla f(x^k)(x - x^k) \\ s.t. & g(x^k) + \nabla g(x^k)(x - x^k) \leq 0 \\ & x \in X \end{bmatrix} \forall x^k \in X \tag{18.105}$$

Let us note that it suffices to include the linearizations of those constraints that are active at (x^k, y^k). This implies that fewer constraints are needed in the Master problem.

By substituting $v(y)$ from Equation (18.104) in the projected problem (Equations (18.100–18.101)), we obtain:

$$\min_x \min_y \quad c^T y + f(x^k) + \nabla f(x^k)(x - x^k) \tag{18.106}$$

subject to:

$$0 \geq g(x^k) + \nabla g(x^k)(x - x^k) + By, \ \forall k \in F \tag{18.107}$$

$$x \in X \tag{18.108}$$

$$y \in Y \cap V \tag{18.109}$$

where $F = \{k : x^k$ is a feasible solution of the primal problem $P(y^k)\}$.

Let us now combine the min operators, and introduce variable μ_{OA} to yield the following equivalent problem:

$$\min_{x,y,\mu_{OA}} \quad c^T y + \mu_{OA} \tag{18.110}$$

subject to:

$$\mu_{OA} \geq f(x^k) + \nabla f(x^k)(x - x^k), \ \forall k \in F \tag{18.111}$$

$$0 \geq g(x^k) + \nabla g(x^k)(x - x^k) + By, \ \forall k \in F \tag{18.112}$$

$$x \in X \tag{18.113}$$

$$y \in Y \cap V \tag{18.114}$$

Furthermore, it is possible to replace $y \in Y \cap V$ only by $y \in Y$ if appropriate integer cuts are made such that the possibility of generating the same integer combinations is excluded. Hence, this problem is equivalent to solving the following:

$$\min_{x,y,\mu_{OA}} \quad c^T y + \mu_{OA} \tag{18.115}$$

subject to:

$$\mu_{OA} \geq f(x^k) + \nabla f(x^k)(x - x^k), \ \forall k \in F \tag{18.116}$$

$$0 \geq g(x^k) + \nabla g(x^k)(x - x^k) + By \ \forall k \in F \tag{18.117}$$

$$x \in X \tag{18.118}$$

$$y \in Y \tag{18.119}$$

$$\sum_{i \in \mathbf{B}^k} y_i^k - \sum_{i \in \mathbf{NB}^k} y_i^k \leq \left| \mathbf{B}^k \right| - 1, \ k \in F \tag{18.120}$$

where $\mathbf{B}^k$ and $\mathbf{NB}^k$ are the sets of binary variables having value 1 and 0, respectively, at iteration k, and $|\cdot|$ denotes the cardinality of a set.

Note that equation $\sum_{i \in \mathbf{B}^k} y_i^k - \sum_{i \in \mathbf{NB}^k} y_i^k \leq |\mathbf{B}^k| - 1, \ k \in F$ will prevent the use of the same combination of integer variables.

The Master problem in Equations (18.115–18.120) is an MILP, and can be solved by MILP algorithms (*e.g.* branch and bound and cutting planes).

In [3] a nice geometric interpretation of this master problem is outlined as follows: To be a valid relaxation of the MINLP, the MILP *underestimates the objective function* and *overestimates the feasible region*. The objective function is underestimated as in Figure 18.1 earlier in this chapter. One can see that the more linearized support functions are added, the better the approximation will be; however, the function is always underestimated (except in the limit where support planes are added for all possible points). On the other hand, the feasible region is overestimated, as in Figure 18.3. In this figure one can appreciate that tangent supports to the feasible region will yield an approximation, and in the limit where support planes are added for all possible points this approximation will be equivalent to the original feasible region.

Hence, from this, the MILP master problem would be equivalent to the original problem if relaxations of the nonlinear functions, for all points x^k that result from fixing all $y \in Y$ to be added. This results in an exhaustive enumeration as we would need to know all feasible x^k points, for which we would need to solve all primal problems $P(y^k)$. Hence, this Master problem is solved only approximately by means of a relaxation.

This relaxation is such that at every iteration, linear supports of the objective and constraints are added only around the current point x^k. As the algorithm proceeds, more linear supports are added with every iteration, and the approximations of the objective function and the feasible region are refined. The Outer Approximation algorithm is described in Algorithm 18.4.

Algorithm 18.4 *OA Algorithm*

1. *Select an initial configuration $y^1 \in Y$ and stopping tolerance ε*
2. *Given a binary configuration y^k solve the resulting Primal problem $P(y^k)$*

   1. *If Primal problem is feasible obtain as its solution x^k. Set the current upper bound*
      $$UBD = P(y^k) = v(y^k)$$
   2. *If Primal problem is infeasible, follow the infeasible Primal approach by minimizing the violation of the constraints and obtain as its solution x^k. Set the current upper bound $UBD = P(y^k) = v(y^k)$*

3. *Solve the relaxed master problem and obtain y^{k+1} and μ_{OA}^{k+1} at the solution. Set $c^T g^{k+1} + \mu_{OA}^{k+1}$ as the new lower bound $c^T g^{k+1} + \mu_{OA}^{k+1}$*

   1. *If $UBD - LBD \le \varepsilon$, terminate, else go to step 2.*
   2. *If the relaxed Master problem is infeasible then terminate and return optimal solution as the current upper bound solution.*

Example

$$\min_{x,y} \quad -y + 3x + \frac{1}{2}x^2 \tag{18.121}$$

subject to:

$$-2.1\,x - \ln\left(\frac{1}{5}x\right) + y - 1 \le 0 \tag{18.122}$$

$$y \in [0, 1]$$

Figure 18.11 illustrates the objective function and constraint values when $y = 1$ and $y = 0$.

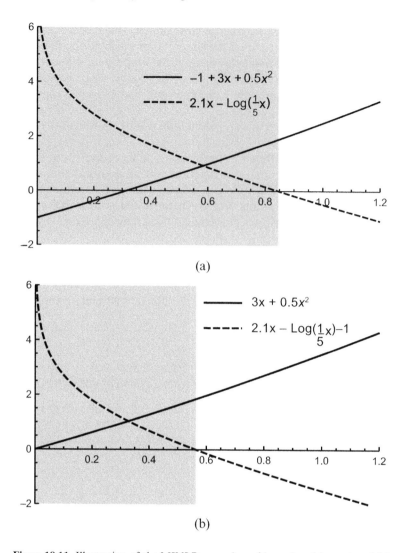

(a)

(b)

Figure 18.11 Illustration of the MINLP example problem when (a) $y = 1$ and (b) $y = 0$. The shaded region is infeasible given the inequality constraint.

Therefore, the optimization that is taking place should choose whether the scenario where $y = 1$ or when $y = 0$ constrains the optimal solution. The OA algorithm would proceed as follows:

Step 1. Let us select $y^1 = 1$

We define the Primal problem at iteration 1 as $P(y^1)$:

$$\min_x \quad -1 + 3x + \frac{1}{2}x^2 \tag{18.123}$$

subject to:

$$-2.1\,x - \ln\left(\frac{1}{5}x\right) + 1 - 1 \le 0 \tag{18.124}$$

which has a solution of $f(x) = 1.8959$, which we set as our current upper bound $UBD = 1.8959$ and a value of $x = 0.8460$.

To construct the Master problem, we linearize nonlinear functions around the primal solution, and obtain:

$$\begin{aligned}
f(x^k) + \nabla f(x^k)(x - x^k) &= (3 \times 0.8460 + 0.5 \times 0.8460^2) \\
&\quad + (3.0 + 0.5 \times 2 \times 0.8460)(x - 0.8460) \\
&= -0.3578 + 3.8460\,x
\end{aligned} \tag{18.125}$$

$$\begin{aligned}
g(x^k) + \nabla g(x^k)(x - x^k) &= (-2.1 \times 0.8460 - \ln(0.2 \times 0.8460)) \\
&\quad + (-2.1 - \tfrac{1}{0.8460})(x - 0.8460) \\
&= 2.77665 - 3.282\,x
\end{aligned} \tag{18.126}$$

Step 2. Solve the master problem

$$\min_{x,y,\mu_{OA}} \quad -y + \mu_{OA} \tag{18.127}$$

subject to:

$$\mu_{OA} \geq -0.3578 + 3.8460\,x \tag{18.128}$$

$$2.77665 - 3.282\,x + y - 1 \leq 0 \tag{18.129}$$

$$y \in [0, 1]$$

which has a solution $y = 0$, $\mu_{OA} = 1.72409$ and we can now set a new lower bound $LBD = 1.72409$.

Figure 18.12 shows the linear approximation of the nonlinear objective and constraint that are used to solve the Master problem.

Step 3. Given that the Master problem had $y = 0$ as itself solution, we have $y^2 = 0$
We define the Primal problem at iteration 2 as $P(y^2)$:

$$\min_{x} \quad 0 + 3x + \frac{1}{2}x^2 \tag{18.130}$$

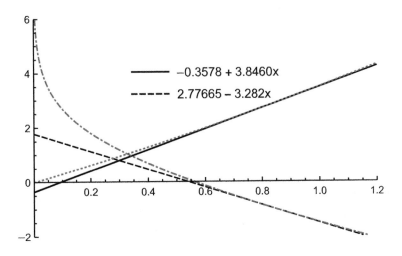

Figure 18.12 Approximation of nonlinear functions by their linear approximations.

subject to:

$$-2.1\,x - \ln\left(\frac{1}{5}x\right) + 0 - 1 \le 0 \qquad (18.131)$$

which has a solution of $f(x) = 1.8489$, which we set as our current upper bound $UBD = 1.8489$ and a value of $x = 0.5634$. Given that we have exhausted all possible y combinations, we have reached the globally optimal solution, and we return the optimal value to be $UBD = 1.8489$ and $x = 0.5634$. If we had more possible values for y, we would continue with the construction of a second Master problem, linearizing around the new optimal point, and adding these new constraints to the previous Master problem.

General Remarks on Outer Approximation

- Problems solved by the OA framework (*i.e.* problems that follow the structure in Equations (18.79–18.82) correspond to a subclass of problems that the GBD approach can address. This is due to the inherent assumptions of separability in x and y, as well as linearity in y. OA is hence a subclass of GBD methods.
- An initial starting point $y^1 \in Y$ can be obtained by relaxing the MINLP to an NLP (setting y variables as continuous) and rounding to the closest integer value.
- For convex problems, a proof presented in [4] shows that OA will terminate in fewer iterations than that of a general GBD. Note that since the relaxed Master problem of OA has many constraints as the iterations proceed, while the relaxed master of GBD adds only one constraint per iteration, fewer iterations does not necessarily mean smaller solution time.
- To explicitly handle some problems with nonlinear equality constraints, [20] proposed the Outer Approximation with Equality Relaxation algorithm, also termed OA/ER.
- A further algorithm termed as Outer Approximation with Equality Relaxation and Augmented Penalty (OA/ER/AP) was developed by [21], which presents milder convexity assumptions than the OA/ER variant to address nonlinear equality constraints and still guarantee global optimality.
- Note that the OA family of methods can be used to solve nonconvex problems; however, in this case it becomes a heuristic method, without guarantee of optimality.
- In practice, outer approximation often works efficiently. However, [5] provides an example where outer approximation takes an exponential number of iterations.

18.3.5 Best Practice for the Solution of MINLPs

In this section we will outline best practices on how mixed integer programming models can be reformulated to facilitate the solution procedures. In many cases, following these suggestions can lead to minor improvements, while in other cases it can have significant effects on multimodality (more than one local minimum) and solution time. These reformulations are based on the fact that convex expressions are preferred over nonconvex ones, while linear formulations are favored over nonlinear ones.

Linearization of Expressions

Given an expression such as $\frac{x_1}{x_2} = \alpha$, where α is some constant, it can be reformulated as $x_1 = \alpha x_2$, hence avoiding the nonlinearity introduced by having a division between variables. This is not particular to mixed integer programming, and is in general advantageous for any optimization model formulation.

Linearizing Bilinear Terms

Given an expression that contains $x_1 y$, where $y \in \{0, 1\}$, it is possible to linearize this expression when $0 \le x_1 \le a$. This is done by introducing a new variable x_2 and replacing the bilinear term $x_1 y$ by x_2 and adding the following constraints:

$$0 \le x_2 \le ya \tag{18.132}$$

$$-a(1-y) \le x_1 - x_2 \le a(1-y) \tag{18.133}$$

This can be done because the product xy is either zero (if $y = 0$) or x_1 (if $y = 1$), and can be conducted for the general case where $b \le x_1 \le a$:

$$yb \le x_2 \le ya \tag{18.134}$$

$$x_1 - a(1-y) \le x_2 \le x_1 - b(1-y) \tag{18.135}$$

This technique can also be used to transform NLPs with bilinear terms, which are nonconvex, to MILPs. MILPs are in general preferred to convonvex NLPs, as there are efficient algorithms to solve this type of problems.

Convexification of Binary Quadratic Programs

Let us consider a binary quadratic program which is not convex:

$$y^T Q y + c^T y \tag{18.136}$$

where $y \in \{0, 1\}^n$. Let us call λ the smallest eigenvalue of Q, where $\lambda < 0$. Hence, we define a new matrix $W = Q - \lambda I$, where I is the identity matrix, and we also define the new gradient $q = c + \lambda e$, where e is the vector of all ones (i.e. $e = (1, \ldots, 1)^T$). We can then define the convex binary quadratic program:

$$y^T W y + q^T y \tag{18.137}$$

where

$$y^T Q y + c^T y = y^T W y + q^T y \tag{18.138}$$

Notice that this is possible due to the binary nature of the y variables, and that $y_i^2 = y_i$.

Use of GBD and OA for Nonconvex Optimization

The GBD and OA methods presented in Sections 18.3.2 and 18.3.3, respectively, are designed to solve convex MINLPs, and have been shown to be very efficient to solve the type of problems they were tailored for. However, if one is to use them to solve nonconvex MINLPs they result in no more than a heuristic, and might not be efficient in practice. On the contrary, branch and bound methods can be slightly modified to yield reasonably good solutions even when confronted with nonconvex MINLPs (the details have been presented in Section 18.3.1).

18.4 REFERENCES

[1] Kannan, R. and Monma, C. "On the computational complexity of integer programming problems." Optimization and Operations Research. 1978;157:p. 161–172.

[2] Belotti, P., Kirches, C., Leyffer, S., Linderoth, J., Luedtke, J., and Mahajan, A. "Mixed-integer nonlinear optimization." Acta Numerica. 2013;22:p. 1–131.

[3] Floudas, A. Nonlinear and Mixed-Integer Optimization: Fundamentals and Applications. Oxford University Press. 1995.

[4] Duran, M. A. and Grossmann, I. E. "An outer approximation algorithm for a class of mixed-integer nonlinear programs." Mathematical Programming. 1986;36(3):p. 307–339.

[5] Hijazi, H., Bonami, P., and Ouorou, A. "An outer-inner approximation for separable MINLPs." Technical report, LIF, Faculté des Sciences de Luminy, Université de Marseille. 2010.

[6] Gomory, R. E. "Outline of an algorithm for integer solutions to linear programs." Bulletin of the American Mathematical Society. 1958;64:p. 275–278.

[7] Cornuejols, G. "Valid inequalities for mixed integer linear programs." Mathematical Programming. 2008;112(1):p. 3–44.

[8] Bonami, P., Biegler, L. T., Conn, A. R., Cornuejols, G., Grossmann, I. E., Laird C. D., Lee, J., Lodi, A., Margot, F., Sawaya, N., and Waechter A. An Algorithmic Framework for Convex Mixed Integer Nonlinear Programs. IBM Research Report RC23771. 2005.

[9] Papoulias, S. A. and Grossmann, I. E. "A structural optimization approach in process synthesis – II. Heat recovery networks." Computers & Chemical Engineering. 1983;7(6):p. 707–721.

[10] Floudas, C. A. and Anastasiadis, S. H. "Synthesis of distillation sequences with several multicomponent feed and product streams." Chemical Enginering Science. 1988;43(9):p. 2407–2419.

[11] Floudas, C. A. and Grossmann, I. E. "Synthesis of flexible heat exchanger networks for multiperiod operation." Computers & Chemical Engineering. 1986;10(2):p. 153–168.

[12] Paules, G. E. and Floudas, C. A. "Synthesis of flexible distillation sequences for multiperiod operation." Computers & Chemical Engineering. 1988;12(4):p. 267–280.

[13] Rich, S. H. and Prokopakis, G. J. "Multiple routings and reaction paths in project scheduling." Industrial & Engineering Chemistry Research. 1940;26(9):p. 1940–1943.

[14] Kondili, E., Pantelides, C. C., and Sargent, R. W. H. "A general algorithm for short term scheduling of batch operations – I. MILP formulation." Computers & Chemical Engineering. 1993;17(2):p. 211–227.

[15] Shah, N., Pantelides, C. C., and Sargent, R. W. H. "A general algorithm for short term scheduling of batch operations – II. Computational issues." Computers & Chemical Engineering. 1993;17(2):p. 229–244.

[16] Voudouris, V. T. and Grossmann, I. E. "Mixed integer linear programming reformulations for batch process design with discrete equipment sizes." Industrial & Engineering Chemistry Research. 1992;31(5):p. 1315–1325.

[17] Voudouris, V. T. and Grossmann, I. E. "Optimal synthesis of multiproduct batch plants with cyclic scheduling and inventory considerations." Industrial & Engineering Chemistry Research. 1993;32(9):p. 1962–1980.

[18] Shah N. and Pantelides, C. C. "Optimal long-term campaign planning and design of batch operations." Industrial & Engineering Chemistry Research. 1991;30(10):p. 2309–2321.

[19] Bixby, R. E., Gu, Z., Rothberg, E., and Wunderling, R. "Mixed integer programming: A progress report. The sharpest cut: The impact of Manfred Padberg and his work." 2004;p. 309–326.

[20] Kocis, G. R. and Grossmann, I. E. "Relaxation strategy for the structural optimization of process flow sheets." Industrial & Engineering Chemistry Research. 1987;26(9): p. 1869–1880.

[21] Viswanathan, J. and Grossmann, I. E. "A combined penalty function and outer approximation for MINLP optimization." Computers & Chemical Engineering. 1990;14(7): p. 769–782.

[22] Eckstein, J. "Parallel branch-and-bound algorithms for general mixed integer programming." SIAM Journal on Optimization. 1994;4(4):p. 794–814.

[23] Achterberg, T. "SCIP – A framework to integrate constraint and mixed integer programming." Technical Report ZIB-Report 04-19, Konrad-Zuse-Zentrum fur Informationstechnik Berlin. 2005.

18.5 FURTHER READING RECOMMENDATIONS

Complexity Analysis of Mixed-Integer Programming

Nemhauser, G. L. and Wolsey, L. A. (1988). Integer and Combinatorial Optimization. John Wiley & Sons. 1988.

Theoretical Discussion on Cutting Planes Method

Cornuejols, G. (2008). "Valid inequalities for mixed integer linear programs." Mathematical Programming. 112(1):p. 3–44.

Feasibility Approach (FA) for MINLPs

Mawengkang, H. and Murtagh, B. A. (1986). "Solving nonlinear integer programs with large scale optimization software." Annals of Operations Research. 1986;5(1):p. 425–437.

Generalized Outer Approximation (GOA) for MINLPs

Fletcher, R. and Leyffer, S. (1994). "Solving mixed integer nonlinear programs by outer approximation." Mathematical Programming. 1994;66(3):p. 327–349.

Generalized Cross Decomposition (GCD) for MINLPs

Gundersen, T. and Grossmann, I. E. (1990). "Improved optimization strategies for automated heat exchanger network synthesis through physical insights." Computers & Chemical Engineering. 1990;14(9):p. 925–944.

Modeling Practices in Integer and Mixed-Integer Programming

Williams, H. P. (1999). Model Building in Mathematical Programming. John Wiley & Sons. 1999.

Applications of Convex MINLP Problems in Chemical Engineering

Floudas, A. (1995). Nonlinear and Mixed-Integer Optimization: Fundamentals and Applications. Oxford University Press. 1995.

18.6 EXERCISES

1. Given the following MILP:

$$\min_{x,y} \quad 2.4x_1 - 2.1y_1 - 5y_2 + 2y_3 \tag{18.139}$$

subject to:

$$3x_1 + y_1 + 2y_2 + 2y_3 \leq 4.5 \tag{18.140}$$

$$x_1 \geq 0 \tag{18.141}$$

$$y_1, y_2, y_3 \in [0, 1] \tag{18.142}$$

(i) Solve the above problem using branch and bound. Use the *integrality relaxation* such that the subproblems are linear programs.
(ii) Do five iterations of the Cutting Planes (using Gomory Cuts).
(iii) Alternate the two previous approaches, such that every iteration consists of a branching, a pruning, and a cut.

2. For the following MINLP:

$$\min \quad -y + 3x + \frac{1}{2}x^2 \tag{18.143}$$

subject to:

$$-2.1\,x - \ln\left(\frac{1}{5}x\right) + y \leq 0 \tag{18.144}$$

$$x_1 \geq 0 \tag{18.145}$$

$$y \in [0, 1, 2] \tag{18.146}$$

(i) Solve the convex MINLP using Outer Approximation.
(ii) Solve the convex MINLP General Benders Decomposition.
(iii) Solve the convex MINLP by branch and bound, by the *integrality relaxation* such that the subproblems are nonlinear programs.

3. For the following nonconvex MINLP:

$$\min \quad y^2 + x^2 - y\,5\cos\left(2\pi x\right)$$

subject to:

$$(x + y)^2 \geq 2$$

$$5 \geq x \geq -5$$

$$y \in [0, 1, 2, 3] \tag{18.147}$$

(i) Solve the nonconvex MINLP by branch and bound, by the *integrability relaxation* such that the subproblems are nonlinear programs.
(ii) Now apply the heuristics mentioned in Section 18.3 and compare both solutions.

19

Global Optimization

19.1 INTRODUCTION

This chapter presents one of the areas where the optimization community has focused significant efforts, namely, finding global solutions to nonconvex optimization problems efficiently. So far in this book we have focused on finding optimal solutions for problems that either a) have only one local (hence global) minimum or b) any "good enough solution" (hence local minimum) is adequate. However, there are many applications where getting a "good enough" solution is not sufficient, and hence global optimization must ensue. Just a few of these examples in chemical engineering are optimizing the Gibbs free energy to find compositions in phase mixtures; calculations on protein folding; safety verification problems where assuming nonglobal solutions of the worst scenario may severely underestimate a true risk; feasibility problems where local solutions give no information as they get stuck in infeasible points; among many others.

Previous chapters of this book have presented linear programs (LPs), convex nonlinear programs (NLPs), mixed-integer linear programs (MILPs) and convex mixed-integer nonlinear programming (MINLPs), all of which can be solved efficiently and for a large number of variables. It is worth noting that the strategies presented for convex NLPs and MINLPs can in most cases (with maybe a few adaptations) obtain local solutions to their nonconvex counterparts. However, as stated previously, there are many problems where being able to find the global minimum and certifying it is paramount. Therefore in the remainder of this chapter we will be discussing global optimization. In general, it should be noted that finding the global optimal solution for a nonconvex problem is much more difficult than simply finding a local solution.

There are different degrees of rigor when it comes to global optimality; we will use Neumaier's classification [1], which is the following:

- **Incomplete methods** use clever intuitive heuristics to search for the global optimum (in the hope of finding it) but have no safeguards if the algorithm gets stuck in a local minimum (the branch and bound heuristics for nonconvex MINLPs presented in Chapter 18 falls into this category).
- **Asymptotically complete methods** reach a global minimum with certainty, or at least with probability one, if allowed to run for an infinite time, but have no means of knowing when a global minimizer has been found.
- **Complete methods** reach a global minimum with certainty, assuming exact computations and indefinitely long run-time, and certify after a finite time that an approximate global minimizer has been found (to within prescribed tolerances).

- **Rigorous methods** reach a global minimum with certainty and within given tolerances even in the presence of rounding errors, except in near degenerate cases, where the tolerances may be exceeded.

The reader must note that the label "deterministic global optimization algorithm" is often used to refer to the last two categories of algorithms mentioned, although many incomplete and asymptotically complete methods are also deterministic.

In this chapter we will be only presenting strategies for complete methods; for the interested reader in incomplete and asymptotically complete methods, Section 19.8 outlines several references. Furthermore, given the complexity, extra computations, and large amounts of safeguards needed to construct a rigorous method, these types of algorithms are slow and have rarely been used in applications with more than 100 variables. Hence our focus on complete methods, which are almost equivalently reliable (except in pathological cases) and much more efficient.

Solving large nonconvex optimization problems to global optimality is one of the main challenges that the optimization community faces. Basic to (almost) all complete/rigorous global optimization algorithms is the branching principle. However, efficient global optimization strategies are a conglomerate of techniques used in conjunction to explore efficiently the solution space and to find the best possible solution. These general concepts and key techniques used by global optimization algorithms are presented in Sections 19.3, 19.4, and 19.5. Before solution techniques are presented, in the following section let us formally define the problem and give the general outline for its solution procedure.

19.2 PROBLEM STATEMENT

Let us define a general nonconvex MINLP as follows:

$$\min_{x,y} \quad f(x, y) \tag{19.1}$$

subject to:

$$h(x, y) = 0 \tag{19.2}$$

$$g(x, y) \leq 0 \tag{19.3}$$

$$x \in X \subseteq \mathbb{R}^n \tag{19.4}$$

$$y \in \mathbb{Z} \tag{19.5}$$

where functions $f(x, y)$, $h(x, y)$, and $g(x, y)$ may be nonconvex. Note that a particular case of the above problem is a nonconvex NLP (where there are no integer variables). Let us define the feasible domain as:

$$\Omega = \left\{ x, y \mid h(x, y) = 0; \ g(x, y) \leq 0; \ x \in X \subseteq \mathbb{R}^n; \ y \in \mathbb{Z} \right\} \tag{19.6}$$

Hence, any point satisfying $x, y \in \Omega$ is a feasible point, while a *solution point* $\hat{x}, \hat{y}$ is characterized by satisfying:

$$f\left(\hat{x}, \hat{y}\right) = \min_{x, y \in \Omega} \quad f(x, y) \tag{19.7}$$

where a *local minimizer* (solution) satisfies $f\left(\hat{x}, \hat{y}\right) \leq f(x, y)$ for all $x, y \in \Omega$ only in some neighborhood of $\hat{x}, \hat{y}$.

At this point, let us mention that there are some cases where global information about the problem is not available. That is, if what we desire to optimize is some kind of black box model where we can only get *local information* about the problem (*e.g.* function values, Hessians, or Jacobians at the current point). These problems must be addressed by particular algorithms, and are not discussed here. The interested reader is referred to the following sources: [1, 6, 7, 8]. From this point onward we will be focusing on global optimization where global information is available.

The general solution procedure to reach a global minimum in a nonconvex problem is as follows. The majority of (if not all) state-of-the-art algorithms for global optimization use a *branching* method combined with some type of *bounding* or *reducing* technique. Branching splits the solution domain into subproblems, while bounding and reducing techniques take care of discarding these subproblems, or trimming their domain until a global solution (or infeasibility of the problem) can be certified to within some tolerance. In order to eliminate subproblems quickly, generating good upper bounds efficiently is important. This is usually done by local optimization with the addition of heuristics to try and find a low-lying local minimum. To create good quality (tight) lower bounds, outer approximation of the feasible domain and underestimation of the objective function are used in order to obtain relaxed problems that are convex and hence have a unique solution. While to efficiently reduce the feasible domain of subproblems, techniques such as constraint propagation and interval arithmetic are employed.

As can be deduced from this, global optimization algorithms have many similarities with those presented in Chapter 18, and they are only more complex given nonconvexities in the integer, continuous, or combined space. To have a better intuition into the framework followed by global optimization algorithms the following section presents a small example.

19.2.1 Qualitative Example Using Branch and Bound

Before we go into any formal detail on the solution procedure for global optimality of nonconvex optimization problems, let us sketch the general solution approach by a branch and bound (B&B) approach.

Assume that we have an optimization problem with many local minima (multimodal), such as the one shown in Figure 19.1.

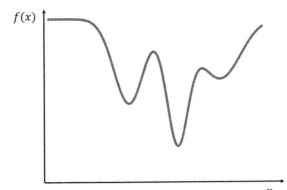

Figure 19.1 Multimodal optimization problem.

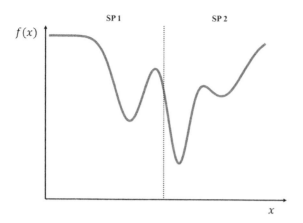

Figure 19.2 Multimodal optimization problem branched into subproblem 1 (SP_1) and subproblem 2 (SP_2).

Notice that Figure 19.1 presents a one-dimensional unconstrained optimization problem, and no complicated optimization algorithm would be needed to find the optimal solution (in fact, plotting it would be more than enough). However, let us use this simple example to walk through the B&B approach.

B&B begins by splitting the solution domain into two or more subproblems; let us assume in this case we create subproblem 1 (SP_1) and subproblem 2 (SP_2), such as shown in Figure 19.2.

We should remember from Chapter 18 that the B&B approach calculates upper and lower bounds for each subproblem, so that these subproblems can be compared, and some of them (the ones we can be sure do not contain the global solution) can be discarded.

To create upper bounds, it suffices to find a local solution to each subproblem (or even a feasible solution). To create lower bounds, as in the case for the solution of MILPs and MINLPs, problems that underestimate the subproblem are solved, and their solution is a lower bound to the original problem. In the case of MILPs and MINLPs, relaxations are easy to construct, and simply by releasing the integrality constraint one could generate a problem that underestimates the original subproblem for the whole domain. In the case of nonconvex NLPs this is not as easy, as one must design such a problem. How to produce these *underestimators* will be discussed later in Section 19.4 of this chapter. For now, let us assume we are able to create reliable lower bounds for both SP_1 and SP_2, which are shown in Figure 19.3.

Once the underestimators have been computed, we can proceed to find a local solution to the original problem, as well as on the underestimating problems. This is shown in Figure 19.4.

The underestimators portrayed in the figure are convex functions; in this way a simple local search algorithm will find their global solution. In this example, we seem to have found the global optimal solution for both SP_1 and SP_2, but this is a mere coincidence, and more importantly, in a problem of multiple dimensions we *would be unable to certify* the solution we found is global until more iterations are carried out. Henceforth, let us assume that although we have already found the global solution we do not know it yet. Finally, we can observe that the lower bound for SP_1 is actually higher (worse) than the upper bound for SP_2, as shown in Figure 19.5.

In Figure 19.5 it is clear that the lower bound of SP_1 is greater than the upper bound of SP_2 (*i.e.* LB1 > UB2); this means that no matter how much we explore SP_1

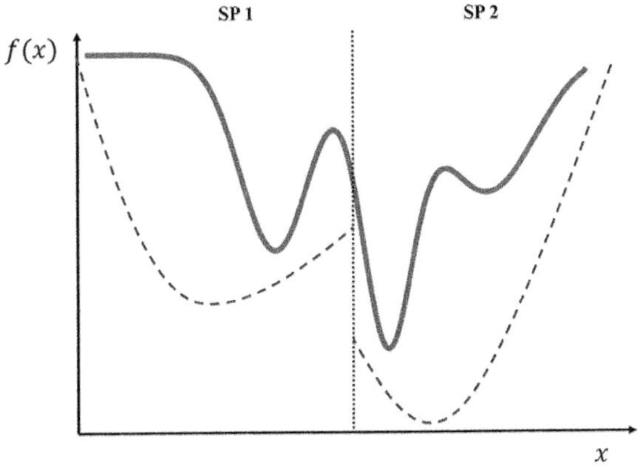

19.2 Problem Statement

Figure 19.3 Multimodal optimization problem where the dashed line represents the underestimating functions for the subproblems.

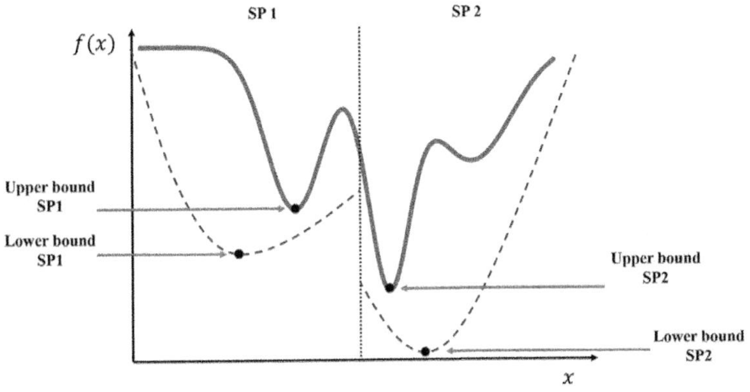

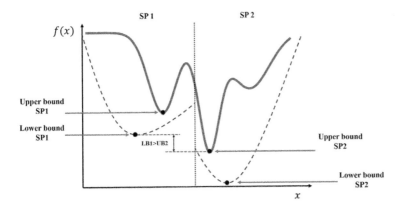

Figure 19.4 Multimodal optimization problem. The dots represent the solutions found by local solvers to both subproblems as well as their underestimators.

Figure 19.5 Multimodal optimization problem. LB1 stands for the lower bound of SP_1 and UB2 means the upper bound of SP_2.

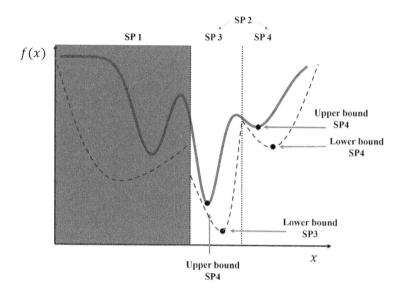

Figure 19.6 Multimodal optimization problem.

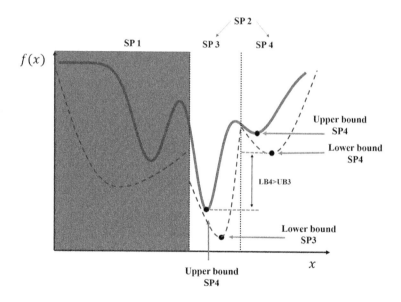

Figure 19.7 Multimodal optimization problem.

we already have a solution that is better than any we could find in this subproblem. Hence, as shown in Figure 19.6, SP_1 is discarded and our effort will now focus on SP_2, once again splitting it into two subproblems (SP_3 and SP_4). It is important to note that the above figure also highlights the necessity of having underestimators whose values are *always* less than or equal to the subproblems (otherwise we could be discarding the subproblem that actually has the global optimum).

We note that once again the upper and lower bounds are computed and analyzed, as shown in Figure 19.7.

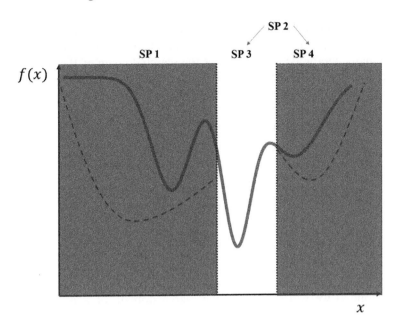

Figure 19.8 Multimodal optimization problem.

Figure 19.7 shows that it is now possible to discard SP_4, as its best value will not match the best known value so far (present in SP3). This then leads to Figure 19.8.

From Figure 19.8 we would again proceed to split SP_3 into SP_5 and SP_6, and follow the same steps already shown until we are left with one subproblem that is small enough so that we would be satisfied, and hence terminate the algorithm and claim that we are at most within ε from the global solution.

Another important aspect to note is that the "closeness" between the under-estimating problem and the real subproblem will have an effect on how quickly subproblems can be discarded. If there is a big gap between the underestimators and the subproblems (*i.e.* the lower bounds produced are not tight), it will not always be clear that a problem has a worse lower bound than another problem's upper bound, and hence the algorithm will need many more iterations to terminate. Consequently, constructing good underestimators is crucial for a global optimization algorithm to be efficient in practice.

The above B&B framework is the "big picture" behind global optimization; however, there are many subtleties that are important to make global optimization algorithms converge faster. These techniques are presented in the subsequent sections of this chapter.

19.3 REDUCING THE DOMAIN FOR A BRANCH AND REDUCE APPROACH

As presented in the qualitative example of the previous section, as well as in Chapter 18, branch and bound is a powerful tool to address global optimization problems. Unfortunately, for problems with continuous variables B&B is not enough, given that searching over a continuous domain is very different than searching over an integer domain. Hence, the approach followed by most (if not all) of the best performing global optimization algorithms is some variant of the branch and reduce (B&R) scheme. In simple words, a B&R scheme is a B&B algorithm but at every iteration (or every so many iterations) it also reduces the search domain of the subproblems by taking advantage of the constraints and the current objective function value, which have some implicit information that can be exploited. This is better shown with an example.

Let us consider the problem:

$$\min_x \quad f(x) = x_1 + 4x_2 \tag{19.8}$$

subject to:

$$x_1^2 + x_2^2 = 12 \tag{19.9}$$

$$-4 \leq x_1 \leq 6 \tag{19.10}$$

$$-5 \leq x_2 \leq 7 \tag{19.11}$$

In what follows we will show how the search domain of a problem can be reduced, hence providing significant increases in performance compared to a pure B&B procedure.

Following the didactic approach presented by Neumaier in his excellent review [1], we will first show the implementation of *constraint propagation* and *interval analysis*. In Sections 19.3.2 and 19.3.1 these two techniques will be formally introduced.

Let us assume that we solve the previously set out optimization problem by a local solver, and the solution obtained is:

$$f(x) = -14.2829 \tag{19.12}$$

$$x_1 = -0.83994, \quad x_2 = -3.3607 \tag{19.13}$$

The *constraint propagation* and *interval arithmetic* approaches will take advantage of the above to reduce the search space, by analyzing the constraints, bounds and their coupled effects. *Constraint propagation* can be applied to this problem as follows.

Given that we have already produced a point with a value $f(x) = -14.2829$, we would like to produce an even better one, and hence we are not interested in any point that does not satisfy the following inequality:

$$x_1 + 4x_2 \leq -14.2829 \tag{19.14}$$

By rearranging this expression, we can obtain an upper bound for x_2:

$$x_2 \leq \frac{-14.2829 - x_1}{4} \tag{19.15}$$

We substitute the value for $x_1 = -4$ that would result in the lowest upper bound (infimum) and hence obtain:

$$x_2 \leq -2.5707 \tag{19.16}$$

which significantly reduces the previously known upper bound for x_2. Following the same procedure for x_1 we obtain:

$$x_1 \leq 5.7171 \tag{19.17}$$

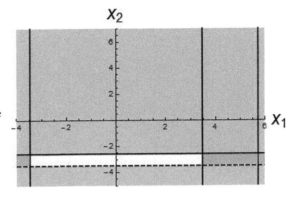

Figure 19.9 Domain reduction by constraint propagation. The figure presents all the feasible space, the shaded area is the infeasible region with the new bounds imposed, while the unshaded area is our remaining feasible region.

This means that by simple propagation of the objective value function we have reduced the search space.

We can now analyze the constraint:

$$x_1^2 + x_2^2 = 12 \tag{19.18}$$

By rearranging this constraint, we obtain:

$$x_1 = \pm\sqrt{12 - x_2^2} \tag{19.19}$$

and hence we obtain the bounds $-3.4641 \leq x_i \leq 3.4641$ for $i = 1, 2$, since for the term inside the square root to give real numbers:

$$12 - x_i^2 \geq 0$$

for $i = 1, 2$.

The procedure just outlined has led to a reduced search space compared to the original domain. This is schematically represented in Figure 19.9.

Let us now use *interval arithmetics* to further reduce the search space.

We will first use a linear relaxation of the constraint to find bounds on the values the variables can attain. Given $h(x) = x_1^2 + x_2^2 = 12$ and the variable bounds $x \in \{[-3.4641, 3.4641], [-3.4641, -2.57]\}$, we can obtain the bounds for the Jacobian $J(x)$ as follows:

$$J(x) = (2x_1, 2x_2) \in \{[-6.9282, 6.9282], [-6.9282, -5.1414]\} \tag{19.20}$$

Through the mean value theorem we know that for any point $\tilde{x} \in x$,

$$F(x) \in F(\tilde{x}) + F'(x)(x - \tilde{x}) \quad \text{when } x \in \mathbf{x} \tag{19.21}$$

where the algorithm Q2 (i) is Algorithm 19.1

In our example we use the point the local solver found as $\tilde{x}$ and obtain:

$$12 \in 12 + [-6.9282, \ 6.9282](x_1 + 0.83994) + [-6.9282, \ -5.1414](x_2 + 3.3607) \tag{19.22}$$

This gives us two equations, one as an upper bound, and the other as a lower bound. The inequality on the lower bound leads to the further domain reduction shown in Figure 19.10.

The inequality on the upper bound reduces the domain as in Figure 19.11.

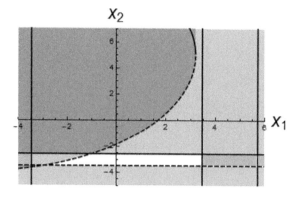

Figure 19.10 Domain reduction by lower bound constraint.

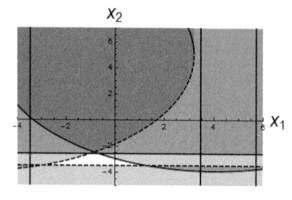

Figure 19.11 Domain reduction by upper bound constraint.

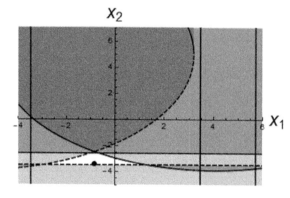

Figure 19.12 Optimal solution shown with domain reduction.

Figure 19.12 also shows the initial solution found by our local solver, which is in fact the global solution.

The important point to make from this example is how much space can be pruned by simple interval arithmetic and constraint propagation. This shows why these are very powerful tools that global optimization algorithms exploit. In the following sections we introduce interval arithmetic and constraint propagation more formally.

19.3.1 Interval Arithmetic

Interval arithmetic is a useful tool to reduce the search space and can be used by propagating the variable bounds. Although, generally, interval arithmetic alone will not yield a solution to a global optimization problem, in conjuction with other techniques it can lead to efficient algorithms. One of its main advantages is that the calculations needed are not computationally intensive.

Let us define $\mathbf{a}$ and $\mathbf{b}$ to be our intervals, $\tilde{a} \in \mathbf{a}$ and $\tilde{b} \in \mathbf{b}$ and $\bullet \in \{+, -, *, /, \}$ to be a binary operator. We now define:

$$\mathbf{a} \bullet \mathbf{b} := \square\{\tilde{a} \bullet \tilde{b}\} \tag{19.23}$$

where the above means that for $\square S = [\inf S, \sup S]$, it is the tightest interval containing S.

For particular cases of the operators these are:

$$\mathbf{a} + \mathbf{b} = [\underline{a} + \underline{b}, \overline{a} + \overline{b}] \tag{19.24}$$

$$\mathbf{a} - \mathbf{b} = [\underline{a} - \overline{b}, \overline{a} - \underline{b}] \tag{19.25}$$

$$\mathbf{a} * \mathbf{b} = \square\{\underline{ab}, a\overline{b}, \overline{a}\underline{b}, \overline{ab}\} \tag{19.26}$$

$$\frac{\mathbf{a}}{\mathbf{b}} = \square\{\underline{a}/\underline{b}, \underline{a}/\overline{b}, \overline{a}/\underline{b}, \overline{a}/\overline{b}\} \quad \text{if } 0 \notin \mathbf{b} \tag{19.27}$$

Furthermore, for the set of elemental univariate functions $\Psi \in \{\log, \exp, \sin, \cos, \text{sqrt}, \ldots\}$, we can also define the intervals containing them as:

$$\Psi(\mathbf{a}) := \square\{\Psi(\tilde{a}) \mid \tilde{a} \in \mathbf{a}\} \tag{19.28}$$

Depending on the monotonicity properties of the function Ψ, the upper and lower bounds on $\Psi(\mathbf{a})$ can be computed at the endpoints of $\mathbf{a}$ and the interior extremal values.

When interval arithmetic is used in *rigorous methods*, the calculations can be rather slow since they must guarantee correctness of outward rounding, to avoid the solution from being lost with the accumulation of round-off errors.

19.3.2 Constraint Propagation

As shown in the example of Section 19.3, constraints in many cases contain implicit information about the feasible region of the optimization problem. Hence, by a general procedure it is possible to extract this information, and reduce the search space (variable bounds) of the problem. This procedure is generally termed *constraint propagation*. We now outline the general formulation to propagate linear constraints as outlined in [9].

We first deal with *linear inequality constraints* of the form:

$$\underline{a} \le \sum_k q_k (x_{D_k}) \tag{19.29}$$

where D is a list of indices, and x_D denotes a subvector of x formed by the components with index in D.

Before we go any further, let us have the following proposition and its proof.

Let q_k be real-valued functions defined on x_{D_k}. The following two statements hold:
(1) If (for a suitable q_k and $\bar{s}$)

$$\bar{q}_k \geq \sup\left\{q_k\left(x_{D_k}\right) \mid x_{D_k} \in \mathbf{x}_{D_k}\right\}, \quad \bar{s} \geq \sum_k \bar{q}_k, \tag{19.30}$$

then, for arbitrary $\underline{a}$,

$$x \in \mathbf{x}, \quad \underline{a} \leq \sum_k q_k\left(x_{D_k}\right) \implies q_k\left(x_{D_k}\right) \geq \underline{a} - \bar{s} + \bar{q}_k \quad \text{for all } k \tag{19.31}$$

(2) If

$$\underline{q}_k \leq \inf\left\{q_k\left(x_{D_k}\right) \mid x_{D_k} \in \mathbf{x}_{D_k}\right\}, \quad \underline{s} \leq \sum_k \underline{q}_k, \tag{19.32}$$

then, for arbitrary $\underline{a}$,

$$x \in \mathbf{x}, \quad \sum_k q_k\left(x_{D_k}\right) \leq \bar{a} \implies q_k\left(x_{D_k}\right) \leq \bar{a} - \underline{s} + \underline{q}_k \quad \text{for all } k \tag{19.33}$$

Proof

The assumptions of part (1) imply:

$$q_k\left(x_{D_k}\right) \geq \underline{a} - \sum_{l \neq k} q_l\left(x_{D_l}\right) \geq \underline{a} - \sum_{l \neq k} \bar{q}_l \geq \underline{a} + \bar{q}_k - \bar{s}, \tag{19.34}$$

hence Equation (19.31) holds. (2) is proven in an equivalent manner.

Given this, for a given domain of the problem we construct Equation (19.30) for constraint (19.29). Then we can check the condition:

$$\underline{a} \leq \bar{s} \tag{19.35}$$

If this condition is violated, it means the current domain is inconsistent with the problem constraints and the search domain can be reduced.

Now, in the case of *two-sided linear inequality constraints* (which include linear equality constraints) of the form:

$$\sum_k q_k\left(x_{D_k}\right) \in \mathbf{a} \tag{19.36}$$

we have:

$$\mathbf{q}_k \supseteq \Box\left\{q_k\left(x_{D_k}\right) \mid x_{D_k} \in \mathbf{x}_{D_k}\right\}, \quad \mathbf{r} \supseteq \mathbf{a} - \sum_k \mathbf{q}_k \tag{19.37}$$

and if $0 \notin \mathbf{r}$ then the domain of x ($x \in \mathbf{x}$) is inconsistent with the constraint (19.36), and hence the search domain can be reduced. This condition must be met due to the fact that $x \in \mathbf{x}$ and constraint (19.36) implies:

$$0 \in \mathbf{a} - \sum_k q_k\left(x_{D_k}\right) \subseteq \mathbf{a} - \sum_k \mathbf{q}_k \tag{19.38}$$

and hence, $0 \in \mathbf{r}$ for the domain and constraint to be consistent.

Let us now analyze the case of a *quadratic constraint* in an interval $\mathbf{c}$, such as $ax^2 + bx \in \mathbf{c}$.

A constraint $ax^2 + bx = \widetilde{c} \in \mathbf{c}$ has two cases regarding the value of a. When $a = 0$ the quadratic constraint becomes $x = \frac{\widetilde{c}}{b}$ and $x = \frac{\left(-b \pm \sqrt{b^2 + 4a\widetilde{c}}\right)}{2a}$ otherwise. When $a \neq 0$, $b^2 + 4a\widetilde{c}$ must be nonnegative. Furthermore, since the varying $\widetilde{c}$ occurs only once in these equations, the range of $\widetilde{c} \in \mathbf{c}$ is given by $\mathbf{c}/b$ when $a = 0$ and by $\frac{\left(-b \pm \sqrt{b^2 + 4a\widetilde{c}}\right)}{2a}$ otherwise. Collecting all these conditions we obtain the interval for the constraint $ax^2 + bx \in \mathbf{c}$:

$$\left\{ x \in \mathbb{R} \mid ax^2 + bx \in \mathbf{c} \right\} = \begin{cases} \emptyset & \text{if } \left(b^2 + 4a\mathbf{c}\right)_+ = \emptyset \\ \emptyset & \text{if } a = b = 0 \notin \mathbf{c} \\ \mathbb{R} & \text{if } a = b = 0 \in \mathbf{c} \\ \frac{\mathbf{c}}{b} & \text{if } a = 0 \\ \frac{-b-w}{2a} \cup \frac{-b+w}{2a} & \text{otherwise} \end{cases} \tag{19.39}$$

where $\cup$ means intersection, $\emptyset$ empty set, and $w = \sqrt{\left(b^2 + 4a\mathbf{c}\right)}$, if $\left(b^2 + 4a\mathbf{c}\right)_+ \neq \emptyset$.

19.4 UNDERESTIMATORS

Designing underestimators is a crucial element in any B&B or B&R framework in global optimization. In this section we outline the most used strategies to create underestimators. First, let us look at ways to underestimate common terms that may appear in optimization problems such as bilinear, trilinear, fractional, fractional trilinear, and concave. Later we look at the two most common ways of relaxing general nonconvex functions: *Factorable Programming Relaxations* and *α-relaxations*.

19.4.1 Underestimating Bilinear Terms

Given a bilinear term xy, its tightest convex lower bounding [12] is obtained by adding a new variable w_{BL}, which takes the value:

$$w_{BL} = \max \left\{ x^L y + y^L x - x^L y^L, \; x^U y + y^U x - x^U y^U \right\} \tag{19.40}$$

This can be included directly into an optimization problem formulation by introducing the following two constraints [13]:

$$w_{BL} \geq x^L y + y^L x - x^L y^L \tag{19.41}$$

$$w_{BL} \geq x^U y + y^U x - x^U y^U \tag{19.42}$$

19.4.2 Underestimating Trilinear Terms

For the case of trilinear terms x, y, z, the new variable w_{TL} along with the following eight inequalities can be added to the optimization problem to create a convex underestimator [14]:

$$w_{TL} \geq xy^L z^L + x^L yz^L + x^L y^L z - 2x^L y^L z^L \tag{19.43}$$

$$w_{TL} \geq xy^U z^U + x^U yz^L + x^U y^L z - x^U y^L z^L - x^U y^U z^U \tag{19.44}$$

$$w_{TL} \geq xy^L z^L + x^L yz^U + x^L y^U z - x^L y^U z^U - x^L y^L z^L \tag{19.45}$$

$$w_{TL} \geq xy^U z^L + x^U yz^U + x^L y^U z - x^L y^U z^L - x^U y^U z^U \tag{19.46}$$

$$w_{TL} \geq xy^L z^U + x^L yz^L + x^U y^L z - x^U y^L z^U - x^L y^L z^L \tag{19.47}$$

$$w_{TL} \geq xy^L z^U + x^L yz^U + x^U y^U z - x^L y^L z^U - x^U y^U z^U \tag{19.48}$$

$$w_{TL} \geq xy^U z^L + x^U yz^L + x^L y^L z - x^U y^U z^L - x^L y^L z^L \tag{19.49}$$

$$w_{TL} \geq xy^U z^U + x^U yz^U + x^U y^U z - 2x^U y^U z^U \tag{19.50}$$

19.4.3 Underestimating Fractional Terms

Underestimators for fractional terms of the form x/y may be introduced by a new variable w_F and the addition of two constraints. These constraints are as follows [14], and depend on the sign of the bounds for x:

$$w_F \geq \begin{cases} \frac{x^L}{y} + \frac{x}{y^U} - \frac{x^L}{y^U} & \text{if } x^L \geq 0 \\ \frac{x}{y^U} - \frac{x^L y}{y^L y^U} + \frac{x^L}{y^L} & \text{if } x^L < 0 \end{cases} \tag{19.51}$$

$$w_F \geq \begin{cases} \frac{x^U}{y} + \frac{x}{y^L} - \frac{x^U}{y^L} & \text{if } x^L \geq 0 \\ \frac{x}{y^L} - \frac{x^U y}{y^L y^U} + \frac{x^U}{y^U} & \text{if } x^U < 0 \end{cases} \tag{19.52}$$

19.4.4 Underestimating Fractional Trilinear Terms

In the case of fractional trilinear term $\frac{xy}{z}$ with bounds $x^L, y^L \geq 0$ and $z^L \geq 0$, the variable w_{FT} is introduced along with the following eight constraints [14]:

$$w_{FT} \geq \frac{xy^L}{z^U} + \frac{x^L y}{z^U} + \frac{x^L y^L}{z} - 2\frac{x^L y^L}{z^U} \tag{19.53}$$

$$w_{FT} \geq \frac{xy^L}{z^U} + \frac{x^L y}{z^L} + \frac{x^L y^U}{z} - \frac{x^L y^U}{z^L} - \frac{x^L y^L}{z^U} \tag{19.54}$$

$$w_{FT} \geq \frac{xy^U}{z^L} + \frac{x^U y}{z^U} + \frac{x^U y^L}{z} - \frac{x^U y^L}{z^U} - \frac{x^U y^U}{z^L} \tag{19.55}$$

$$w_{FT} \geq \frac{xy^U}{z^U} + \frac{x^U y}{z^L} + \frac{x^L y^U}{z} - \frac{x^L y^U}{z^U} - \frac{x^U y^U}{z^L} \tag{19.56}$$

$$w_{FT} \geq \frac{xy^L}{z^U} + \frac{x^L y}{z^L} + \frac{x^U y^L}{z} - \frac{x^U y^L}{z^L} - \frac{x^L y^L}{z^U} \tag{19.57}$$

$$w_{FT} \geq \frac{xy^U}{z^U} + \frac{x^U y}{z^L} + \frac{x^L y^U}{z} - \frac{x^L y^U}{z^U} - \frac{x^U y^U}{z^L} \tag{19.58}$$

$$w_{FT} \geq \frac{xy^L}{z^U} + \frac{x^L y}{z^L} + \frac{x^U y^L}{z} - \frac{x^U y^L}{z^L} - \frac{x^L y^L}{z^U} \tag{19.59}$$

$$w_{FT} \geq \frac{xy^U}{z^L} + \frac{x^U y}{z^L} + \frac{x^U y^U}{z} - 2\frac{x^U y^U}{z^L} \tag{19.60}$$

19.4.5 Underestimating Concave Univariate Terms

The convex envelope of a univariate concave function $f_{u,conc}(x)$ is obtained by linearizing the function at the lower bound of its variable range as follows:

$$f_{u,conc}(x^L) + \frac{f_{u,conc}(x^U) - f_{u,conc}(x^L)}{x^U - x^L}\left(x - x^L\right) \tag{19.61}$$

19.4.6 Factorable Programming Relaxations

When nonconvex functions cannot be directly underestimated by these approaches, factorable programming relaxations can be used to underestimate nonconvex functions f_N as well as to create new constraints that relax nonconvex inequality constraints g_N. As its name suggests, factorable programming is used to relax functions that are sums and products of univariate functions (*i.e.* factorable functions). The idea of this approach is that it is difficult to create lower-bounding functions from scratch for complex functions. However, we do know lower and upper boundings for functions such as logarithms, fractional and integer powers and exponentials (which are either convex or concave), and for operators $+, *, /$ (which have been presented above), and hence this approach proposes to factor composite functions into functions for which lower and upper boundings are known. Let us exemplify this with the following function: $x_1 + \ln\left(x_2 + \sqrt{x_3} \exp(x_1 \, x_4)\right)$:

$$
\begin{aligned}
x_5 &= x_1 \, x_4 & x_9 &= x_2 + x_8 \\
x_6 &= \exp(x_5) & x_{10} &= \ln(x_9) \\
x_7 &= x_3 \, x_6 & x_{11} &= x_1 + x_{10} \\
x_8 &= \sqrt{x_7}
\end{aligned}
$$

It is now straightforward to create convex relaxations on each of the variables as they include only one operation. This approach, however, has two major drawbacks. The first is that given we are doing relaxations over relaxations, these bounding problems often exhibit a large relaxation gap. Furthermore, we can see that in our test function we have x_1 twice in our formulation; however, when the problem is factored this is not taken into account, and this may alone contribute to a larger gap between the relaxation and the original function.

19.4.7 α-relaxations

This is yet another approach to creating underestimators of general nonconvex functions. A function $f_N(\mathbf{x})$ is underestimated in the domain $\left[\mathbf{x}^L, \mathbf{x}^U\right]$ by the function $f_{u,\alpha}(\mathbf{x})$ as follows [2]:

$$f_{u,\alpha}(\mathbf{x}) = f_N(\mathbf{x}) + \sum_{i=1}^{n} \alpha_i \left(x_i^L - x_i\right)\left(x_i^U - x_i\right) \tag{19.62}$$

where the α_i are positive scalars.

Equation (19.62) possesses the two crucial properties of an underestimator: 1) $f_{u,\alpha}(\mathbf{x}) < f_N(\mathbf{x})$, given that the summation term is negative and 2) since the summation term is quadratic, it is convex, and hence with big enough α_i the whole function $f_{u,\alpha}(\mathbf{x})$ will be convex. The question now is how large must α_i be so that $f_{u,\alpha}(\mathbf{x})$ is convex?

One approach would simply be to make the α_i very large, such that you can ensure that the function is convex; however, the larger the α_i the greater the difference will be between the true function and the underestimator. As was mentioned earlier, the greater the difference between the two, the lower the quality of the underestimator, and the more inefficient the algorithm will be. It then becomes crucial to make the α_i just big enough so that the underestimating function is convex. Unfortunately, there is another inconvenience, as it turns out that calculating the optimal α_i is very computationally intensive, and given that this value will be computed every time we want to construct an underestimating function this would greatly diminish the performance of a global optimization algorithm. Therefore, as a tradeoff in an efficient optimization algorithm, a good enough value of the α_i must be computed, while at the same time not incurring too much computational time. We will now present the strategies proposed in [2] to compute such α_i.

As mentioned in earlier chapters of this book, if the Hessian matrix $H_{f_{u,\alpha}}(\mathbf{x})$ of the function $f_{u,\alpha}(\mathbf{x})$ is positive semi-definite, then the function will be convex. Given the Hessian matrix H_{f_N} we can compute:

$$H_{f_{u,\alpha}} = H_{f_N} + 2I_\alpha \tag{19.63}$$

where I_α is a diagonal matrix whose diagonal elements are the α_i.

Let us first start with the case where we only compute one α for all the elements in $\mathbf{x}$, such that:

$$f_{u,\alpha}(\mathbf{x}) = f_N(\mathbf{x}) + \alpha \sum_{i=1}^{n} \left(x_i^L - x_i \right) \left(x_i^U - x_i \right) \tag{19.64}$$

For this case, it would be preferable for α to be:

$$\alpha \geq \max \left\{ 0, -1/2 \min_{i, \mathbf{x}^L \leq \mathbf{x} \leq \mathbf{x}^U} \lambda_i(\mathbf{x}) \right\} \tag{19.65}$$

where λ_i are the eigenvalues of the Hessian matrix H_{f_N}. Hence, we can formulate the following optimization problem:

$$\min_{\mathbf{x}, \lambda} \quad \lambda \tag{19.66}$$

subject to:

$$H_{f_N}(\mathbf{x}) - \lambda I = 0 \tag{19.67}$$

$$\mathbf{x} \in \left[\mathbf{x}^L, \mathbf{x}^U \right] \tag{19.68}$$

where I is the identity matrix. Unfortunately, given that the elements of the Hessian matrix H_{f_N} are likely to be nonlinear functions themselves, computing a α that is valid throughout the underestimation domain is a difficult problem in itself. Furthermore, the problem in Equations (19.66–19.68) is a challenging nonconvex optimization problem; hence, using the Hessian matrix H_{f_N} is not a sensible approach to compute

α (or α_i). Fortunately, a valid lower bound on the minimum eigenvalue of H_{f_N} can be easily approximated when the interval Hessian matrix $[H_{f_N}] \supseteq H_{f_N}$ is introduced. A simple example of $[H_{f_N}]$ is as follows.

For the function $f_1 = x_1^2 x_2 + x_1^2 x_2^3$, we can compute its Hessian matrix:

$$
H_1 = \begin{bmatrix} 2x_2 + 2x_2^3 & 2x_1 + 6x_1 x_2^2 \\ 2x_1 + 6x_1 x_2^2 & 6x_1^2 x_2 \end{bmatrix}
\tag{19.69}
$$

Assuming the domain interval is $1 \le x_1 \le 2, 0 \le x_2 \le 1$, then the interval Hessian matrix is:

$$
[H_1] = \begin{bmatrix} [1, 5] & [1, 16] \\ [1, 16] & [0, 24] \end{bmatrix}
\tag{19.70}
$$

Having shown an example of the interval Hessian matrix $[H_{f_N}]$ we now proceed to obtain the value of α by solving:

$$
\alpha \ge \max \{0, -1/2 \quad \lambda_{\min} ([H_{f_N}])\}
\tag{19.71}
$$

where $\lambda_{\min} ([H_{f_N}])$ is the minimum eigenvalue of $[H_{f_N}]$.

Solving the problem in Equation (19.71) is a difficult task, but an approximate value can be computed with a modest effort. The method presented to approximate $\lambda_{\min} ([H_{f_N}])$ has a complexity of $\mathcal{O}(n^2)$, where n is the dimension of $\mathbf{x}$, and comes from *Gerschgorin's theorem* for interval matrices. This theorem states that for a real matrix $\mathbf{A}$ the eigenvalues are bounded below by $\check{\lambda}_{\min}$:

$$
\check{\lambda}_{\min} = \min_i \left(a_{ii} - \sum_{j \ne i} |a_{ij}| \right)
\tag{19.72}
$$

where a_{ij} are the elements of matrix $\mathbf{A}$. [2] extended the theorem for the case of interval matrices as follows.

For an interval matrix $[\mathbf{A}] = ([\underline{a}_{ij}, \overline{a}_{ij}])$ a lower bound for the eigenvalues is given by:

$$
\lambda_{\min} \ge \min_i \left[\underline{a}_{ij} - \sum_{j \ne i} \max \left(|\underline{a}_{ij}|, |\overline{a}_{ij}| \right) \right]
\tag{19.73}
$$

With Equation (19.73) we are now able to compute in reasonable time an α such that we can convexify our nonconvex function.

Another approach is to compute a different α_i for each variable such as the following:

$$
f_{u,\alpha}(\mathbf{x}) = f_N(\mathbf{x}) + \sum_{i=1}^n \alpha_i \left(x_i^L - x_i \right) \left(x_i^U - x_i \right)
\tag{19.74}
$$

In this case, we cannot use the minimum eigenvalue of the interval Hessian matrix to get ideal values for α_i, and hence a different approach is required. Let present a few definitions before introducing how to calculate the α_i values. [2] A square matrix A is an *M-matrix* if all its off-diagonal elements are nonpositive and a real positive vector u exists such that $Au > 0$. Inequalities are understood component-wise.

The comparison matrix $\langle A \rangle$ of the interval matrix $[\mathbf{A}] = \left(\left[\underline{a}_{ij}, \overline{a}_{ij} \right] \right)$ is [2]

$$\langle A \rangle_{ij} = \begin{cases} 0 & \text{if } i = j \text{ and } 0 \in \left[\underline{a}_{ij}, \overline{a}_{ij} \right] \\ \min \left\{ \left| \underline{a}_{ij} \right|, \left| \overline{a}_{ij} \right| \right\} & \text{if } i = j \text{ and } 0 \notin \left[\underline{a}_{ij}, \overline{a}_{ij} \right] \\ -|a|_{ij} & \text{if } i \neq j \end{cases} \quad (19.75)$$

where $|a|_{ij} = \max \left\{ \left| \underline{a}_{ij} \right|, \left| \overline{a}_{ij} \right| \right\}$

A square interval matrix $[A]$ is an *H-matrix* if its comparison matrix $\langle A \rangle$ is an M-matrix. It can be shown that if $[A]$ is an H-matrix, it is regular and 0 is not an eigenvalue of $[A]$ [3].

The method we present here to calculate the α_i values also presents a complexity of $\mathcal{O}\left(n^2\right)$, where n is the dimension of $\mathbf{x}$, the method relied on in the *Scaled Gerschgorin Theorem*.

For any vector $d > 0$ and a symmetric interval matrix $[A]$, define the vector α as:

$$\alpha_i = \max \left\{ 0, -\frac{1}{2} \left(\underline{a}_{ii} - \sum_{j \neq i} |a|_{ij} \frac{d_j}{d_i} \right) \right\} \quad (19.76)$$

where $|a|_{ij} = \max \left\{ \left| \underline{a}_{ij} \right|, \left| \overline{a}_{ij} \right| \right\}$. Then, for all $A \in [A]$, the matrix $A_{\mathcal{L}} = A + 2\Delta$, where $\Delta = \text{diag}(\alpha_i)$ is positive semidefinite. The proof of this theorem is beyond the scope of this presentation, but the interested reader can review it in [2]. The choice of $d > 0$ is arbitrary; if one were to set all the entries of $d = 0$, then α becomes:

$$\alpha_i = \max \left\{ 0, -\frac{1}{2} \left(\underline{a}_{ii} - \sum_{j \neq i} |a|_{ij} \right) \right\} \quad (19.77)$$

where the second term inside the max operator is the same one used in the method from *Gerschgorin's theorem*.

Another good choice for d can be $d_i = x_i^U - x_i^L$; this intuitively reflects that variables which can take a wide range of values have a larger effect on the quality of the underestimator than those that take a smaller range of values.

19.4.8 General Remarks on Underestimators

Other approaches to compute α-relaxations exist that have a complexity of $\mathcal{O}\left(n^3\right)$. This will yield more accurate relaxations with a better gap; however, there is a tradeoff between quality of the lower bound and efficiency in its computation. In many cases, the possibility of getting a slightly better lower bound does not make up for the computational time lost in its computing. The interested reader is referred to [2] for greater detail.

In practice, several relaxations are computed in parallel and the one that presents the smallest gap is chosen as an underestimator in the underlying iteration.

19.4.9 Equality Constraint Relaxation

As it has been defined over the course of this book, any equality that does not have a linear structure will cause the optimization problem to be nonconvex. Dealing with terms that are bilinear, trilinear, fractional, and fractional trilinear is straightforward,

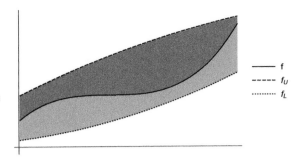

Figure 19.13 Equality relaxation, where f is the original nonlinear equality, f_L is a convex underestimator and f_U is a concave overestimator.

as complicating variable terms are replaced by new variables that participate linearly in the problem. Let us assume we have the following equality with bilinear, trilinear, fractional, and fractional trilinear terms:

$$\sum_{i=1}^{bt} b_i \, x_{BL_i,1} \, x_{BL_i,2} + \sum_{i=1}^{tt} t_i \, x_{TL_i,1} \, x_{TL_i,2} \, x_{TL_i,3}$$

$$+ \sum_{i=1}^{ft} f_i \frac{x_{F_i,1}}{x_{F_i,2}} + \sum_{i=1}^{ftt} f t_i \frac{x_{FT_i,1} \, x_{FT_i,2}}{x_{F_i,3}} = 0 \tag{19.78}$$

Equation (19.78) can be replaced by:

$$\sum_{i=1}^{bt} b_i \, w_{BL_i} + \sum_{i=1}^{tt} t_i \, w_{TL_i} + \sum_{i=1}^{ft} f_i \, w_{F_i} + \sum_{i=1}^{ftt} f t_i \, w_{FT_i} = 0 \tag{19.79}$$

where inequalities for the w variables corresponding to the bilinear, trilinear, fractional, and fractional trilinear terms are added in conjunction with Equation (19.79). On the other hand, if we have equality constraints that are nonlinear (general nonconvex or univariate concave terms), any such equality constraint $h(x) = 0$ must be rewritten as two inequalities of opposite signs:

$$h(x) \leq 0 \tag{19.80}$$

$$-h(x) \leq 0 \tag{19.81}$$

and these two inequalities must now be underestimated independently. The use of these inequalities will bound the equality constraint inside a feasible region as shown in Figure 19.13.

In this way, the two inequalities are added to the inequality set and the branch and reduce procedure can ensue.

General Remarks on Equality Constraint Relaxation

Another approach to deal with nonlinear equality constraints is to construct piecewise linear approximations of the equality constraints, and append these approximations into the original problem. Although this approach has been shown to be useful in practice, many authors deem the equality relaxation in Section 19.4.9 as yielding higher performance. Details as to how piecewise linear approximations are constructed can be found in [4].

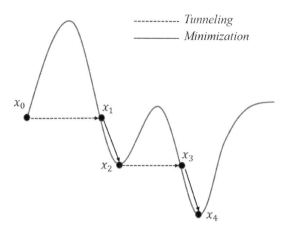

Figure 19.14 Tunneling framework for unconstrained optimization.

19.5 TUNNELING

In previous sections we have talked about how to create good lower bounds, and although simply finding a local minimum to our optimization problem is frequently used to obtain an upper bound, tunneling is a slightly more complex way that may result in a better solution, and hence a better upper bound.

19.5.1 Unconstrained Tunneling

The tunneling technique of an unconstrained optimization problem is schematically represented in Figure 19.14.

The general framework for tunneling in an unconstrained optimization problem consists of two phases: 1) Minimization phase, 2) tunneling phase. In the minimization phase a local search algorithm is used to obtain a local solution to the optimization problem $\min_x f(x)$, let us say $f(x_{local}^*)$. In the tunneling phase we try to find a new point whose objective function is equal to $f(x_{local}^*)$, and hence we find a new starting point to conduct a minimization phase again. This is done by finding a zero (through a "zero seeking algorithm") to the tunneling function:

$$T\left(x,\ f(x_{local}^*)\right) = f(x) - f(x_{local}^*) \tag{19.82}$$

This procedure is repeated until we are unable to find a new point $x \neq x_{local}^*$ that has the same objective value as our current local minimum. If we are unable to find a zero for Equation (19.82), it becomes subjective to decide when to stop, as ensuring that there is no zero for this equation can become extremely expensive and hence a stopping criterion should be put in place. Stopping criteria are quite subjective and as we are only seeking to provide a good upper bound, not too much computational time should be devoted to tunneling. Some authors, as in [5], suggest the following as a stopping criterion:

It is very likely that the first tunneling step is relatively easy compared to the subsequent ones; let us say we take N iterations to find our first tunneling point. Then we once again go through a minimization phase, and when we reach the second tunneling phase, we give a limit of αN iterations, where α, being in itself subjective, can be of values around 10–100. Hence, every subsequent tunneling phase can be given an iteration limit which is in terms of the previous number of iterations.

However, we can come across the problem that Equation (19.82) presents a zero at our current solution x^*_{local}. To avoid our zero seeking algorithm from getting stuck in the solution we already have, we can introduce a pole at x^*_{local}, with pole strength θ, such that our new tunneling function is now:

$$T\left(x,\ f(x^*_{local}),\ x^*_{local}, \theta\right) = \frac{f(x) - f(x^*_{local})}{\left[(x - x^*_{local})^T (x - x^*_{local})\right]^\theta} \tag{19.83}$$

At the beginning of each tunneling phase θ is given the value $\theta = 0$, and small increments (say $\theta = 1.0$) are added to θ until the pole is strong enough to cancel the zero at x^*_{local}.

19.5.2 Constrained Tunneling

In the case of constrained global optimization where we have a problem:

$$\min_x\ f(x) \tag{19.84}$$

subject to:

$$h(x) = 0 \tag{19.85}$$

$$g(x) \le 0 \tag{19.86}$$

$$x \in X \subseteq \mathbb{R}^n \tag{19.87}$$

The constrained tunneling framework, defines $\Phi(x)$ as the active set of constraints given by:

$$\Phi(x) = \left\{h(x);\ g(x) \mid g(x)_j - \varepsilon \le 0,\ j = 1, \ldots, v\right\} \tag{19.88}$$

and proceeds to solve the system:

$$\Phi(x) \le 0 \tag{19.89}$$

$$T_{con}\left(x,\ f(x^*_{local}),\ x^*_{local}, \theta\right) \le 0 \tag{19.90}$$

Let us remark that $\Phi(x)$ will contain all equality constraints; however, it will only contain the v inequalities that satisfy $g(x) \le \varepsilon$, and will hence change at every iteration (depending on which inequality constraints are active with some tolerance ε).

It is now possible to use $T_{con}(\cdot)$ in the same framework as $T(\cdot)$ was used in the unconstrained case; however, the point x^* which minimizes $T_{con}(\cdot)$ can be:

1. An unconstrained minimum where $\nabla_x f(x^*) = 0$, x^* being an interior point to the constraint set.
2. An unconstrained minimum where $\nabla_x f(x^*) = 0$, x^* being on a boundary to the constraint set.
3. A constrained minimum where $\nabla_x f(x^*) \ne 0$.

In cases 1 and 2, the tunneling function $T(\cdot)$ is the same as in the unconstrained case. In case 3, however, there are feasible points x that are close to x^* which satisfy

$f(x) - f(x^*) \leq 0$; this would cause the minimizing stage to be drawn to the same region of attraction of x^*. To avoid this, the following equation is used for $T(\cdot)$:

$$T_{con}\left(x,\ f(x_{local}^*),\ x_{local}^*, x^C, \theta_C\right) = \frac{f(x) - f(x_{local}^*)}{\left[\left(x - x^C\right)^T \left(x - x^C\right)\right]^{\theta_C}} \tag{19.91}$$

where in this case, the pole x^C is positioned along the line joining x^* and the current x.

The constrained tunneling algorithm proceeds in a similar fashion to the unconstrained case. It starts at a feasible point x_0, and conducts a local contained minimization (Minimization phase) to find $x_{local}^{1,*}$ (where 1 denotes the current iteration). Then, starting from this optimal point $x_{local}^{1,*}$, it proceeds to the tunneling phase, where it tries to satisfy Equations (19.89) and (19.90). If this new point is found, it proceeds with another minimization phase and so on, until a stopping criterion is met.

General Remarks on Tunneling

In contrast with purely using constraint propagation or interval arithmetic, specific implementations of a (almost purely) tunneling framework have been applied to solve global optimization problems. The interested reader is referred to [4, 10, 11] for further detail.

Another way to conduct tunneling is by finding points with a better objective function value that are *nearly feasible*, and then use feasibility restoration strategies to try to obtain a feasible point. These strategies can be reviewed in detail in [1].

19.6 PSEUDOCODE FOR A GLOBAL OPTIMIZATION ALGORITHM

In summary, global optimization algorithms are a combination of a variety of techniques for a nonconvex optimization problem of the following structure:

$$\min_x \quad f(x) \tag{19.92}$$

subject to:

$$h(x) = 0 \tag{19.93}$$

$$g(x) \leq 0 \tag{19.94}$$

$$x \in X \subseteq \mathbb{R}^n \tag{19.95}$$

A general framework for a B&R algorithm used for global optimization can be the following:

Algorithm 19.1 *Branch and Reduce for Nonconvex Optimization*

1. *Initialization: Set the best value so far $f^* = \infty$, initialize the subproblem candidate list by appending the original problem.*
2. *If the subproblem candidate list is empty return the best solution found so far f^*, if this solution is $f^* = \infty$ declare the problem as infeasible. Else:*

    1. *Select one of the subproblems in the candidate list (SP_i)*
    2. *Carry out constraint propagation and interval arithmetic techniques to reduce the search space of SP_i. The resulting problem is your new SP_i from this point onward.*

3. *Conduct a local optimization for a number of times to determine a low-lying optimum.*
4. *Conduct tunneling in one or more of the most promising low-lying optimums identified.*
5. *Choose the best local optimum and declare this as $f^U_{SP_i}$*
6. *Create convex underestimators $\check{SP_i}$ for* SP$_i$ *(e.g. $\alpha-$relaxations and factorable programming)*

   1. *Optimize the different $\check{SP_i}s$*
   2. *Choose the underestimator which presents the least gap compared to $f^U_{SP_i}$ and declare this as the lower bound $f^L_{SP_i}$ for subproblem* SP$_i$

7. *Apply fathoming*

   1. *If the solution $f^U_{SP_i}$ is infeasible, then* SP$_i$ *is also infeasible, hence go back to step 2. There is an optional feasibility restoration phase at this point, where the algorithm tries to find at least one feasible point for* SP$_i$; *however, this can be time consuming.*
   2. *If $f^L_{SP_i} \geq f^*$ then* SP$_i$ *has no feasible solution which is better than the best solution found so far, hence go back to 2.*
   3. *If the optimal solution $f^U_{SP_i}$ is feasible for* SP$_i$ *this is a feasible solution for the original problem. If $f^U_{SP_i} \leq f^*$ this is the new best found solution (i.e. $f^* = f^U_{SP_i}$).*

8. *If there is no other subproblem in the list, evaluate if the feasible domain X_i is small enough to be within some tolerance ε, if so, return f^* as the global optimum and terminate algorithm. Else, continue to (i).*
9. *Branch the current subproblem* SP$_i$ *and add the smaller subproblems to the candidate subproblem list. Go to step 2.*

State-of-the-art algorithms for global optimization follow a similar structure to that shown in Algorithm 19.1.

19.7 REFERENCES

[1] Neumaier, A. "Complete search in continuous global optimization and constraint satisfaction." Acta Numerica. 2004;13(1):p. 271–369.

[2] Adjiman, C. S., Dallwig, S., Floudas, C. A., and Neumaier, A. "A global optimization method, αBB, for general twice – Differentiable constrained NLPs." Computers & Chemical Engineering. 1998;22(9):p. 1137–1158.

[3] Neumaier, A. "An optimality criterion for global quadratic optimization." Journal of Global Optimization. 1992;2(2):p. 201–208.

[4] Belotti, P., Kirches, C., Leyffer, S., Linderoth, J., Luedtke J., and Mahajan, A. "Mixed-integer nonlinear optimization." Acta Numerica. 2013;22:p. 1–131.

[5] Levy, A. V. and Gómez, S. Numerical Optimization. SIAM, Chapter: "The tunneling method applied to global optimization." 1985.

[6] Jones, D. R., Perttunen, C. D., and Stuckman B. E. "Lipschitzian optimization without the Lipschitz constant." Journal of Optimization Theory and Applications. 1993;79(1): p. 157–181.

[7] Huyer, W. and Neumaier, A. "Global optimization by multilevel coordinate search." Journal of Global Optimization. 1999;14(4):p. 331–355.

[8] Pintér, J. D. Global Optimization in Action. Kluwer. 1996.

[9] Dallwig, S., Neumaier, A., Schichl H. GLOPT – A Program for Constrained Global Optimization. Developments in Global Optimization. 1997.

[10] Yao, Y. "Dynamic tunneling algorithm for global optimization." IEEE Transactions on Systems, Man, ad Cybernetics. 1989;19(5):p. 1222–1230.

[11] Yasuda, K. and Kanazawa, T. "Multitrajectory dynamic tunneling algorithm for global optimization." Electrical Engineering in Japan. 2005;154(1):p. 47–58.

[12] Al-Khayyal, F. A. and Falk J. E. "Jointly constrained biconvex programming." Mathematics of Operations Research. 1983;8(2):p. 273–286.

[13] McCormick, G. P. "Computability of global solutions to factorable nonconvex programs: Part I – Convex underestimating problems." Mathematical Programming. 1976;10(1): p. 147–175.

[14] Maranas, C. D. and Floudas C. A. "Finding all solutions of nonlinearly constrained systems of equations." Journal of Global Optimization. 1995;7(2):p. 143–183.

19.8   FURTHER READING RECOMMENDATIONS

More Extensive Treatment on Complete Methods for Global Optimization

Neumaier, A. (2004). "Complete search in continuous global optimization and constraint satisfaction." Acta Numerica. 13(1):p. 271–369.

Belotti, P., Kirches, C., Leyffer, S., Linderoth, J., Luedtke, J., and Mahajan, A. (2013). "Mixed-integer nonlinear optimization." Acta Numerica. 22:p. 1–131.

Interval Arithmetic from a Theoretical and Practical Perspective

Neumaier, A. (1991). Interval Methods for Systems of Equations. Cambridge University Press.

Adjiman, C. S., Dallwig, S., Floudas, C. A., and Neumaier, A. (1998). "A global optimization method, αBB, for general twice–differentiable constrained NLPs." Computers & Chemical Engineering. 22(9):p. 1137–1158.

Epperly, T. G. W. and Pistikopoulos, E. N. (1997). "A reduced space branch and bound algorithm for global optimization." Journal of Global Optimization. 11(3): p. 287–311.

Deterministic Global Optimization from the Constraint Propagation Perspective

Van Hentenryck, P., Michel, L., and Deville, Y. (1997). Numerica: A Modeling Language for Global Optimization. MIT Press.

Global Optimization with Discrete Variables

Parker, R.G. and Rardin, R. L. (1988). Discrete Optimization. Academic Press.

Heuristics on Branching Rules for Bound-Constrained Problems

Csendes, T. and Ratz, D. (1997). "Subdivision direction selection in interval methods for global optimization." SIAM Journal on Numerical Analysis. 34(3):p. 922–938.

Ratz, D. (1996). "On branching rules in second-order branch-and-bound methods for global optimization." Scientific Computation and Validated Numerics. p. 221–227.

Ratz, D. and Csendes, T. (1995). "On the selection of subdivision directions in interval branch-and-bound methods for global optimization." Journal of Global Optimization. 7(2):pp. 183–207.

Recent Survey on Constraint Propagation

Bliek, C., Spellucci, P., Vicente, L. N., Neumaier, A., Granvilliers, L., Monfroy, E., Benhamou, F., Huens, E., Van Hentenryck, P., Sam-Haroud, D., and Faltings, B. (2001). "Algorithms for Solving Nonlinear Constrained and Optimization Problems: The State of the Art." A progress report of the COCONUT project: www.mat.univie.ac.at/~neum/glopt/coconut/StArt.html.

Extensive Treatment on Local Descent Techniques and Their Heuristics

Boender, C. G. E., Rinnooy Kan, A. H. G., Timmer, G. T. and Stougie, L. (1982). "A stochastic method for global optimization." Developments in Global Optimization. 22(1): pp. 125–140.

Levy, A. V. and Gómez, S. (1985). Numerical Optimization. SIAM, Chapter: "The tunneling method applied to global optimization."

Review on Nonconvex MINLP Problems

Tawarmalani, M. and Sahinidis, N. V. (2002). Convexification and Global Optimization in Continuous and Mixed-Integer Nonlinear Programming: Theory, Algorithms, Software, and Applications. Kluwer.

19.9 EXERCISES

1. Let us assume that we wish to find underestimators for the following function:

$$\frac{\exp(x_1\,x_2)}{(x_1\,x_2)} + \cos\left(x_2 + \sqrt{x_1}\right) - \sin\left(\ln\left(x_1 + x_2^2\right)\right)\frac{x_1}{x_2}$$

for $x_1, x_2 \in [1, 3]$

   (i) Use factorable programming to determine a lower bounding function for the above expression.

   (ii) Use α-relaxations to determine a lower bounding function for the above expression.

   (iii) Plot and compare the three functions.

2. The *Rastrigin function* is a nonconvex function designed to trap local solvers in its many local solutions. Its 2D formulation is as follows:

$$R(x) = A2 + \left(x_1^2 - A\cos(2\pi x_1)\right) + \left(x_2^2 - A\cos(2\pi x_2)\right)$$

   (i) For $A = 10$ and $x_1, x_2 \in [-5.12, 5.12]$, find its global minimum by implementing the uncosntrained version of Algorithm 19.1.

   (ii) Add an additional bilinear term to the objective function:

$$R'(x) = A2 + \left(x_1^2 - A\cos(2\pi x_1)\right) + \left(x_2^2 - A\cos(2\pi x_2)\right) + x_1 x_2$$

   Find the global minimum to $R'(x)$.

3. We now have a constrained version of the optimization problem:

$$\min \quad R(x)$$

subject to:

$$(x_1 + x_2)^2 \geq 2$$

For $A = 10$ and $x_1, x_2 \in [-5.12, 5.12]$, find its global minimum by implementing the constrained version of Algorithm 19.1.

20

Optimal Control Problems (Dynamic Optimization)

20.1 PROBLEM STATEMENT, SINGLE-STAGE AND MULTISTAGE PROBLEMS

The solution of Optimal Control Problems (OCPs) may be defined as the determination of time-dependent controls and time-invariant parameters that optimize a performance index associated with a dynamical system. In other words, optimal control relates to the optimization of dynamical systems, where dynamical systems refers to processes that evolve with time, and are characterized by *states* $x(t)$ and can be controlled by inputs $u(t)$, which are called *controls*.

Given that OCPs examine the performance of systems that operate under transient conditions, they play a crucial role in chemical engineering. Some examples are set point changes in process plants, process plant start-ups, and processes that are inherently non-steady state, such as batch processes. It is therefore important to understand and to be able to solve these problems efficiently. The OCP formulation is an optimization problem that includes an objective function, and is constrained by a set of differential equations (or differential-algebraic equations) that describe the dynamics of the system. Its solution describes the trajectories of control variables $u(t)$ that minimize the objective function subject to the dynamical behavior of the system under study.

The general form of a dynamical system is given by:

$$\dot{x}(t) = f\left(x(t)\right) \tag{20.1a}$$

$$x(t_0) = x_0 \tag{20.1b}$$

$$t \in \left[t_0, t_f\right] \tag{20.1c}$$

with final time t_f and $x\,(t)$, $\dot{x}\,(t)$, $f\,(\cdot) \in \mathbb{R}^n$.

Furthermore, more generally, the ordinary differential equation (ODE) system in Equations (20.1a–20.1c) can include algebraic variables and constraints. This yields the following differential-algebraic equation (DAE) system:

$$\dot{x}(t) = f\left(x(t), y(t)\right) \tag{20.2a}$$

$$0 = g(x(t), y(t)) \tag{20.2b}$$

$$x(t_0) = x_0 \tag{20.2c}$$

$$t \in \left[t_0, t_f\right] \tag{20.2d}$$

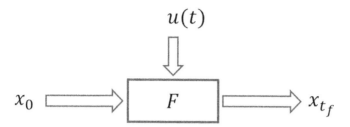

Figure 20.1 Single-stage system representation.

with algebraic variable vector $y(t) \in \mathbb{R}^m$. When this system is to be optimized with respect to some performance index, the resulting optimization problem is an OCP. OCPs can be single-stage or multistage systems, depending on the problem under consideration. Consider first an OCP formulation of a single-stage system as shown in Figure 20.1.

A single-stage system is initially known at an n-dimensional state vector $x(0)$. This initial state, subject to the system dynamics, and a q-dimensional control vector u determines the state vector at the end of the stage, namely, $x(t_f)$. The relation between the initial and final states in a single-stage system can be written as in Equation (20.3).

$$x(t_f) = F(x(0), u(t)) \tag{20.3}$$

Notice that the final state in Equation (20.3) is not only a function of the initial state $x(0)$ but also of the control input trajectory $u(t)$. The specific example model in Equations (20.4a–20.4d) constitutes a semi-explicit index-1 DAE system. Hence, a single-stage OCP can be defined as: Performance index:

$$\min_{u(\cdot),\, t_f} \quad J = c(x(t_f), y(t_f)) + \int_{t_0}^{t_f} L(x(t), y(t), u(t))\, dt \tag{20.4a}$$

subject to:
System dynamics:

$$\dot{x}(t) = f(x(t), y(t), u(t)) \tag{20.4b}$$

$$0 = g(x(t), y(t), u(t)) \tag{20.4c}$$

$$x(t_0) = x_0 \tag{20.4d}$$

$$t \in \left[t_0, t_f\right] \tag{20.4e}$$

with final time t_f, state variable vector $x(t)$, $\dot{x}(t)$, $f(\cdot) \in \mathbb{R}^n$, algebraic variable vector $y(t) \in \mathbb{R}^m$, and control variable vector $u(t) \in \mathbb{R}^q$. Note that the performance index J includes two terms, the first term comprises the cost at the final time, while the integral term defines the costs incurred during the system's trajectory.

Example 1: Single-Stage Optimal Control Problem

A simple example of a single-stage OCP can be that of thinking of a car that starts at an initial point $x(t_0) = x_0$, let us say with initial velocity zero $\dot{x}(t_0) = 0$, and needs to reach a certain destination $x(t_f)$ also with final velocity being zero $\dot{x}(t_f) = 0$. Assuming the car goes in a straight line it can have as control variable the acceleration (positive or negative). An objective would be to minimize the time it takes the car to go from its initial position to its final destination, while satisfying all constraints. This problem can be formulated as follows:

$$\min_{u(\cdot),t_f} \quad t_f$$

subject to:
ODEs: laws of motion:

$$\dot{x}_1(t) = x_2(t)$$

$$\dot{x}_2(t) = u(t)$$

$$0 \leq t \leq t_f$$

Initial conditions: car stationary at origin:

$$x_1(0) = 0 \qquad x_2(0) = 0$$

Final conditions: car stationary at destination:

$$x_1(t_f) = x_{t_f} \qquad x_2(t_f) = 0$$

This problem can also be schematically represented as in Figure 20.2.

A more general class of OCPs are multistage problems. In this special class of problems the objective function and the system dynamics can change discontinuously, which defines the end of one stage and the beginning of another.

Let us now consider a multistage system over $N - 1$ stages defined as:

$$x^i\left(t_0^{(i)}\right) = J^i\left(x^i, u^i\right), \qquad i = 2, 3, \ldots, N \qquad (20.5)$$

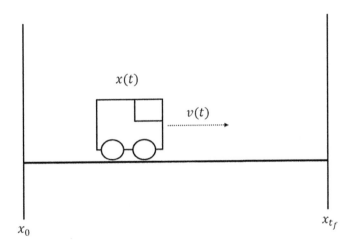

Figure 20.2 Single-stage car example.

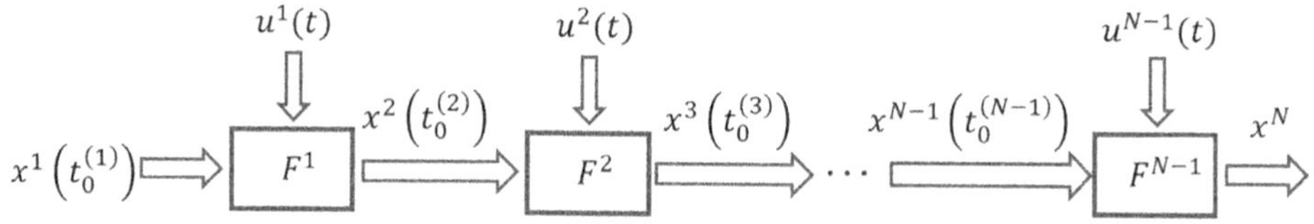

Figure 20.3 Multi-stage system representation.

where $x^{(1)}(t_0^{(1)})$ is given as the initial condition, and J^i is the junction condition at stage i. Equation (20.5) implies that the final state of stage i is the initial state of stage $i+1$, which is shown schematically in Figure 20.3.

It is then possible to define a general multistage OCP as:
Performance index:

$$\min_{u^i(\cdot),\, t_f^i;\, i=1,2,\dots,N} \quad J = \sum_{i=1}^{N} c^i(x^i(t_f^i), y^i(t_f^i)) + \sum_{i=1}^{N} \int_{t_0^i}^{t_f^i} L^i(x^i(t), y^i(t), u^i(t))\, dt$$

$$(20.6a)$$

subject to:
System dynamics for each stage:

$$f^i\left(\dot{x}^i(t), x^i(t), y^i(t), u^i(t)\right) = 0 \qquad i = 0, 1, \dots, N \qquad (20.6b)$$

Initial conditions for first stage:

$$J_0\left(\dot{x}^1(t_0), x^1(t_0), y^1(t_0), u^1(t_0)\right) = 0 \qquad (20.6c)$$

Stage junction conditions:

$$J_i\left(\dot{x}^{i+1}(t_0^{i+1}), x^{i+1}(t_0^{i+1}), y^{i+1}(t_0^{i+1}), u^{i+1}(t_0^{i+1}), \dot{x}^i(t_f^i), x^i(t_f^i), y^i(t_f^i), u^i(t_f^i)\right) = 0$$
$$i = 2, \dots, N$$

$$(20.6d)$$

Equality and inequality path constraints for each stage:

$$g_{path}^i(\dot{x}^i(t), x^i(t), y^i(t), u^i(t)) \leq 0 \qquad i = 1, 2, \dots, N \qquad (20.6e)$$

$$h_{path}^i(\dot{x}^i(t), x^i(t), y^i(t), u^i(t)) = 0 \qquad i = 1, 2, \dots, N \qquad (20.6f)$$

$$t^i \leq t \leq t^{i+1} \qquad i = 1, 2, \dots, N-1 \qquad (20.6g)$$

Equality and inequality point constraints for stages:

$$g_{point}^i(\dot{x}^i(t_f^i), x^i(t_f^i), y^i(t_f^i), u^i(t_f^i)) \leq 0 \qquad i = 2, \dots, N-1 \qquad (20.6h)$$

$$h_{point}^i(\dot{x}^i(t_f^i), x^i(t_f^i), y^i(t_f^i), u^i(t_f^i)) = 0 \qquad i = 2, \dots, N-1 \qquad (20.6i)$$

Equality and inequality point constraints for first and final stage:

$$g_{point}^0(\dot{x}^1(t_0), x^1(t_0), y^1(t_0), u^1(t_0)) \leq 0 \qquad (20.6j)$$

$$h_{point}^0(\dot{x}^1(t_0), x^1(t_0), y^1(t_0), u^1(t_0)) = 0 \qquad (20.6k)$$

$$g_{point}^N(\dot{x}^N(t_f^N), x^N(t_f^N), y^N(t_f^N), u^N(t_f^N)) \le 0 \tag{20.6l}$$

$$h_{point}^N(\dot{x}^N(t_f^N), x^N(t_f^N), y^N(t_f^N), u^N(t_f^N)) = 0 \tag{20.6m}$$

Control bounds:

$$u^{i,L} \le u^i(t) \le u^{i,U} \qquad i = 1, 2, \ldots, N \tag{20.6n}$$

Stage sequencing constraints:

$$t^{i-1} < t^i \qquad i = 2, 3, \ldots, N \tag{20.6o}$$

with final time t_f^N which states the final time of the last stage N, state vector variables $x^i(t)$, $\dot{x}^i$, $f^i \in \mathbb{R}^{n_i}$, algebraic vector variables $y^i(t) \in \mathbb{R}^{m_i}$, and control variable vector $u^i(t) \in \mathbb{R}^{q_i}$ for all $i = 1, \ldots, N$ stages. Let us raise a few comments on this problem:

The performance index J includes the sum of the costs at the final time t_f^i for each stage, as well as the sum of the costs of the system's trajectory for all stages.

- Equation (20.6b) gives the differential equations that govern each stage (it may be the case that the equations of each stage are the same).
- Equation (20.6c) dictates the initial conditions of the overall multistage problem.
- Equations (20.6d) give the junction conditions that link each stage.
- Equations (20.6e–20.6f) present the equality and inequality constraints that must be satisfied all throughout the dynamics of the system.
- Equations (20.6h–20.6i) present the equality and inequality constraints that must be satisfied at the boundaries of intermediate stages.
- Equations (20.6j–20.6m) present the equality and inequality constraints that must be satisfied at the first and final stages.
- Equations (20.6n–20.6o) show the bounds for the controls and stage times, respectively.

The following section introduces the principles on which the solution strategies for these problems are based.

Example 2: Multistage Optimal Control Problem

Example 1 in this chapter is again considered here. However, in this case, at given distances the car will be instantaneously loaded with extra mass, thus defining a multistage system with the stages being defined by the given distances between loading stations [3], as shown in Figure 20.4.

This multistage optimal control problem can be defined as follows:

$$\min_{u^i(\cdot),\, t_f^i;\, i=1,2,3} t_f^3 \tag{20.7}$$

subject to:
ODEs: laws of motion:

$$\dot{x}_1^i(t) = x_2^i(t) \qquad i = 1, 2, 3 \tag{20.8}$$

$$\dot{x}_2^i(t) = u^i(t) \qquad i = 1, 2, 3 \tag{20.9}$$

Initial conditions: car stationary at origin:

$$x_1^1(0) = 0 \qquad x_2^1(0) = 0 \tag{20.10}$$

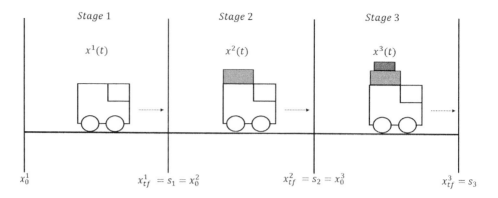

Figure 20.4 Multi-stage car example (after [3]).

Junction conditions: momentum conservation:

$$m^i x_2^i \left(t_f^i \right) - m^{i+1} x_2^{i+1} \left(t_0^{i+1} \right) = 0 \qquad i = 1, 2 \tag{20.11}$$

Final conditions for each stage:

$$x_1^i(t_f^i) = s_i \qquad i = 1, 2, 3 \tag{20.12}$$

Final stage condition for velocity:

$$x_2^3(t_f^3) = 0 \tag{20.13}$$

Now that we have shown the problem statement for an OCP let us start with the solution procedures in the following section.

20.2 PONTRYAGIN'S MINIMUM (MAXIMUM) TO SOLVE SINGLE-STAGE DYNAMICS

Optimal control problems are not too different from optimization problems introduced in Chapter 7 in the sense that conditions for optimality must also be stated. To derive the necessary conditions for optimality (also termed as Pontryagin Minimum Principle) for OCPs the single-stage system in Equations (20.4a–20.4e) will be analyzed. These conditions can then be generalized for multistage problems. To aid in this analysis it is useful to first define a Hamiltonian function as:

$$H(x, y, u, \lambda, v) = L(x, y, u) + \lambda^T f(x, y, u) + v^T g(x, y, u) \tag{20.14}$$

where $\lambda(t)$ and $v(t)$ are Euler–Lagrange vectors of multipliers. Time as argument has been removed from Equation (20.14) and from the remainder of this analysis for simplicity of presentation. This derivation is conducted through a perturbation analysis as follows.

Consider perturbations $\delta u \equiv \delta u(t)$, which lead to a feasible control, given that the base value $u(t)$ is also feasible. Perturbing the DAE system in Equations (20.4b) and (20.4c) would have state perturbations satisfying Equations (20.15a) and (20.15b), in order to describe the new feasible trajectories $(x(t) + \delta x(t), y(t) + \delta y(t))$:

$$\delta \dot{x} = f_x \delta x + f_y \delta y + f_u \delta u \tag{20.15a}$$

$$0 = g_x \delta x + g_y \delta y + g_u \delta u \tag{20.15b}$$

Consider now a perturbation of the performance index in Equation (20.4a) and append the perturbation of the DAE system to the integral term via the use of the multipliers $\lambda(t)$ and $v(t)$ for the differential and algebraic equations, respectively. With this the following is obtained:

$$\delta J = \left(c_x(x, y)dx + c_y(x, y)dy \right)\big|_{t_f}$$
$$+ \int_{t_0}^{t_f} \left[\left(-\lambda^T \delta \dot{x} + \lambda^T f_x \delta x + \lambda^T f_y \delta y + \lambda^T f_u \delta u \right) \right.$$
$$+ \left(v^T g_x \delta x + v^T g_y \delta y + v^T g_u \delta u \right)$$
$$+ \left. \left(L_x \delta x + L_y \delta y + L_u \delta u \right) \right] dt \tag{20.16}$$

An optimum with respect to $u(\cdot)$ must be such that it is a stationary point for the performance index. Thus the sum of the terms multiplied by each of the perturbations (*i.e.* terms multiplied by $\delta \dot{x}$, δu, δx and δy) in Equation (20.16) will be zero individually. The perturbation $\delta \dot{x}$ is problematic and integration by parts is used; as a first step the following is noted:

$$\int_{t_0}^{t_f} \lambda^T \delta \dot{x} \, dt = \int_{t_0}^{t_f} \lambda^T d(\delta x) \tag{20.17}$$

The integration by parts rule is given by:

$$\int_{t_0}^{t_f} a \, db = [ab]_{t_0}^{t_f} - \int_{t_0}^{t_f} b \, da \tag{20.18}$$

Setting $a = \lambda^T$, $b = \delta x$ and noticing that $da = \dot{\lambda}^T dt$ yields that:

$$\int_{t_0}^{t_f} \lambda^T \delta \dot{x} \, dt = \left[\lambda^T \delta x \right]_{t_0}^{t_f} - \int_{t_0}^{t_f} \dot{\lambda}^T \delta x \, dt \tag{20.19}$$

By collecting terms from the integral form in the respective perturbations δx, δy, δu, the following is obtained:

$$\dot{\lambda}^T + \lambda^T f_x + v^T g_x + L_x = 0$$
$$\lambda^T f_y + v^T g_y + L_y = 0$$
$$\lambda^T f_u + v^T g_u + L_u = 0$$

The final condition for the system (also referred to as an adjoint system) is derived by considering that $(\delta y)_{t_f} = -(g_y^{-1} g_x \delta x)_{t_f}$ [4] from the algebraic equation part of the DAE system, and is found to be:

$$\lambda^T(t_f) = \left[c_x - c_y g_y^{-1} g_x \right]_{t_f} \tag{20.21}$$

Putting all of this together into one system of equations results in the following Two-Point Boundary Value Problem (TPBVP):

$$\dot{x}^T = f(x, y, u) \tag{20.22a}$$

$$0 = g(x, y, u) \tag{20.22b}$$

$$\dot{\lambda}^T = -(\lambda^T f_x + v^T g_x + L_x) = -\dot{H}_x \tag{20.22c}$$

$$0 = \lambda^T f_y + v^T g_y + L_y = \dot{H}_y \tag{20.22d}$$

$$0 = \lambda^T f_u + v^T g_u + L_u = \dot{H}_u \tag{20.22e}$$

with boundary conditions :

$$x^T(t_0) = x_0$$

$$\lambda^T(t_f) = \left[c_x - c_y g_y^{-1} g_x \right]_{t_f} \tag{20.22f}$$

where we use the expression for the Hamiltonian H derived in Equation (20.14).

By solving the problem in Equations (20.22a–20.22f) the solution of the original OCP is obtained. This is possible because u^* is determined by the minimum principle, while x^*, y^*, λ^*, and v^* are obtained by the ODE system and the adjoint equations. In this way, the number and type of conditions in the TPBVP matches the number and type of unknowns needed to solve the OCP. The requirement for this formulation to be solvable is for the Hamiltonian to be an explicit function of the control variables, otherwise higher derivatives of the Hamiltonian have to be considered [4]. Let us note that this solution procedure is also termed the *indirect (Euler–Lagrange) method.*

In the following example, a reformulation of an OCP will be conducted by transcribing it into a TPBVP to portray the indirect approach.

Dynamic Optimization by Pontryagin's Minimum Principle

Consider the OCP of a batch reactor [5]:

$$\max_{u(\cdot)} \phi = x_2(1.0) \tag{20.23}$$

subject to:

$$\dot{x}_1(t) = -\left(u(t) + \frac{u(t)^2}{2} \right) x_1(t) \tag{20.24}$$

$$\dot{x}_2(t) = u(t) x_1(t) \tag{20.25}$$

$$x_1(0) = 1, \quad x_2(0) = 0 \tag{20.26}$$

$$0.0 \le u(t) \le 5.0, \qquad 0 \le t \le 5. \tag{20.27}$$

The Hamiltonian (dropping time arguments) $H(x, y, u, \lambda, v)$ is defined as follows:

$$H = -\lambda_1 \left(u + \frac{u^2}{2} \right) x_1 + \lambda_2 u x_1 \tag{20.28}$$

In order to solve this system by the indirect (Euler–Lagrange) method, equations for the Lagrange multipliers λ_1 and λ_2 and an equation for the control profile u are introduced:

$$\frac{d\lambda_1}{dt} = -\frac{\partial H}{\partial x_1} \tag{20.29}$$

$$\frac{d\lambda_2}{dt} = -\frac{\partial H}{\partial x_2} \tag{20.30}$$

$$\frac{\partial H}{\partial u} = 0 \tag{20.31}$$

The initial conditions are given for the state variables x_1 and x_2 by the system, and the final conditions are given as follows:

$$\lambda_1(t_f) = \frac{\partial \phi}{\partial x_1} \tag{20.32}$$

$$\lambda_2(t_f) = \frac{\partial \phi}{\partial x_2} \tag{20.33}$$

$$\left.\frac{\partial H}{\partial u}\right|_{t_0} = 0 \tag{20.34}$$

There are five equations with five boundary conditions. The system equations and the boundary conditions are summarized as follows:

System equations:

$$\dot{x}_1 = -\left(u + \frac{u^2}{2}\right) x_1 \tag{20.35}$$

$$\dot{x}_2 = u\, x_1 \tag{20.36}$$

Boundary conditions:

$$x_1(0) = 1, \quad x_2(0) = 0 \tag{20.37}$$

$$\lambda_1(t_f) = \frac{\partial \phi}{\partial x_1} \quad \lambda_2(t_f) = \frac{\partial \phi}{\partial x_2} \tag{20.38}$$

$$\left.\frac{\partial H}{\partial u}\right|_{t_0} = 0 \tag{20.39}$$

It can now be observed that two out of the five boundary conditions, namely the ones for the Lagrange multipliers λ_1 and λ_2, are specified at final time t_f. Therefore this constitutes a Two-Point Boundary Value Problem (TPBVP), which can be solved by readily available software and algorithms.

20.2.1 General Comments and Remarks on Indirect Methods

Methods that solve OCPs by directly solving the resulting TPBVP are termed as *indirect methods*. The most common indirect methods include the shooting method, multiple shooting method, and collocation methods. The main difficulty is finding a first estimate of the adjoint variables that produces a solution reasonably close to the true final time conditions. Hence, trajectories in the state/adjoint space frequently diverge if poor initial conditions are provided. Furthermore, the problem with such approaches is that while integrating in the forward direction of the original dynamical system may be stable, that of the adjoint system may not be so. A further comlication with the indirect approach is the handling of control bounds, and general (path) constraints.

20.3 TRANSCRIPTION TO NLP PROBLEMS VIA DISCRETIZATION

The difficulties presented by indirect methods to efficiently solve optimal control problems motivated the reformulation of OCPs into finite dimensional optimization problems, termed *direct methods*. In direct methods, the control (and possibly state) variables of the OCP are discretized over time elements using piecewise continuous functions, and thus the problem is transcribed to an NLP. Within this approach methods can be separated into two subgroups: direct sequential methods and direct simultaneous methods, both of which will be introduced in subsequent sections.

20.3.1 Direct Sequential Methods: Control Vector Parameterization

Direct sequential methods only discretize the control variables $u(t)$ and the resulting approach is termed as Control Vector Parametrization (CVP). In this formulation a DAE model is solved in an inner loop while the parameters discretizing the control variables are updated. Gradients of the objective function are calculated by direct sensitivity equations of the DAE system or by integration of adjoint equations, or even finite differences.

Consider the multistage optimization problem for a DAE system:
Performance index:

$$\min_{u^i(\cdot),\, t_f^i;\, i=1,2,\dots,N} J = \sum_{i=1}^{N} c^i(x^i(t_f^i), y^i(t_f^i)) + \sum_{i=1}^{N} \int_{t_0^i}^{t_f^i} L^i(x^i(t), y^i(t), u^i(t))\, dt$$

$$(20.40a)$$

subject to:
System dynamics:

$$\dot{x}^i(t) = f^i\left(x^i(t), y^i(t), u^i(t)\right) \qquad i = 0, 1, \dots, N \qquad (20.40b)$$

$$x^1(t_0^1) = x_0 \qquad (20.40c)$$

$$h_{path}^i(x^i(t), y^i(t), u^i(t)) = 0 \qquad i = 1, 2, \dots, N \qquad (20.40d)$$

Stage junction conditions:

$$\dot{x}^{i+1}(t_0^{i+1}) = \dot{x}^i(t_f^i)$$
$$x^{i+1}(t_0^{i+1}) = x^i(t_f^i)$$
$$y^{i+1}(t_0^{i+1}) = y^i(t_f^i) \qquad (20.40e)$$
$$u^{i+1}(t_0^{i+1}) = u^i(t_f^i)$$
$$i = 1, \dots, N - 1$$

Bounds:

$$t^i \le t \le t^{i+1} \qquad i = 1, 2, \ldots, N - 1 \tag{20.40f}$$

$$u^{i,L} \le u^i\,(t) \le u^{i,U} \qquad i = 1, 2, \ldots, N \tag{20.40g}$$

$$t_{i-1} < t_i \qquad i = 1, 2, \ldots, N \tag{20.40h}$$

with final time t_f^N which states the final time of the last stage N, state vector variables $x^i\,(t), \dot{x}^i, f^i \in \mathbb{R}^{n_i}$, algebraic vector variables $y^i\,(t) \in \mathbb{R}^{m_i}$, and control variable vector $u^i\,(t) \in \mathbb{R}^{q_i}$ for all $i = 0, 1, \ldots, N$ stages. Equations (20.40a–20.40h) are in semi-implicit form; furthermore to be index-1 such a system must have $\mathbf{h_{path}}(\cdot)$ (where by this we mean the whole $h^i_{path}(x^i(t), y^i(t), u^i(t)) = 0 \quad i = 1, 2, \ldots, N$ system) solvable for $y(t)$; hence det $\left[\mathbf{h_{path}}(\cdot)_y\right] \ne 0 \forall t$, where $\mathbf{h_{path}}(\cdot)_y$ is the derivative of $\mathbf{h_{path}}(\cdot)$ with respect to $y(t)$. Allowing us to reduce the DAE system to an equivalent ODE system.

To solve OCPs by the sequential method the control variables $u(t)$ are parameterized by piecewise continuous functions with coefficients that determine their profile, and hence the decision-making process depends only on this finite set of parameters.

The DAEs in Equations (20.40a–20.40h) are solved internally, while an outer loop optimizes over the control vector $\mathbf{u}$.

The sequential method is hence the following:

1. Problem initialization, provide initial guess for control vector parameters.
2. Until convergence criterion is met do:

   1. Given the specification of the control variables $u^i(t)$, the DAE system in Equations (20.40a–20.40h) is treated as an initial value problem (IVP). The solution of this IVP computes the state variables.
   2. Gradients of the objective function and the constraints with respect to the parameters parameterizing $u^i(t)$ are computed. This information is passed on to the NLP solver.
   3. The NLP solver adjusts the parameter values for the controls on the DAE system. The new parameter values are passed on to step 2(a), or if the termination criterion is met, the algorithm continues to step 3.

3. After convergence of step 2, the latest computed states and controls comprise the solution to the OCP.

Concerning step 2(b) in the framework outlined, there are several ways to compute the gradients of the objective function and the constraints with respect to the parameters of the control variables $u^i(t)$, namely, finite differences, direct sensitivity equations, or adjoint equations. These different approaches are presented in the subsequent sections.

Derivative Evaluation via Finite Differences

Obtaining derivatives through finite differences (also known as perturbation) is the most straightforward way to obtain gradients for the parameters of the control variables. Although this framework is easy to implement, the gradients obtained are prone to truncation and round-off errors, in addition to being computationally expensive.

Let us denote δe_j as a small perturbation on the j-th element, such that δ is a small value, and e_j represents a vector of all zeroes, except for the j-th entry, where the value is one. It is then possible to define a perturbation vector $\delta \mathbf{p}_j$ (where $\mathbf{p}$ is the vector of parameters that parameterize the control vectors) of the $\mathbf{p}$ vector, with a small perturbation on the j-th entry:

$$\delta \mathbf{p}_j = \mathbf{p} + \delta e_j \tag{20.41}$$

Then we can define the perturbation vector of the objective function and constraints of our DAE system as:

$$\psi(\delta \mathbf{p}_j) = \left[\Phi(\delta \mathbf{p}_j), \ E_{const}(\delta \mathbf{p}_j)^T, \ I_{const}(\delta \mathbf{p}_j)^T \right] \tag{20.42}$$

where E_{const} and I_{const} stand for the equality and inequality constraints, respectively. It is then possible to calculate the derivative for the jth element as:

$$\nabla_{\delta \mathbf{p}_j} \psi(\mathbf{p}) = \frac{\psi(\mathbf{p}) - \psi(\delta \mathbf{p}_j)}{\delta} \tag{20.43}$$

This approach is derived via forward divided differences, and has an error of order $\mathcal{O}(\delta)$. Furthermore, note that the evaluation of Equation (20.43) for all gradient components necessitates $n_p + 1$ integrations of the DAE system (where n_p is the total number of parameters parametrizing the $u^i(t)$ control variables). This must be done at every iteration of the sequential method algorithm, which makes this approach computationally expensive. If a more accurate derivative approximation was desired, a centered divided difference approach could be used. This would reduce the error to order $\mathcal{O}(\delta^2)$ and would require two integrations of the DAE system per parameter, that is $2n_p + 1$ DAE integrations overall (per iteration).

As alternatives to approximate derivatives through finite differences, the direct sensitivity and adjoint methods have been developed. Both of these approaches offer higher accuracy and efficiency in derivative calculations.

Derivative Evaluation via Direct Sensitivity Equations

A different approach to computing the derivatives of a dynamic system is through the calculation of the direct sensitivity equations. For this, consider a single-stage dynamic system, expressed as a semi-explicit index-1 DAE, where we have the parameterized controls $u(p)$. By dropping the time as arguments, we have:

$$\dot{x} = f(x, y, u(p), p) \tag{20.44a}$$
$$0 = g(x, y, u(p), p) \tag{20.44b}$$
$$x(t_0) = x_0 \tag{20.44c}$$

The sensitivity equations of this DAE system with respect to the parameters p that parameterize the controls u are:

$$\dot{x}_p = \frac{\partial f}{\partial x} x_p + \frac{\partial f}{\partial y} y_p + \frac{\partial f}{\partial u} \frac{\partial u}{\partial p} + \frac{\partial f}{\partial p} \tag{20.45a}$$

$$0 = \frac{\partial g}{\partial x} x_p + \frac{\partial g}{\partial y} y_p + \frac{\partial g}{\partial u} \frac{\partial u}{\partial p} + \frac{\partial g}{\partial p} \tag{20.45b}$$

$$x_p(t_0) = \frac{\partial x_0}{\partial p} \tag{20.45c}$$

where $\dot{x}_p = \frac{\partial \dot{x}}{\partial p}$, $x_p = \frac{\partial x}{\partial p}$, $y_p = \frac{\partial y}{\partial p}$.

Note that Equations (20.45a–20.45c) comprise a linear time varying (LTV) index-1 DAE system, and the solution of the state profile must be known to compute them. Once the solution to the direct sensitivity equations is obtained, the gradients for the objective and the constraints in Equations (20.44a–20.44c) are computed through the chain rule as follows:

$$\nabla_p \psi = \dot{x}_p(t_f)\frac{\partial \psi}{\partial \dot{x}} + \frac{\partial \psi}{\partial p} \tag{20.46}$$

To illustrate how direct sensitivity equations are obtained, let us look at the following example.

Sensitivity Calculations by Direct Sensitivities

Consider the following DAE system:

$$\dot{x}_1 = u_1 (10 x_2 - x_1) + y_1^2 \tag{20.47}$$
$$\dot{x}_2 = u_1 (x_1 - 10 x_2) - (1 - u_2) x_2 \tag{20.48}$$
$$0 = (1 - x_1 - x_2) u_3 - y_1 \tag{20.49}$$
$$x_1(0) = u_3, \qquad x_2(0) = 0 \tag{20.50}$$

Here, given this simple system, let us assume that each control variable is parameterized by a single parameter, where $p_{1,2,3} = u_{1,2,3}$ (simply replacing the u_i by p_i in this system). The system has $n_x = 2$ differential states, $n_y = 1$ algebraic states, and $n_p = 3$ number or parameters that paremeterize the controls. The resulting sensitivity equation system results in a DAE system of $n_p (n_x + n_y) = 9$ equations, which consists of $n_p n_x = 6$ differential equations, and $n_p n_y = 3$ algebraic equations.

Notice that $\frac{\partial u}{\partial p} = I$ given the fact that we are parameterizing directly each u_i by p_i and $\frac{\partial f}{\partial p} = \mathbf{0}$ given that p does not appear explicitly in Equations (20.51–20.50), where I is the identity matrix and $\mathbf{0}$ is a vector of all zeroes of appropriate dimensions, respectively.

Computing the differential sensitivity equation matrix $\dot{x}_p$ in Equation (20.45a) with entries $\dot{x}_{p,ij}$ for state variable x_i and parameter p_j is as follows:

$$\begin{bmatrix} \dot{x}_{p,11} & \dot{x}_{p,12} & \dot{x}_{p,13} \\ \dot{x}_{p,21} & \dot{x}_{p,22} & \dot{x}_{p,23} \end{bmatrix} = \begin{bmatrix} \frac{\partial f_1}{\partial x_1} & \frac{\partial f_1}{\partial x_2} \\ \frac{\partial f_2}{\partial x_1} & \frac{\partial f_2}{\partial x_2} \end{bmatrix} x_p + \begin{bmatrix} \frac{\partial f_1}{\partial y_1} \\ \frac{\partial f_2}{\partial y_1} \end{bmatrix} y_p$$

$$+ \begin{bmatrix} \frac{\partial f_1}{\partial u_1} & \frac{\partial f_1}{\partial u_2} & \frac{\partial f_1}{\partial u_3} \\ \frac{\partial f_2}{\partial u_1} & \frac{\partial f_2}{\partial u_2} & \frac{\partial f_2}{\partial u_3} \end{bmatrix} \begin{bmatrix} 1 & 0 & 0 \\ 0 & 1 & 0 \\ 0 & 0 & 1 \end{bmatrix} + \begin{bmatrix} 0 & 0 & 0 \\ 0 & 0 & 0 \end{bmatrix} \tag{20.51}$$

$$= \begin{bmatrix} -u_1 & 10 u_1 \\ u_1 & -10 u_1 - (1 - u_2) \end{bmatrix} x_p + \begin{bmatrix} 2 y_1 \\ 0 \end{bmatrix} y_p$$

$$+ \begin{bmatrix} (10 x_2 - x_1) & 0 & 0 \\ (x_1 - 10 x_2) & x_2 & 0 \end{bmatrix} \begin{bmatrix} 1 & 0 & 0 \\ 0 & 1 & 0 \\ 0 & 0 & 1 \end{bmatrix} \tag{20.52}$$

The algebraic sensitivity equation matrix is computed in the following way as per Equation (20.45b):

$$0 = \left[\begin{array}{cc} \frac{\partial g_1}{\partial x_1} & \frac{\partial g_1}{\partial x_2} \end{array} \right] x_p + \left[\frac{\partial g_1}{\partial y_1} \right] y_p$$

$$+ \left[\begin{array}{ccc} \frac{\partial g_1}{\partial u_1} & \frac{\partial g_1}{\partial u_2} & \frac{\partial g_1}{\partial u_3} \end{array} \right] \left[\begin{array}{ccc} 1 & 0 & 0 \\ 0 & 1 & 0 \\ 0 & 0 & 1 \end{array} \right] + \left[\begin{array}{ccc} 0 & 0 & 0 \\ 0 & 0 & 0 \end{array} \right] \tag{20.53}$$

$$0 = \left[\begin{array}{cc} -u_3 & -u_3 \end{array} \right] x_p + [-1] y_p$$

$$+ \left[\begin{array}{ccc} 0 & 0 & (1 - x_1 - x_2) \end{array} \right] \left[\begin{array}{ccc} 1 & 0 & 0 \\ 0 & 1 & 0 \\ 0 & 0 & 1 \end{array} \right] + \left[\begin{array}{ccc} 0 & 0 & 0 \\ 0 & 0 & 0 \end{array} \right] \tag{20.54}$$

The initial conditions are provided as in Equation (20.45b):

$$x_p(t_0) = \left[\begin{array}{ccc} x_{p11}(0) & x_{p12}(0) & x_{p13}(0) \\ x_{p21}(0) & x_{p22}(0) & x_{p23}(0) \end{array} \right] = \left[\begin{array}{ccc} 0 & 0 & 1 \\ 0 & 0 & 0 \end{array} \right] \tag{20.55}$$

By integrating the DAE system in Equations (20.47–20.50) forward in time for some value of the controls, the solution for all states is obtained. Then with the state trajectory information it is possible to solve the sensitivity Equations (20.51–20.53), and (20.55).

In most implementations, state and state-sensitivity equations are solved simultaneously, so that a local error control can be performed on both systems and to avoid the need to store the state profiles (to later solve the sensitivity system). Furthermore, given that the joint state-sensitivity system grows in proportion to $n_p (n_x + n_y)$ efficient DAE solvers for these systems have been developed. These solvers exploit the structure (*e.g.* sparsity, bandedness) of the system, and the fact that both sensitivity and state equations share the same Jacobian and other properties. Given the complexity of solving the state-sensitivity system, usually implicit multistep integration algorithms are employed. The three most popular methods used to compute the solution for these systems differ in the way the corrector formula is solved while sharing the same predictor step. These three variations are staggered direct, simultaneous corrector, and staggered corrector methods.

Derivative Evaluation via Adjoint Equations

In some cases, a dynamic optimization problem will have a large number of parameters, and if in particular the number of states is also large, the direct sensitivity approach is inefficient to compute derivatives. In such cases, the adjoint equations are the best method to compute the problem's gradients. For this, let us define ψ_{t_f} (which is either the objective function or a constraint at final time t_f) as an element of ψ in Equation (20.42) at the end of the integration horizon. Now to ψ_{t_f} we adjoin the DAEs and define the following:

$$\hat{\psi}\left(x(t), y(t), u(t), p\right) = \psi_{t_f}\left(x(t_f), p\right)$$

$$+ \int_0^{t_f} \left[\lambda(t)^T \left(f\left(x(t), y(t), u(t), p\right) - \dot{x}(t)\right) \right.$$

$$\left. + v(t)^T g(x(t), y(t), u(t), p) \right] dt \tag{20.56}$$

Given n_p parameters that parameterize the controls, the gradient of $\hat{\psi}$ with respect to p, that is $\hat{\psi}_p$, $j = 1, \ldots, n_p$ is obtained by applying the chain rule:

$$
\hat{\psi}_j \left(x(t), y(t), u \right) = \left(\psi_{t_f} \right)_p \left(x(t_f), p \right) + \left(\psi_{t_f} \right)_x \left(x(t_f), p \right) x_p \left(t_f \right)
$$
$$
+ \int_0^{t_f} \left[\lambda(t)^T \left\{ f_p \left(x(t), y(t), u(t), p \right) \right. \right.
$$
$$
+ f_x \left(x(t), y(t), u(t), p \right) x_p(t) - \dot{x}_p(t)
$$
$$
+ f_y \left(x(t), y(t), u(t), p \right) y_p(t) \right\} + v(t)^T \left\{ g_p \left(x(t), y(t), u(t), p \right) y_p(t) \right.
$$
$$
\left. \left. + g_x \left(x(t), y(t), u(t), p \right) x_p(t) \right\} \right] dt \tag{20.57}
$$

Integrating $\int_0^{t_f} \lambda(t)^T \dot{x}(t) \, dt$ by parts:

$$
\int_0^{t_f} \lambda(t)^T \dot{x}(t) \, dt = \left[\lambda(t)^T x(t) \right]_{t_0}^{t_f} - \int_{t_0}^{t_f} \left\{ \dot{\lambda}(t)^T x_p(t) \right\} dt
$$

thus, Equation (20.57) becomes:

$$
\hat{\psi}_j \left(x(t), y(t), u \right) = \left(\psi_{t_f} \right)_p \left(x(t_f), p \right) + \left\{ \left(\psi_{t_f} \right)_x \left(x(t_f), p \right) - \lambda(t_f)^T \right\} x_p \left(t_f \right)
$$
$$
+ \lambda(t_0)^T x(t_0) + \int_0^{t_f} \left[\lambda(t)^T f_p \left(x(t), y(t), u(t), p \right) \right.
$$
$$
+ \left\{ f_x \left(x(t), y(t), u(t), p \right)^T \lambda(t)^T + \dot{\lambda}(t)^T \right\}^T x_p(t)
$$
$$
+ \lambda(t)^T f_y \left(x(t), y(t), u(t), p \right) y_p(t)
$$
$$
+ v(t)^T \left\{ g_p \left(x(t), y(t), u(t), p \right) y_p(t) + g_x \left(x(t), y(t), u(t), p \right) x_p(t) \right.
$$
$$
\left. \left. + g_y \left(x(t), y(t), u(t), p \right) y_p(t) \right\} \right] dt \tag{20.58}
$$

Through perturbations as conducted in Section 20.2 we perturb $\delta x(t), \delta y(t), \delta p$, dropping the x, u, y, t and p dependencies to obtain

$$
\delta \hat{\psi}_j = \left[\left(\psi_{t_f} \right)_x - \lambda(t_f)^T \right]^T \delta x_p(t_f) + \lambda(t_0)^T \delta x(t_0) + \left[\left(\psi_{t_f} \right)_p \right]^T \delta p(t)
$$
$$
+ \int_0^{t_f} \left\{ \left[f_x^T \lambda^T - \dot{\lambda}^T + v g_x \right]^T \delta x(t) \right. \tag{20.59}
$$
$$
+ \left[f_y^T \lambda + g_y v \right]^T \delta y(t)
$$
$$
\left. + \left[f_p \lambda + g_p v \right]^T \delta p(t) \right\} dt \tag{20.60}
$$

We now wish for $\delta \hat{\psi}_j$ to be only affected by $\delta p(t)$; however, since the perturbations $\delta x(t)$, $\delta y(t)$, $\delta x(t_0)$ can take arbitrary values, the terms multiplied by these perturbations must be zero; as such, from the final state perturbation $\delta x(t_f)$ we obtain the boundary condition:

$$
\left(\psi_{t_f} \right)_x = \lambda(t_f)^T \tag{20.61}
$$

From the state perturbation $\delta x(t)$ we obtain the differential equation for the adjoint variables:

$$\dot{\lambda}^T = -v\, g_x - f_x{}^T \lambda^T \tag{20.62}$$

The perturbation of the algebraic variables $\delta y(t)$ leads to the algebraic equation:

$$\lambda^T f_y + v^T g_p^T = 0 \tag{20.63}$$

We then assume that the initial state is actually a function of the control variable (therefore of p), hence $x(t_0) = x_0(p)$, which makes the perturbation $\delta x(t_0)$ a function of p, and hence:

$$\lambda(t_0)^T \delta x(t_0) = \lambda(t_0)^T \left[\frac{\partial x_0}{\partial p} \right]^T \delta p(t) \tag{20.64}$$

With this assumption, Equation (20.59) becomes:

$$\delta \psi_j = \left[\left((\psi_{t_f})_p + \frac{\partial x_0}{\partial p} \lambda(t_0) \right)^T + \int_0^{t_f} (f_p\, \lambda + g_p\, v)^T \, dt \right] \delta p(t) \tag{20.65}$$

To compute these derivatives, the state equations are computed by a forward pass in time, while the adjoint equations are computed with the use of the state equations by a backward pass in time.

The following example will show how the adjoint sensitivity equations are computed.

Sensitivity Calculations by Calculating Adjoint Sensitivities

Consider the following DAE system:

$$\dot{x}_1 = u_1 (10\, x_2 - x_1) + y_1^2 \tag{20.66}$$

$$\dot{x}_2 = p_1 (x_1 - 10\, x_2) - (1 - p_2)\, x_2 \tag{20.67}$$

$$0 = (1 - x_1 - x_2)\, p_3 - y_1 \tag{20.68}$$

$$x_1(0) = u_3, \qquad x_2(0) = 0 \tag{20.69}$$

Note that here we assume the control variables u have already been parameterized by parameters p.

As mentioned in the direct sensitivity example, this system has $n_x = 2$ differential states, $n_y = 1$ algebraic states, and $n_p = 3$ control variables. Furthermore, depending on the number of constraints at the final time, and the objective function, the resulting adjoint DAE system results in $n_\psi (n_x + n_y)$ equations.

Given this system, we have:

$$\lambda(t)^T f(x, y, p) + v(t)\, g(x, y, p) = \lambda_1 \left(p_1 (10\, x_2 - x_1) + y_1^2 \right)$$
$$+ \lambda_2 (p_1 (x_1 - 10\, x_2) - (1 - p_2)\, x_2) \tag{20.70}$$
$$+ v_1 ((1 - x_1 - x_2)\, p_3 - y_1)$$

with Equation 20.70, and following from Equation (20.62) we can define the differential adjoint equations as follows:

$$\dot{\lambda}_1 = v_1\, p_3 + \lambda_1\, p_1 - \lambda_2\, p_1 \tag{20.71}$$

$$\dot{\lambda}_2 = v_1 \, p_3 - 10 \, \lambda_1 \, p_1 + \lambda_2 \, (10 \, p_1 + (1 - p_2)) \tag{20.72}$$

$$\lambda_1(t_f) = \frac{\partial \psi_{t_f}}{\partial x_1}, \qquad \lambda_2(t_f) = \frac{\partial \psi_{t_f}}{\partial x_2} \tag{20.73}$$

The algebraic constraint is obtained by Equation (20.63):

$$2 \, y_1 \, \lambda_1 - v_1 = 0 \tag{20.74}$$

By solving the DAE system in Equations (20.71–20.74), the profiles for $\lambda(t)$ and $v(t)$ are obtained. Following Equation (20.65) we obtain the sensitivities for the control parameters p:

$$\frac{\partial \hat{\psi}_j}{\partial p_1} = \frac{\partial \psi_j}{\partial p_1} + \int_0^{t_f} [\lambda_1 \, (10 \, x_2 - x_1) + \lambda_2 \, (x_1 - 10 \, x_2)] \, dt$$

$$\frac{\partial \hat{\psi}_j}{\partial p_2} = \frac{\partial \psi_j}{\partial p_2} + \int_0^{t_f} [\lambda_2 \, x_2] \, dt$$

$$\frac{\partial \hat{\psi}_j}{\partial p_3} = \frac{\partial \psi_j}{\partial p_3} + \lambda_1(0) + \int_0^{t_f} [v_1 \, (1 - x_1 - x_2)] \, dt$$

As can be observed by comparing Equations 20.47–20.50 with the current example, if $n_\psi < 3$ the current example would have a smaller amount and less complex equations than the system for the direct sensitivities. Therefore, this adjoint system could be expected to be easier to solve than the one resulting from the direct sensitivity approach given that $n_\psi < n_p$.

General Remarks on Derivative Calculations

Three approaches have been presented to calculate the sensitivities to address OCPs through the sequential approach. From these three approaches the finite differences approach is in general the easiest to implement; however, it is also the most inefficient in terms of computational time, and the one most prone to have errors. Comparing between calculating the direct sensitivity equations and the adjoint sensitivity equations, on one hand, the cost of computing the direct sensitivity equations is proportional to the number of states and controls. On the other hand, the cost of computing the adjoint sensitivity equations is proportional to the number of functions $\hat{\psi}_j$ (objective and constraint functions). Furthermore, both approaches are not prone to errors, and offer stability as long as the state equations are stable. In general, if the number of functions $\hat{\psi}_j$ is less than the number of parameterizing parameters p, then the adjoint sensitivity approach is preferred over computing the direct sensitivity equations, due to its lower computational cost.

20.3.2 General Comments and Remarks on the Direct Sequential Approach

Since the DAE system is to be solved at each iteration of the sequential approach, direct sequential methods are said to be feasible path methods. If for some reason the algorithm exits before its full convergence, it can return a fully feasible solution with the best objective found so far.

Another inherent advantage over the direct transcription approach described in Section 20.4 is that the accuracy of the state variables is handled via the error control mechanism of a numerical integration solver.

A disadvantage compared to direct transcription, however, is that finding an optimal solution, or even a feasible solution, may prove difficult when the system is unstable or the differential equations do not have a solution for certain control values; in other words, only stable or mildly unstable dynamic systems can be handled by the sequential approach [2].

Moreover, a large amount of computational effort may be wasted in obtaining accurate state values when the controls are far away from their optimal values, as the method must undergo many integrations of the DAE system.

20.4 CONTROL AND STATE PARAMETERIZATION VIA ORTHOGONAL COLLOCATION

Another important class of methods to solve optimal control problems are those that transcribe the problem into an NLP by dicretizing both the state and control variables. The most popular terms for this approach are *direct transcription* and *full discretization*. Fully discretizing both controls and states avoids the need to solve the dynamic system in an inner loop, contrary to control vector parameterization. Furthermore, if the method converges, the system becomes feasible only at the optimal solution. This mitigates possible convergence issues that may arise when a DAE system is solved multiple times. However, if the system is not dealt with carefully, convergence problems might still exist when solving the resulting (possibly very large-scale) NLP.

There are a number of methods that can be used to discretize a time-varying system. However, given that this is not simply the integration of a system, but at the same time an optimization procedure is taking place, not all methods designed for initial value problems (IVP) are suitable to address the current problem. The most prominent candidate to deal with these discretization schemes are *orthogonal collocation methods*. This approach is equivalent to discretizing through an implicit Runge–Kutta method with high order accuracy and excellent stability properties [6]. These methods are computationally too expensive as initial value solvers, but for boundary value problems and optimal control problems, which require implicit solutions anyway, this discretization is a less expensive way to obtain accurate solutions [8].

20.4.1 Orthogonal Collocation

Orthogonal collocation methods are a subclass of collocation methods in dynamic optimization. Orthogonal collocation offers advantages over other direct transcription methods: It is a single step method, which is preferred as it does not rely on smooth profiles that extend over previous time steps. As the NLP formulation needs to deal with potential discontinuities in control profiles this is an essential property. Orthogonal collocation is a high-order implicit discretization method which provides accurate profiles with relatively few finite elements, as well as good stability characteristics when facing stiff systems. Another advantage is that the discretization

points are selected optimally from the approximation error point of view at the values
of the roots of orthogonal polynomials.

Let us define the following single-stage OCP:
Objective function, evaluated at the final time t_f:

$$\min_{u(\cdot),t_f} \quad \phi\left(x(t_f), y(t_f), u(t_f)\right) \tag{20.75a}$$

subject to:
Process dynamics, described by a set of index-1 DAEs:

$$\dot{x}(t) = f\left(x(t), y(t), u(t)\right), \quad t_0 \leq t \leq t_f \tag{20.75b}$$

Coupled algebraic equations:

$$g\left(x(t), y(t), u(t)\right) = 0 \tag{20.75c}$$

Initial conditions for differential process variables:

$$x(t_0) = x(t_0) \tag{20.75d}$$

Trajectory inequality constraints (path constraints), imposed at all times t:

$$h\left(x(t), y(t), u(t)\right) \leq 0, \quad t_0 \leq t \leq t_f \tag{20.75e}$$

Simple bounds on control functions, $u(t)$, and final time t_f for each stage:

$$u^L \leq u(t) \leq u^U, \quad t_0 \leq t \leq t_f \tag{20.75f}$$

where x is the differential variables of the stage, $\dot{x}$ the time derivative of differential
variables x, y the algebraic variables of the stage, and u the control variables.

It is noted that each stage is considered to have its own time argument and the
integration limits are between t_0 and t_f, which is to be viewed as the duration of the
stage dynamics. The time itself is not included as an argument in the specification
of the DAE's considered. Should this appear explicitly, then an additional ordinary
differential equation may be defined to carry its value for the stage requiring it. In
what follows we will outline the general procedure to discretize dynamic optimization
problems through collocation, and although many researchers have been working in
this field of research, the specific framework presented here follows from Biegler and
his research [6, 7, 8].

To transform a DAE optimization problem into an NLP, one can approximate
state and control variables through a family of polynomials on finite elements
$(t_0 < t_1 < \dots < t_N = t_f)$. The differential profile for the state variables can be
discretized by a Runge–Kutta type scheme. There are a number of polynomials from
which the roots can be obtained to get the collocation points. From the different
options, Radau roots have been deemed to be the most effective in practice due to
their stability and high precision [7]. In the case of Radau polynomials, the final point
of each element is also a collocation point [6].

This is explicitly embedded in the following discretization equations so as to avoid
duplication of variables:

$$x(t) = x_{i-1} + h_i \sum_{j=1}^{K} \Phi_j(\tau)\,\dot{x}_{ij} \tag{20.76}$$

The differential variables appearing are $\dot{x}_{ij} = \frac{dx(t_{ij})}{dt}$ where $t_{ij} = t_{0,i} + h_i \tau_j$, with $j = 1, 2, \ldots, K$, where K is the number of collocation points in the element i and τ_j are normalized collocation points $0 \leq \tau_1 < \tau_2 < \cdots < \tau_k \leq 1$ comprising the values of the roots of orthogonal polynomials. Continuity between elements is enforced by matching the state values and gradients at each step, as follows:

$$x_i = x_{i,0} + h_i \sum_{j=1}^{K} \Phi_j(\tau)\, \dot{x}_{ij} \tag{20.77}$$

where continuity between elements is enforced by:

$$x_i = x_{i-1,K} \tag{20.78}$$

By differentiating Equation (20.77) with respect to time, we get:

$$\frac{dx_i(t)}{dt} = \dot{x}_i(t) = h_i \sum_{j=1}^{K} \frac{d\Phi_j(\tau)}{d\tau} \frac{d\tau}{dt} \dot{x}_{ij}$$

$$= \sum_{j=1}^{K} \Phi'_j(\tau)\dot{x}_{ij} \tag{20.79}$$

where $\Phi'_j(\tau) = \frac{d\Phi_j(\tau)}{d\tau}$.

This polynomial must satisfy the conditions:

$$x(t_{0,i}) = x_{i,0} \Rightarrow \Phi_j(0) = 0, \quad j = 1, 2, \ldots, K$$
$$\dot{x}_i(t_{ij}) = \dot{x}_{ij} \Rightarrow \Phi'_l(\tau_j) = \delta_{l,j}, \quad l, j = 1, 2, \ldots, K \tag{20.80}$$

where $\delta_{l,j}$ is the Kronecker delta.

Thus by choosing:

$$\Phi_j(\tau) = \int_0^\tau \phi_j(\hat{\tau}) d\hat{\tau}, \quad j = 1, 2, \ldots, K \tag{20.81}$$

the desired conditions are satisfied. $\phi_j(\tau)$ are Lagrange interpolation basis polynomials of degree $K - 1$ (for interpolation over $K - 1$ points). $\Phi_j(\tau)$ are polynomials of degree $K - 1$ (with also K coefficients in number).

The algebraic and control profiles are represented as:

$$y_i(t) = \sum_{j=1}^{K} \Phi_j(\tau)\, y_{ij} \tag{20.82}$$

$$u_i(t) = \sum_{j=1}^{K} \Phi_j(\tau)\, u_{ij} \tag{20.83}$$

The discretization of the DAEs has collocation order K, and number of elements N. The discretization involves the following modeling aspects.

1. Control value extrapolation at stage initial time $t = 0$ (element 1 at $\tau = 0$) so as to enable the initialization problem:

$$u_0 = \sum_{j=1}^{K} \phi_j(0) \cdot u_{1,j} \tag{20.84}$$

2. Stage initialization problem at stage initial time $t = 0$ (given the boundary conditions, it calculates the full initial state for $\dot{x}^0, x^0, y^0$):

$$f\left(\dot{x}_0, x_0, y_0, u_0\right) = 0 \tag{20.85}$$

3. Interior points discretization for elements $i = 1, 2, 3, \ldots, N$, with collocation points in each element $j = 1, 2, \ldots, K$ (element i contains the element initial value symbol $x^{(i,0)}$ for variables x):

$$f\left(\dot{x}_{i,j}, x_{i,j}, y_{i,j}, u_{i,j}\right) = 0 \tag{20.86}$$

$$x_{i,j} = x_{i-1,K} + h_i \sum_{l=1}^{K} \Phi_l(\tau_j)\dot{x}_{i,l} \tag{20.87}$$

noting that $x_{i-1,K} = x_{i,0}$.

4. Continuity of differential state variable at initial points of elements, $i = 2, 3, \ldots, N$: Continuity is explicitly included in the state variable relationships by using the end value of the previous element for all elements. All of this assumes Radau roots as collocation points so that the element final points are always collocation points.

5. The Lagrange interpolation polynomials appearing in the formulation are defined as follows:

$$\phi_j(\tau) = \prod_{\substack{k=1 \\ k \neq j}}^{K} \frac{(\tau - \tau_k)}{(\tau_j - \tau_k)}, \quad j = 1, 2, \ldots, K \tag{20.88}$$

$$\Phi_j(\tau) = \int_0^{\tau} \phi_j(\hat{\tau})d\hat{\tau}, \quad j = 1, 2, \ldots, K \tag{20.89}$$

6. Constraints and bounds for the discretization parameters:

$$t = \sum_{i=1}^{N} h_i \tag{20.90}$$

7. Constraints and relationships involving the stage parameters t can be added as the user requires. The bounds on the element sizes are as follows:

$$h^L \leq h_i \leq h^U, \quad i = 1, 2, \ldots, N \tag{20.91}$$

8. The bounds on the time derivatives, differential variables, algebraic variables, and control variables can generically be implemented as follows for a generic variable z:

$$z^L \leq z_{i,j} \leq z^U \tag{20.92}$$

$$j = 0, 1, 2, \ldots, K, \quad i = 1, 2, \ldots, N \tag{20.93}$$

With the above, the original OCP (20.75a–20.75f) can be formulated as follows:
Objective function:

$$\min_{x_{i,j},\, y_{i,j},\, u_{i,j}} \Psi(x_N) \tag{20.94a}$$

subject to:
Process dynamics:

$$\dot{x}_{i,j} = f(x_{i,j},\, y_{i,j},\, u_{i,j}) \tag{20.94b}$$

Collocation Constraints:

$$x_{i,j} = x_{i-1,K} + h_i \sum_{l=1}^{K} \Phi_l(\tau_j)\dot{x}_{i,l} \tag{20.94c}$$

Continuity Constraints:

$$x_{i,0} = x_{i-1,K} \tag{20.94d}$$

Coupled algebraic equations:

$$g(x_{i,j},\, y_{i,j},\, u_{i,j}) = 0 \tag{20.94e}$$

Initial conditions:

$$x_{1,0} = x(t_0) \tag{20.94f}$$

Path constraints (inequality state constraints):

$$h\left(x_{i,j},\, y_{i,j},\, u_{i,j}\right) \leq 0 \tag{20.94g}$$

Simple bounds:

$$y^L \leq y_{i,j} \leq y^U, \quad u^L \leq u_{i,j} \leq u^U, \quad t_0 \leq t \leq t_f \tag{20.94h}$$

$$t_f^L \leq t \leq t_f^U \tag{20.94i}$$

The following example will show how the formulation of the direct transcription approach is carried out.

OCP Reformulation by Collocation over Finite Elements in Time

Consider the following OCP.
Objective function:

$$\min \quad \left(x_1(t_f)\right)^2 + \left(x_2(t_f)\right)^2 \tag{20.95a}$$

subject to:
Dynamics of the system:

$$\dot{x}_1(t) = u_1(t)\,(10\,x_2(t) - x_1(t)) + y_1^2(t) \tag{20.95b}$$
$$\dot{x}_2(t) = u_1(t)\,(x_1(t) - 10\,x_2(t)) - (1 - u_2(t))\,x_2(t) \tag{20.95c}$$
$$0 = (1 - x_1(t) - x_2(t))\,u_3(t) - y_1(t) \tag{20.95d}$$

Initial conditions:

$$x_1(0) = u_3(0), \quad x_2(0) = 0 \tag{20.95e}$$

Terminal constraint:

$$0 = u_2(t_f)x_2(t_f) - 5 \tag{20.95f}$$

Bounds:

$$-1,000 \le x_2(t) \le 1,000$$

$$-50 \le u_j(t) \le 100$$

A direct transcription formulation using Radau roots for a discretization into N elements and K collocation points per element is as follows. We denote the elements and collocation points as superscripts in this example to avoid confusion between variable number labels and exponents:

$$\min_{x^{i,j},y^{i,j},u^{i,j}} \quad \left(x_1^{N,K}\right)^2 + \left(x_2^{N,K}\right)^2 \tag{20.96a}$$

subject to:
Dynamics of the system:

$$\dot{x}_1^{i,j} = u_1^{i,j}\left(10x_2^{i,j} - x_1^{i,j}\right) + \left(y_1^{i,j}\right)^2 \tag{20.96b}$$

$$\dot{x}_2^{i,j} = u_1^{i,j}\left(x_1^{i,j} - 10x_2^{i,j}\right) - \left(1 - u_2^{i,j}\right)x_2^{i,j} \tag{20.96c}$$

$$0 = \left(1 - x_1^{i,j} - x_2^{i,j}\right)u_3^{i,j} - y_1^{i,j} \tag{20.96d}$$

Collocation constraints:

$$x_1^{i,j} = x_1^{i-1,K} + h^i \sum_{l=0}^{K} \Phi_l(\tau_j)\dot{x}_1^{i,l} \tag{20.96e}$$

$$x_2^{i,j} = x_2^{i-1,K} + h^i \sum_{l=0}^{K} \Phi_l(\tau_j)\dot{x}_2^{i,l} \tag{20.96f}$$

Continuity constraints:

$$x_1^{i,0} = x_1^{i-1,K} \tag{20.96g}$$

$$x_2^{i,0} = x_2^{i-1,K} \tag{20.96h}$$

Bounds:

$$-1,000 \le x_m^{i,j} \le 1,000 \tag{20.96i}$$

$$-50 \le u_m^{i,j} \le 100 \tag{20.96j}$$

$$for \quad i = 1,2,\ldots,N, \quad j = 1,2,\ldots,K, \quad m = 1,2$$

Initial conditions:

$$x_1^{1,0} = u_3^{1,0}, \qquad x_2^{1,0} = 0 \tag{20.96k}$$

Terminal constraint:

$$0 = u_2^{N,K}x_2^{N,K} - 5 \tag{20.96l}$$

This is now a full NLP: Notice that we optimize for the controls u but also for the differential and algebraic variables x and y. This approach discretizes the problem into an NLP and it optimizes for all variables regardless of their original class.

20.4.2 General Remarks on the Direct Transcription Approach

When a dynamic optimization problem is large, the direct transcription approach gives rise to large-scale NLP problems. Because these NLPs originate from dynamical systems, they are sparse and structured, and hence structure-exploiting solvers are required if large systems are to be solved. Furthermore, other lines of research have focused on decomposition techniques that are tailored to exploit the structure of these NLPs [1].

A difficulty with this approach is that the numbers of elements and collocation points must be chosen *a priori*; while sufficiently many elements and points need to be considered for obtaining an accurate solution to the DAE system, too many elements and points may give rise to NLPs of intractable size.

A common characteristic of direct transcription methods is that the differential equations are satisfied at the converged solution of the NLP problem only. The disadvantage is that intermediate NLP iterates have no physical significance in general. An advantage, on the other hand, is that no computational effort is wasted in obtaining feasible solutions to the DAEs away from the converged solution of the NLP problem.

Given that the (inequality and equality) path constraints are discretized into a finite number of inequality constraints, which must hold at every collocation point in every element, path constraints are more easily handled than in the direct sequential approach. However, there is no guarantee that the path constraints will be satisfied between collocation points.

A variant of orthogonal collocation methods that is often called the pseudospectral optimal control method uses only one collocation interval, but on this interval it uses a very high order polynomial. State constraints are then typically enforced at all collocation points [2].

20.5 REFERENCES

[1] Fu, J., Faust, J. M. M., Chachuat, B., and Mitsos A. "Local optimization of dynamic programs with guaranteed satisfaction of path constraints." Automatica. 2015;62:pp. 184–192.

[2] Rawlings, J. B., Mayne, D. Q., and Diehl, M. M. Model Predictive Control: Theory, Computation, and Design. Nob Hill Publishing. 2019.

[3] Vassiliadis, V. S. Computational Solution of Dynamic Optimization Problems with General Differential-Algebraic Constraint. PhD thesis. Imperial College London. 1993.

[4] Bryson E. B. and Ho, Y.-C. Applied Optimal Control. John Wiley & Sons. 1975.

[5] Ray, W. H. Advanced Process Control. McGraw-Hill. 1981.

[6] Biegler, L. T. "An overview of simultaneous strategies for dynamic optimization." Chemical Engineering and Processing: Process Intensification. 2007;46(11):pp. 1043–1053.

[7] Kameswaram, S. and Biegler L. T. "Convergence rates for direct transcription of optimal control problems using collocation at Radau points." Computational Optimization and Applications. 2008;41(1):pp. 81–126.

[8] Biegler, L. T. Nonlinear Programming. MOS-SIAM Series on Optimization. 2010.

20.6 FURTHER READING RECOMMENDATIONS

More Extensive Treatment on Optimal Control Problems

Bryson, E. B. and Ho, Y.-C. (1975). Applied Optimal Control. John Wiley & Sons.

Optimal Control Problem Formulation for Model Predictive Control

Rawlings, J. B., Mayne, D. Q., and Diehl, M. M. (2019). Model Predictive Control: Theory, Computation, and Design. Nob Hill Publishing.

20.7 EXERCISES

1. A high speed train is travelling between stations A and B. The route is 1,000 L, where L is a unit measure of length. The trip consists of three distinct stages: from distance 0 to 250 L the speed limit is 15 L T^{-1}, where T is a unit measure of time. From distance 250 L to 750 L the speed limit is 25 L T^{-1}. From distance 750 L to the final destination at 1,000 L the speed limit is 10 L T^{-1}. The train is capable of accelerating at $+ 2$ L T^{-2}, and decelerating at 3 L T^{-2}.

   (i) Formulate an optimal control problem with the goal of minimizing the total travel time.
   (ii) Use an implicit Euler discretization scheme, allowing different numbers of discrete elements over the different stages of the problem, to render it into a finite dimensional optimization problem. Treat the element sizes as unequal free variables.
   (iii) Identify the source of nonlinearity in the resulting discretized optimal control problem. Justify whether this reformulated problem is convex or nonconvex.

2. Plug flow reactors are commonly used in industry for efficient reaction processes. Often they are carried out with a fixed bed containing a catalyst, enhancing certain chemical reactions. In this problem it is sought to find the optimal policy of mixing two catalysts along the length of such a reactor. The maximization of the yield in product 3 is formulated in the following manner:

$$\max_{u(\cdot)}$$

subject to:

$$\frac{dy_1}{dx} = u\,(15y_2 - y_1)$$

$$\frac{dy_2}{dx} = u\,(y_2 - 5y_1) - (2 - u)\,y_3$$

$$y_3 = 1 - y_1 - y_2$$

$$y_1(0) = 1.0$$

$$y_2(0) = 0.0$$

$$0 \le u(x) \le 1.0$$

$$0 \le x \le 2.0$$

where x is the distance along the reactor. Reformulate the above problem into a finite dimensional NLP.

21

System Identification and Model Predictive Control

21.1 INTRODUCTION

Control theory is a field of engineering and mathematics that deals with the behavior of dynamical systems with inputs, and how their behavior is modified by output through feedback. In this chapter, we shall briefly introduce two important fields in control theory, *i.e.*, identification and control, and show how these two problems relate to optimization theory introduced in this book.

The objective of control theory is to control a dynamical system so that its output tracks a desired reference signal. To do this, two steps are essential:

1. A model (mathematical description) of the dynamical system of interest is needed. This step uses experimental data to estimate the unknown parameters in the system. This step is called *System identification (SI)*, which will be introduced in Section 21.3.
2. Once a model is obtained from data, a controller is designed that monitors the output and compares it to a reference signal. The difference between actual and desired output is fed back to the input of the system in order to correct this discrepency. In particular, we shall introduce the widely used *Model Predictive Control* (MPC), which will be discussed in Section 21.4.

SI uses statistical methods to build mathematical models of dynamical systems from measurement data. The approaches enable efficient inference of the model parameters, which quantifies the interactions between model state variables. From the derived model, predictions can be made about the dynamical behavior in different conditions (different input signals) using simulation techniques, offering the means to investigate a previously uncharacterized behavior, and to generate testable hypotheses about the functionality of the system. The model can be used to predict future outputs and design controllers.

Since the 1980s [1], MPC is an advanced process control that has been used in the process industries, such as chemical plants and oil refineries. Model predictive controller design relies on the knowledge of underlying dynamical models of the chemical process, most often linear state-space models obtained by SI. MPC uses the current plant measurements, the current dynamical states of the process, the MPC models, and the variable targets and limits of the process to calculate future changes in the control variables. These changes are calculated to make the dependent variables close to targets while satisfying constraints on both control and dependent variables. The MPC typically implement only the first change in each control variable, and repeats the calculation at the next sampling period.

21.2 DYNAMICAL SYSTEMS

Many systems in the areas of chemical engineering, electrical engineering, mechanical engineering, and process engineering show the dynamical behavior over time. The output of a dynamical system at any time instant depends on its history, and not just on the present input, that is, a dynamical system has memory. For identification and control of dynamical systems it is of great importance that the systems are expressed in a convenient way. Based on the system under study, different mathematical models can be used for analyzing the dynamical behavior of the system. Next, we will give a simple illustration of the concepts for dynamical systems and models by analyzing a biological system and introduce some important models used for identification and control.

21.2.1 Modeling: A Simple Example

We now consider a well-characterized biochemical system described in Figure 21.1 as the production of progesterone from cholesterol. Cholesterol is a fat molecule that can be ingested and synthesized by the cell, while progesterone is an essential female hormone involved in supporting gestation.

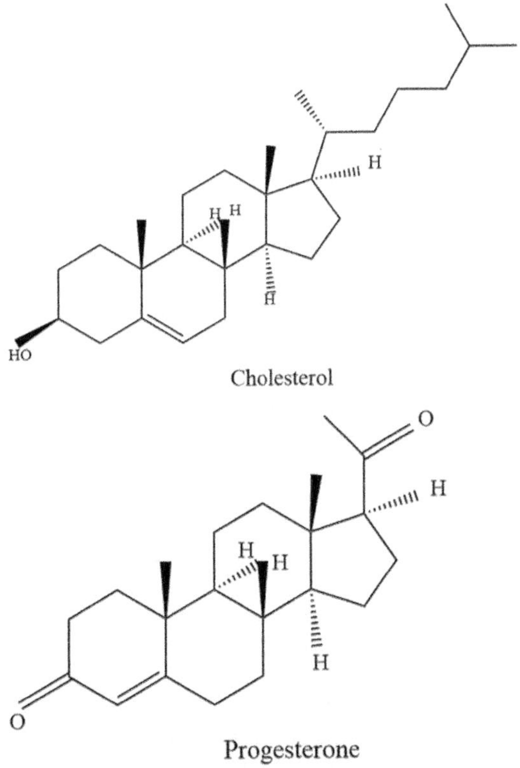

Figure 21.1 Chemical representation of cholesterol and progesterone molecules.

Cholesterol

Progesterone

Let x_1 and x_2 represent the concentrations of cholesterol and progesterone, respectively. The chemical reaction can be formulated as follows:

$$X_1 \underset{k_2}{\overset{k_1}{\rightleftharpoons}} X_2 \tag{21.1}$$

where X_1 and X_2 represent cholesterol and progesterone, respectively. Assuming that the chemical reaction (21.1) follows the mass-action kinetics, the mathematical equations that describe the abundance evolution of progesterone and cholesterol can be related in terms of the forward and reverse reaction rate constants k_1 and k_2, respectively.

The reaction in Equation (21.1) implies that X_1 and X_2 in (21.1), but here x_1 and following x_2 increases at a rate $k_2 x_2$ and decreases at rate $k_1 x_1$, while the opposite can be said for x_2. Therefore, the rate of change of x_1 with respect to time, $\frac{dx_1}{dt}$, is equal to $k_2 x_2 - k_1 x_1$. Hence, the material balances of the reaction in Equation (21.1) can be described by a set of differential equations:

$$\frac{dx_1}{dt} = -k_1 x_1 + k_2 x_2 \tag{21.2}$$

$$\frac{dx_2}{dt} = k_1 x_1 - k_2 x_2. \tag{21.3}$$

Note that discrete-time data is collected in practice, but it is desired to model the systems in continuous time. The fundamental reason lies in the fact that biochemical systems evolve in continuous time and hence their discrete-time models may result in big difference of parameter values, which in turn can lead to wrong conclusions.

Control is a common concept, since there are always some quantities, such as temperature, altitude, or speed, which are to be controlled in a certain manner. Control theory is concerned with how the systems respond to external force and how to guide the response to the desired state. Core in control systems is the study of dynamical systems. In the case of controlling the biochemical systems, control decisions, *i.e.*, control variables, are expected to be derived and implemented over real time. Control theory is based on firm mathematical foundations. The behavior of the system variables to be controlled is typically described by a set of differential or difference equations. We take the biochemical system shown here as an example. For simplicity, we assume there is a linear relationship between the amount of food consumed and the amount of cholesterol produced, *i.e.*, x_1 can be controlled by the amount of consumed food u. With the control variable u, the dynamical relationship becomes:

$$\frac{dx_1}{dt} = -k_1 x_1 + k_2 x_2 + k_3 u \tag{21.4a}$$

$$\frac{dx_2}{dt} = k_1 x_1 - k_2 x_2, \tag{21.4b}$$

where k_3 is the correlation parameter between consumed food and cholesterol increase. Equations (21.4a) and (21.4b) model the biochemical process under investigation with three parameters to be identified.

21.2.2 Mathematical Model

Based on this example, we illustrate the mathematical modeling process for a biological system with differential equations. In general, a dynamical system can be described by a difference or differential equation. Here some important mathematical model structures used for dynamical systems are introduced.

Differential Equation for Continuous-Time

Considering continuous processes, the relation between the input and the output variables of the system can be described by the ordinary differential equation:

$$a_n y^{(n)}(t) + a_{n-1} y^{(n-1)}(t) + \cdots + a_1 \dot{y}(t) + a_0 y(t)$$
$$= b_0 u(t) + b_1 \dot{u}(t) + \cdots + b_m u^{(m)}(t) \tag{21.5}$$

where t is the time index and $u(t)$, $y(t)$ are the input and output variables, respectively. $u^{(n)}$ and $y^{(n)}$ represent the n-th time derivative with respect to the input and output variables, respectively. For the linear dynamical system, the coefficients $a_0, a_1, a_2, \cdots, a_n$ and $b_0, b_1, b_2, \cdots, b_m$ are constant.

When the system involves spatiotemporal dynamics, the partial differential equation is considered to describe a distributed parameter system as follows:

$$F\left(x, u, \frac{\partial u}{\partial t}, \frac{\partial u}{\partial x}, \frac{\partial^2 u}{\partial x \partial t}, \frac{\partial^2 u}{\partial x^2} u, \frac{\partial^3 u}{\partial x^3} u, \frac{\partial^3 u}{\partial x^2 \partial t} u, \cdots \right) = 0 \tag{21.6}$$

where $u = u(x, t)$ is a function of variables x and t.

It is worth noting that the model in Equation (21.5) is usually chosen to describe finite-dimensional systems. For a specific set of differential equations, an analytic solution can be found for both types of models given in Equations (21.5) and (21.6).

Example 1

We consider the following first-order ordinary differential equation:

$$\frac{dy}{dt} + p(t)y = q(t), \quad y(t_0) = 0. \tag{21.7}$$

Its analytical solution is given by:

$$y(t) = \int_{t_0}^{t} q(s) e^{-\int_s^t p(\tau) d\tau} ds.$$

Example 2

Consider the homogeneous heat conduction equations in the n-dimensional case:

$$u_t - \Delta u = 0, \quad u \in \mathbb{R}^n, \quad 0 < t \leq T$$
$$u(x, 0) = \varphi(x), \quad x \in \mathbb{R}^n.$$

By taking the Fourier transform [2] with respect to the state variable x, the solution can be given as follows:

$$u(x, t) = (4\pi t)^{-n/2} \int_{\mathbb{R}^n} \varphi(y) e^{-\frac{\|x-y\|^2}{4t}} dy. \tag{21.8}$$

Difference Equations for Discrete-Time

By discretizing the continuous-time model in Equation (21.5), the difference equation is derived to describe the process in the discrete-time domain as follows:

$$y(k) + a_1 y(k-1) + \cdots + a_n y(k-n) = b_1 u(k-1) + \cdots + b_m u(k-m) \quad (21.9)$$

where the notation k is used instead of kT for simplicity, with T being the sampling period. The coefficients of this difference equation are usually different from the coefficients of the differential equation given in (21.5).

In an analogous way, Equation (21.6) can be discretized by substituting the difference scheme into the derivative terms $u_x, u_{xx}, \ldots$, such as the central difference scheme, or high-order difference scheme [3]. The processing methods involve various choices and result in different schemes. The basic concepts are given in the next sections.

State-Space Representation

When all the past information of the system is closely relevant to the future evolution of the system, they have to be taken into account in the model. The state-space model is a good choice to fulfill this purpose. We can formulate a general state-space model with linear time invariant (LTI) parameters as:

$$\dot{x}(t) = Ax(t) + Bu(t) \quad (21.10a)$$

$$y(t) = Cx(t) \quad (21.10b)$$

where state variable $x(\cdot) \in \mathbb{R}^n$, input $u(\cdot) \in \mathbb{R}^m$, output $y(\cdot) \in \mathbb{R}^p$, and (A, B, C) are real matrices with appropriate dimension. For the example given in Equations (21.10a) and (21.10b), the real matrices (A, B, C) can be uniquely defined by:

$$A = \begin{bmatrix} -k_1 & k_2 \\ k_1 & -k_2 \end{bmatrix}, B = \begin{bmatrix} k_3 \\ 0 \end{bmatrix}, C = \begin{bmatrix} 0 & 1 \end{bmatrix}.$$

Equations (21.10) are called a continuous-time state-space model. Herein, the relationship between the input and output variables is written as a first-order differential equation in the auxiliary state vector $x(\cdot) \in \mathbb{R}^n$. Inspired by the example in Equation (21.7), we calculate the solution from t to $t + T$:

$$y(t + T) = Ce^{AT}x(t) + C \int_t^{t+T} e^{A(t-\tau)} Bu(\tau) d\tau. \quad (21.11)$$

Alternatively, without loss of generality, assuming that the input is constant over the sampling interval and $T = 1$, we can discretize the continuous time system to give:

$$x(k + 1) = A_d x(k) + B_d u(k) \quad (21.12a)$$

$$y(k) = Cx(k) \quad (21.12b)$$

where $A_d = e^A$ and $B_d = \int_0^1 e^{A(1-\tau)} B d\tau$ from Equation (21.11). We then have the discrete-time state-space form to describe the system behavior. One of the important factors for the success of MPC is the type of process model used. Once the system is obtained, we can introduce MPC.

Mutual Transformation

Based on the different requirements, different mathematical models can be used for describing linear dynamical systems. To execute the optimization algorithm, or understand the system accurately, the models introduced previously can be mutually transformed. By the Laplace transform [7], we can transform Equation (21.5) into the transfer function mode:

$$Y(s) = \frac{b_1 s^{m-1} + b_2 s^{m-2} + \cdots + b_{m-1} s + b_m}{s^n + a_1 s^{n-1} + \cdots + a_{n-1} s + a_n} U(s) := G_1(q)U(s).$$

By the z transform [7], the transfer function model representation for the difference equation can be obtained:

$$y(k) = \frac{b_1 z^{-1} + b_2 z^{-2} + \cdots + b_{m-1} z^{-m+1} + b_m z^{-m}}{1 + a_1 z^{-1} + \cdots + a_{n-1} z^{-n+1} + a_n z^{-n}} u(k) := G_2(q)u(k).$$

In the same way, the discrete-time state-space model can be deduced:

$$y(k) = C(qI - A)^{-1} Bu(k) := G_3(q)u(k).$$

Thus the transfer function model for continuous/discrete cases is usually given by the following representation:

$$y(t) = G(q) u(t). \tag{21.13}$$

Once the transfer functions are determined, the impulse response model representation for dynamical systems is determined which can be explained by the solutions in Equations (21.7) and (21.8). By the transfer function representation, the state-space model can be transformed into the difference/differential equation.

Stochastic Model

In real life, however, noise has effect on experimental data at different levels and makes SI challenging. Rewrite in the measurement devices, stochastic process variations (intrinsic), or environmental fluctuations (extrinsic). Noise is incorporated into the model given by Equation (21.13):

$$y(k) = G(q)u(k) + H(q)e(k). \tag{21.14}$$

In particular, the noise term in state-space models is divided into two parts, leading to the modified state-space equations:

$$\tilde{x}(k + 1) = A\tilde{x}(k) + Bu(k) + Ke(k) \tag{21.15a}$$

$$y(k) = C\tilde{x}(k) + e(k) \tag{21.15b}$$

where the variable $e(\cdot)$ is a random signal, typically assumed to be white (Gaussian) noise with the variance σ^2 and zero mean, i.e., K is a vector with the same number of columns as the dimension of states.

To estimate the state of the stochastic model given by Equations (21.15a) and (21.15b), Kalman filtering is efficiently used. Kalman filtering [4, 5], also known as linear quadratic estimation (LQE), is an algorithm that uses a series of measurements observed over time containing noise and other uncertainties. According to the Kalman filtering theory, we introduce its principles with the LTI discrete-time stochastic model. The Kalman filtering model assumes the true state at time $k + 1$ is evolved from the state at k according to:

$$x(k) = Ax(k - 1) + Bu(k) + w(k) \tag{21.16}$$

where A is the state transition matrix applied to the previous state $x(k - 1)$, B is the control-input matrix applied to the control vector $u(k)$, and $w(k)$ is the process noise assumed to be drawn from a zero mean multivariate normal distribution $\mathcal{N}$, with covariance matrix $Q(k)$: $w(k) \sim \mathcal{N}(0, Q(k))$.

At time k an observation (or measurement) $y(k)$ of the true state $x(k)$ is made according to:

$$y(k) = Cx(k) + v(k) \tag{21.17}$$

where C is the observation matrix which maps the true state space into the observed space, and $v(k)$ is the observation noise which is assumed to be zero mean Gaussian white noise with covariance matrix $R(k)$: $v(k) \sim \mathcal{N}(0, R(k))$.

The Kalman filtering is a recursive estimator. The notation $\hat{x}(k + 1 \mid k)$ represents the estimate of x at time $k+1$ given observations up to and including time k. The state of the filter is represented by two variables: $\hat{x}(k \mid k)$, the *a posteriori* state estimate at time k given observations up to and including time k, and $P(k \mid k)$, the *a posteriori* error covariance matrix (a measure of the estimated accuracy of the state estimate).

The Kalman filtering progresses in two phases: "predict" and "update." The "predict" phase uses the state estimation from the previous time step to produce an estimate of the state at the current time step. In the "update" phase, the current *a priori* prediction is combined with current observation information to refine the state estimate. This improved estimation is termed the *a posteriori* state estimation. The detailed recursive estimation is given as follows:

Predict

- Predicted (*a priori*) state estimation : $\hat{x}(k \mid k - 1) = A\hat{x}(k - 1 \mid k - 1) + Bu(k)$
- Predicted (*a priori*) estimation covariance: $P(k \mid k-1) = AP(k-1 \mid k-1)A^T + Q(k)$

Update

- Innovation or measurement residual term: $\tilde{z}(k + 1) = y(k) - C\hat{x}(k \mid k - 1)$
- Innovation covariance: $S(k) = R(k) + CP(k \mid k - 1)C^T$
- Kalman gain: $K(k) = P(k \mid k - 1)C^T S^{-1}(k)$
- Updated (*a posteriori*) state estimate: $\hat{x}(k \mid k) = \hat{x}(k \mid k - 1) + K(k)\tilde{z}(k)$
- Updated (*a posteriori*) estimate covariance: $P(k \mid k) = [\mathbf{I} - K(k)C]P(k \mid k - 1)[\mathbf{I} - K(k)C]^T + K(k)R(k)K^T(k)$
- Measurement residual: $\tilde{z}(k|k) = y(k) - C\hat{x}(k - 1 \mid k - 1)$

Typically, the two phases alternation, with the prediction advancing the state until the next predefined observation, and the update incorporated in the observation.

However, this is not necessary; if an observation is unavailable for some reason, the update may be skipped and multiple prediction steps may be performed. Likewise, if multiple independent observations are available at the same time, multiple update steps may be performed (usually with different observation matrix C).

21.3 INTRODUCTION TO SYSTEM IDENTIFICATION

The goal of SI is to identify a mathematical description (model) of the dynamical system from a set of inputs and corresponding outputs for a specific application, such as satisfying scientific curiosity, prediction and control, state estimation, fault diagnosis, simulation, and so on. The procedure of SI involves the three basic entities:

1. Objectives, prior knowledge, data: based on the prior system knowledge and modeling objectives, collecting the relevant data is necessary for building a model of the dynamical systems.
2. Model set: in order to describe the dynamical system, the choice of the most suitable model from among all possible mathematical models should be made. Many candidates exist, including parametric and nonparametric models, deterministic and stochastic models. In general, we shall also neglect some less important aspects for practical purposes.
3. Model structure: how to best match the model with the available information of the dynamical system (*e.g.* state-space model structure).

In this subsection, our goal is to develop basic expressions for the SI problem together with the error function. We identify parameters from data by translating the task into a nonlinear optimization problem. In what follows, some important identification methods, including the least-squares (LS) method, maximum likelihood methods, as well as the gradient modified method, are discussed, which depend on different basic principles. The state-space model described in Equations (21.12a–21.12b) as well as its transfer function representation in Equation (21.13) is selected as the basic model representation for analysis.

21.3.1 Mathematical Statement for SI

Suppose that we have recorded inputs and outputs over time interval $1 \leq k \leq N$, that is

$$\mathcal{D}^N = \{u(1), y(1), \ldots, u(N), y(N)\}, \tag{21.18}$$

and that we want to model a LTI dynamical system using the discrete-time state-space model, given by:

$$x(k+1) = Ax(k) + Bu(k) + Ke(k) \tag{21.19a}$$

$$y(k) = Cx(k) + e(k) \tag{21.19b}$$

where $y(k)$ represents the output of the dynamical system, system matrix $A \in \mathbb{R}^{n \times n}$, input matrix $B \in \mathbb{R}^{n \times m}$, and output matrix $C \in \mathbb{R}^{n \times n}$ are unknown, and $e(k)$ represents the white noise that is statistically independent from the input $u(k)$. The internal state and the input of the dynamical system are given by

$$x(k) = [x_1(k)\, x_2(k)\, \cdots\, x_n(k)]^T$$
$$u(k) = [u_1(k)\, u_2(k)\, \cdots\, u_m(k)]^T.$$

To identify the state-space model that describes the relationship between input and output data, we first parameterize the dynamical system by using all components of the system matrices. This processing procedure amounts to using the unknown values of physical coefficients in a mathematical model derived from the laws of physics. Further, the resulting parameterized predictor modeled by Kalman filtering theory is given by:

$$\hat{x}(k+1|\theta) = A(\theta)\hat{x}(k|\theta) + B(\theta)\hat{u}(k) + K(\theta)[y(k) - C(\theta)\hat{x}(k|\theta)] \qquad (21.20a)$$

$$\hat{y}(k|\theta) = C(\theta)\hat{x}(k|\theta). \qquad (21.20b)$$

This formulation is natural when considering the noise term to be unimportant or when dealing with deterministic models. The process of developing basic expressions for the problem of the parameter-estimation, *e.g.*, identifying this parameter vector θ for Equations (21.20a–21.20b) can thus be summarized in the following way:

Step 1. Determine the parameter and predictor
Step 2. Choose the objective function for the parameter-estimation problem
Step 3. Solve the parameter-estimation optimization problem
Step 4. Analyze the statistical properties of the estimated parameters

21.3.2 How to Parameterize the LTI State-Space Model

In the specific state-space model such as shown in Section 21.2.1, we choose the forward and reverse reaction rates k_1 and k_2 as the parameters. Whether a parameterized model can be identified uniquely from the input-output data or not is an important problem arising in the parameterization schemes. Some properties for obeying prior knowledge about the dynamical system, such as the stability and the conservative property, will have to be considered. For this, we first present some concepts that will be used to define the identifiability.

The model set: let the parameter vector θ range over the set $\mathcal{M} \subset \mathbb{R}^d$ such that state-space models with different parameters are generated. We consider the set of state-space models with different parameters.

It is worth noting that the model set is usually noncountable. However, to execute a linesearch method [6], the model needs to be indexed by the parameter. As in Equation (21.13), the state-space predictor model in Equations (21.20a–21.20b) can be rewritten into the transfer function form:

$$\hat{y}(k|\theta) = C(\theta)[qI - A(\theta) + K(\theta)C(\theta)]^{-1}B(\theta)u(k)$$
$$+ C(\theta)[qI - A(\theta) + K(\theta)C(\theta)]^{-1}K(\theta)y(k) := G(q,\theta)\hat{x}(k|\theta). \qquad (21.21)$$

Let $\mathcal{R}_n^{n \times m}$ denote the model set of all $n \times m$ proper rational functions with a degree of at most n. In the following we introduce a parametrization of the state-space predictor model in Equations (21.20a–21.20b).

Definition A parameterization is a differentiable mapping from a parameter open subset $\mathcal{M} \subset \mathbb{R}^d$ to the space of rational transfer functions $\mathcal{R}_n^{n \times m}$, such that the gradients of the predictor functions are stable. The mapping $\mathcal{S}$ is also called the state-space model structure.

A state-space model structure $\mathcal{S}$ is globally identifiable at θ^* if the condition holds:

$$\mathcal{M}(\theta) = \mathcal{M}(\theta^*), \theta \in \Omega \Rightarrow \theta = \theta^*. \tag{21.22}$$

We emphasize some remarks when parameterizing state-space models:

1. Parameterizing Equation (21.21) directly is advantageous because of the simple form with respect to parameter θ;
2. The parameterization needs to consider the size of the parameter vector θ and the surjective/injective property of the mapping $\mathcal{M}(\theta)$;
3. The optimization problem should be easy to execute after parameterization;
4. The numerical sensitivity of the model structure $\mathcal{M}(\theta)$ with respect to θ must be considered.

Parameterization Case Study

Consider a parameterization of the simple example in Section 21.2.1, in which the system matrix is given by assigning every element in the system matrix:

$$A = \begin{bmatrix} -k_1 & k_2 \\ k_1 & -k_2 \end{bmatrix}, B = \begin{bmatrix} k_3 \\ 0 \end{bmatrix}, C = \begin{bmatrix} 0 & 1 \end{bmatrix}.$$

We parameterize the system by using all components in the system matrix, which results in

$$\hat{x}(k+1|\theta) = \begin{bmatrix} \theta_1 & \theta_2 \\ \theta_3 & \theta_4 \end{bmatrix} \hat{x}(k|\theta) + \begin{bmatrix} k_3 \\ 0 \end{bmatrix} \hat{u}(k|\theta) \tag{21.23}$$

$$\hat{y}(k|\theta) = \begin{bmatrix} 0 & 1 \end{bmatrix} \hat{x}(k|\theta). \tag{21.24}$$

But this parameterization cannot uniquely identify a state-space model.

21.3.3 Objective Function for Parameter-Estimation Problem

Given the parameterized predictor model in Equations (21.20a)–(21.20b), the forward problem consists of predicting the future and then designing the controller, provided the parameter vector θ are known. Provided we have a measurement $y(k)$ corresponding to the input $u(k)$, the inverse problem of SI aims at selecting a "best" estimator $\hat{\theta}$ such that the resulting predicted output from the predictor model in Equations (21.20a–21.20b) matches the measured output as closely as possible. In essence, it is an optimization problem that searches for the best parameter values.

Let $\hat{y}(k|\theta)$ be the output of the predictor model given a choice of θ. For this, an error function is defined, which compares the measurement and an estimate given a parameter set $\theta(A, B, C)$ as in the state-space model:

$$\epsilon(k|\theta) = y(k) - \hat{y}(k|\theta), \forall k$$

for every sampling time k. Next the magnitude of error is quantified and an overall mismatch function is defined:

$$V_N(\theta) = \frac{1}{N} \sum_{k=1}^{N} l(\epsilon(k|\theta))$$

Let $V_N(\theta)$ be an objective function, that indicates how large the error between the predicted output and the measured output is. The process of identifying the parameter vector θ that minimizes $V_N(\theta)$ then becomes the optimization problem with constraints, that is:

$$\hat{\theta} = \arg\min_{\theta} \; V_N(\theta) \tag{21.25}$$

subject to $\theta \in \mathcal{M}(\theta) \subset \mathbb{R}^P$ and Equations (21.20a–21.20b).

A numerical solution to Equation (21.25) is usually obtained by iterative methods such as Newton's method or quasi-Newton method. The main difference in these numerical minimization methods is the estimate of the derivatives of the function $V_N(\theta)$.

Objective Function for State-Space Model

Consider the LS function given as follows:

$$V_N(\theta) = \frac{1}{N} \sum_{k=1}^{N} \| y(k) - \hat{y}(k|\theta) \|_2^2. \tag{21.26}$$

For simplicity, assume that the state-space model in Equation (21.21) has been described by the observable normal for a single-input-single-output (SISO) process with:

$$A = \begin{bmatrix} 0 & 1 & \cdots & 0 \\ 0 & 0 & I_{n-3} & 0 \\ \vdots & 0 & & 1 \\ & & \cdots & \\ -\alpha_n & -\alpha_{n-1} & \cdots & -\alpha_1 \end{bmatrix} \text{ and } \begin{cases} B = [\beta_1 \; \beta_2 \; \cdots \; \beta_m]^T \\ C = [1\,0\,\cdots\,0]. \end{cases}$$

The defined expression can be calculated $V_N(\theta)$. The discrete state equation results in the transfer function:

$$G(z,\theta) = \frac{\beta_1 z^{-1} + \beta_2 z^{-2} + \cdots + \beta_{m-1} z^{-m+1} + \beta_m z^{-m}}{1 + \alpha_1 z^{-1} + \cdots + \alpha_{n-1} z^{-n+1} + \alpha_n z^{-n}}.$$

Thus Equation (21.21) can be rewritten into the regression model:

$$\hat{y}(k|\theta) = \Phi^T(k)\theta \tag{21.27}$$

with

$$\Phi(k) = \left[y(k-1) \; \cdots \; y(k-n) \, u(k-1) \; \cdots \; u(k-m) \right]^T$$
$$\theta = [-\alpha_1 \; -\alpha_2 \; \cdots \; -\alpha_n \, \beta_1 \; \cdots \; \beta_m]^T .$$

Inserting Equation (21.27) into Equation (21.26), the objective function can be rewritten as:

$$V_N(\theta) = \frac{1}{N} \sum_{k=1}^{N} \| y(k) - \Phi^T(k)\theta \|_2. \tag{21.28}$$

Next, we shall see that this form is convenient for estimating the parameters and one can obtain an analytic solution to the parameter estimation problem.

Derivative Formulas for the LS function

Because of the difficulty of obtaining the analytical solution of the LS problem given in Equation (21.26), finding good approximate solutions using numerical methods will be helpful. Based on the specific problem, iterative search methods can be divided into many types depending on how to use the derivative of the LS function. Here we suppose that the derivative emerging in the following exists. Then we deduce the Jacobian matrix as well as Hessian matrix of $V_N(\theta)$ and introduce the search direction corresponding to different numerical algorithms. We can rewrite Equation (21.26) into matrix form:

$$V_N(\theta) = \frac{1}{N} H_N^T(\theta) H_N(\theta)$$

with the error vector:

$$H_N(\theta) = \begin{bmatrix} \epsilon(1|\theta) & \epsilon(2|\theta) & \cdots & \epsilon(N|\theta) \end{bmatrix}^T.$$

The first-order and second-order derivative of $V_N(\theta)$ at some $\theta \in \Omega$ can be expressed by use of the chain rule:

$$V_N'(\theta) = \frac{\partial V_N(\theta)}{\partial \theta} = \frac{1}{N} \frac{\partial H_N^T(\theta)}{\partial \theta} H_N(\theta) + \frac{1}{N}(I_p \otimes H_N^T(\theta)) \frac{\partial H_N(\theta)}{\partial \theta}$$

$$= \frac{2}{N} \left(\frac{\partial H_N(\theta)}{\partial \theta^T} \right)^T H_N(\theta)$$

$$V_N''(\theta) = \frac{\partial^2 V_N(\theta)}{\partial \theta \partial \theta^T} = \frac{2}{N} \frac{\partial^2 H_N^T(\theta)}{\partial \theta^T \partial \theta}(I_p \otimes H_N(\theta)) + \frac{2}{N} \left(\frac{\partial H_N(\theta)}{\partial \theta^T} \right)^T \frac{\partial H_N(\theta)}{\partial \theta^T}.$$

21.3.4 Optimization Method for Parameter Estimation

When carrying out the search for minimizing $V_N(\theta)$, the iterative formula is given by:

$$\hat{\theta}^{(i+1)} = \hat{\theta}^{(i)} + \lambda_i S^{(i)}$$

where $S^{(i)}$ is a search direction based on information about $V_N(\theta)$ and λ_i is the step size. Using a different search direction will result in different numerical methods. In general, we use the Taylor series expansion to approximate the objective function, such as the second-order Taylor expansion around the given point $\theta^{(i)}$:

$$V_N(\theta) \approx V_N(\theta^{(i)}) + (V_N'(\theta^{(i)}))^T (\theta - \theta^{(i)}) + \frac{1}{2}(\theta - \theta^{(i)})^T V_N''(\theta^{(i)})(\theta - \theta^{(i)}). \quad (21.29)$$

To make the objective function $V_N(\theta)$ reach a minimum in the next iteration point amounts to minimizing the approximate Taylor expansion. The necessary condition for taking the extremum becomes:

$$V_N'(\theta^{(i)}) + V_N''(\theta^{(i)})(\theta - \theta^{(i)}) = 0.$$

Further, under the condition that the Hessian matrix at $\theta^{(i)}$ is positive definite, it holds that:

$$\theta^{(i+1)} = \theta^{(i)} - (V_N''(\theta^{(i)}))^{-1} V_N'(\theta^{(i)}).$$

In this case, we select the search direction $S^{(i)} = (V_N''(\theta^{(i)}))^{-1} V_N'(\theta^{(i)})$, which is known as Newton's method, as discussed in Chapter 5. Some other classical methods are:

- **the steepest descent method:** $S^{(i)} = -V_N'(\theta^{(i)})$;
- **the conjugate gradient method:** $S^{(i)} = -V_N'(\theta^{(i)}) + \beta_i S^{(i-1)}$, where β_i is a scalar value parameter that makes $\{S^{(i)}\}$ mutually conjugate;
- **the Gauss–Newton method:** $S^{(i)} = \frac{2}{N} \left(\frac{\partial H_N(\theta)}{\partial \theta^T}\right)^T$, where $\left(\frac{2}{N} \left(\frac{\partial H_N(\theta)}{\partial \theta^T}\right)^T \frac{\partial H_N(\theta)}{\partial \theta^T}\right)^{-1}$ is the approximation of Hessian matrix $V_N(\theta^{(i)})$.

21.3.5 The Statistical Properties of $\hat{\theta}$

As seen from the previous introduction, there are many choices for objective function and the resulting estimator has different stochastic properties. In general, the unbiasedness, efficiency, and consistency (or weak consistency) can be considered for analyzing the accuracy of the estimator $\hat{\theta}$, and are defined as follows:

Unbiasedness: If $E(\hat{\theta}) = \theta$;

Efficiency: If an estimator $\hat{\theta}$ minimizes the covariance $\text{cov}(\hat{\theta}) = E[(\hat{\theta} - E(\hat{\theta}))(\hat{\theta} - E(\hat{\theta}))^T]$;

Consistency: If for any $\delta > 0$, there holds $\lim_{N \to \infty} P(\|\hat{\theta}_N - \theta\| \geq \delta) = 0$.

For the special case in Equation (21.28), we can calculate the parameter vector $\hat{\theta}$ by the derivative formulas and optimization method in Sections 21.3.3 and 21.3.4, given by:

$$\hat{\theta} = \left[\frac{1}{N} \sum_{k=1}^{N} \Phi(k) \Phi^T(k)\right]^{-1} \left[\frac{1}{N} \sum_{k=1}^{N} \Phi(k) y(k)\right]$$

where the measurement output $y(k)$ is generated by the system with true parameter value θ_0 and white noise sequence $e(k)$:

$$y(k) = \Phi^T(k) \theta_0 + e(k).$$

Thus the estimator $\hat{\theta}$ is unbiased for the special case, resulting from $e(k)$ being independent of $\Phi(k)$.

21.3.6 Other Parameter Estimation Methods

Here, two different estimation methods are presented, in which the parameters and/or the outputs will be given as a series of random variables in the stochastic frame. Unlike the LS method, where the considered noise term is used to analyze estimation error, the parameter, and the observed quantity have the property of stochastic from the probability point of view. For estimating the parameter θ in Equation (21.21), we suppose that Equation (21.18) is a set of stochastic process under the condition of independent observation.

Maximum Likelihood

Let $p(y(k)|\theta)$ be the PDF of random variable $y(k)$ given parameters θ, which result in the joint probability density function for N independent observations $Y_N = [y(1), y(2), \cdots, y(N)]$:

$$p(Y_N|\theta) = \Pi_{k=1}^{N}p(y(k)|\theta).$$

For a set of deterministic variables Y_N, this form is a function of θ, which is known as the likelihood function with:

$$L(Y_N|\theta) = p(Y_N|\theta)$$

Thus, if $L(Y_N|\hat{\theta})$ takes a maximum value for any $\theta \in \Omega$, the estimator $\hat{\theta}$ is the nearest to the exact parameters. Here the objective function for estimating parameter θ is given by:

$$\hat{\theta} = \arg\max_{\theta} L(Y_N|\theta).$$

Based on the fact that $\ln L(Y_N|\theta)$ as well as $L(Y_N|\theta)$ have a maximum simultaneously at $\hat{\theta}$, we can obtain the estimator by using the extremum principle:

$$\left.\frac{\partial \ln L(Y_N|\theta)}{\partial \theta}\right|_{\theta=\hat{\theta}} = 0.$$

Bayesian Estimation

Unlike the Maximum Likelihood method aiming to maximize the conditional PDF, the Bayesian method seeks the *a posteriori* PDF of the parameters. Assume the conditional PDF of the output $y(k)$ given parameter θ and the input-output $D_{k-1} = [u(1) \; y(1) \; \cdots \; u(k-1) \; y(k-1)]$, denoted by p, is known. The parameters can be regarded as a random variable with *the priori* PDF such that the parameter can be estimated by the Bayesian formula:

$$p(\theta|D_k) = p(\theta|u(k), y(k), D_k) = p(\theta|y(k), D_k) = \frac{p(y(k)|\theta, D_k)p(\theta|D_k)}{\int_{-\infty}^{\infty} p(y(k)|\theta, D_k)p(\theta|D_k)d\theta}.$$

To achieve optimality, there are two different approaches. One way is to maximize the *a posteriori* PDF:

$$\hat{\theta} = \arg\max_{\theta} p(\theta|D_k). \tag{21.30}$$

Another way is taking the following objective function

$$\hat{\theta} = \arg\min_{\theta} \int w(\hat{\theta}, \theta)p(\theta|D_k)d\theta. \tag{21.31}$$

In particular, the optimal criterion $w(\hat{\theta}, \theta)$ is a quadratic function $||\hat{\theta} - \theta||_2^2$. To find the minimum, the extremum principle will be used again. In addition, the optimal estimator $\hat{\theta}$ for some particular cases is the expectation of the parameter [7], that is:

$$\hat{\theta} = E(\theta|D_k). \tag{21.32}$$

General Remarks on Identification Methods

In practice, many different identification methods have been developed (see [7] and Chapter 7). For example, the LS method selects a LS cost function as the objective function, which is easier to implement and can yield the global minimum efficiently. Some other LS methods arise, including the weighted LS method, the repeated LS method, and the instrumental-variable methods for different model structures [7].

- **Regularization:** because of the potential ill-posedness of the LS method, a penalty term is introduced to make the problem better conditioned. It can be used to reduce the variance and improve the performance of the LS solution for regression problems [8, 9, 10]. Inspired by the kernel learning in machine learning [10], kernels are also used in system identification [11, 12, 13, 14], which can encode prior information on the exponential stability of the system. In addition, see also [15] for sparse regularization and [16] for model reduction.
- **Nonlinear System Identification:** although nonlinear models can describe complex systems with various natural phenomena, dealing with nonlinearity causes difficulty. In general, the nonlinear term can be expanded on the basis function such that the LS method can be used, *e.g.* the overparameterization method [17, 18] for the Hammerstein model. Some other methods include subspace methods [19], frequency domain methods [20], maximum-likelihood methods [21], and nonparametric Bayesian methods [22].

21.4 INTRODUCTION TO MODEL PREDICTIVE CONTROL (MPC)

MPC is a control strategy that takes full advantage of an online model (in the *control computer*) to calculate predictions of the dynamical system output and to optimize future control actions. The MPC strategy is visualized as shown in Figure 21.2. As a matter of fact, MPC is not a single specific control strategy but rather a family of control methods that have been developed based on certain key principles, three of which are:

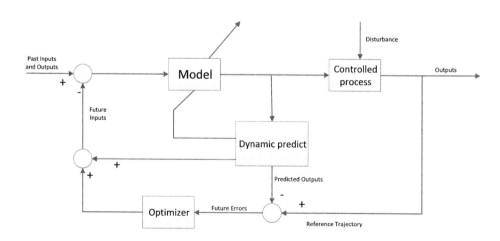

Figure 21.2 MPC strategy as a block scheme.

1. **Prediction Model:** choosing the appropriate prediction model is the core of MPC. An appropriate model captures the full dynamical behavior and hence predicts the future state/output well. In a deeper level, it should allow for theoretical analysis.
2. **Calculating the control law by optimizing the objective function(s):** MPC aims to make the future output track the required reference trajectory within the considered horizon, while satisfying all constraints on the control variables and state variables.
3. **Receding horizon control strategy for rectification:** when applying the first control signal to each step, the future output under the control can be calculated and used as the initial predictor values at the next step. However, plant model mismatch and environmental disturbance may cause the drift-off between the predicted output and measured values, such that feedback compensation is necessary.

Therefore, the MPC strategy is intuitive and makes good practical sense:

(i) use a model to predict the evolution of the process output as a function of future (desired) control actions
(ii) Rewrite, which incorporates the errors between intended and predicted system outputs, and possibly also the required control efforts

MPC is of an open generic nature with intuitive principles and performance-oriented design parameters. It can handle nonlinear multi-variable control problems in a straightforward way. Moreover, it is currently the most natural approach to solve constrained control problems, a topic that plays a significant role in everyday industrial practice, *e.g.,* quality constraints, actuator limitations, and safety restrictions. The inherent feedforward compensation of measured disturbances makes MPC able to control dynamical systems with unusual behaviors such as non-minimum phase systems and unstable systems.

In this section, we use the state-space model for LTI systems as a benchmark to illustrate basic principles of MPC. We also introduce computational and numerical aspects for minimizing the objective function, which will present the relationship between MPC and optimization methods. Further, we give some remarks on key advantages/disadvantages of MPC and some other MPC methods with different models used.

21.4.1 Mathematical Statement for MPC

The various algorithms of the extensive MPC family differ mainly in (i) the type of model used to represent the system and its disturbances and (ii) the objective function(s) to be optimized, with or without constraints. The basic principles of the MPC strategy are characterized in Figure 21.3.

We consider the stabilization problem for LTI systems described by the state-space model given in Equations (21.12a)–(21.12b) with $B = I$ and $C = I$:

$$x(k+1) = Ax(k) + u(k) \tag{21.33a}$$

$$y(k) = x(k) \tag{21.33b}$$

where $u(k) = [u_1(k) \cdots u_N(k)]^T \in \mathbb{R}^N$ represent the MPC decision values for the N states $x(k) \in \mathbb{R}^N$. The MPC element $u(k)$ for doing so will be calculated by solving a moving horizon optimization problem as described in the following section.

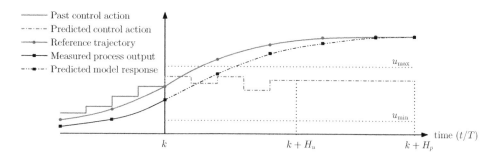

Figure 21.3 The philosophy of the MPC strategy.

The model output $y(k)$ represents the effect of the system input $u(k)$ on the actual system output $\bar{y}(k)$. This is a non-measurable signal because only the combined effect of system input and the disturbances is measurable at the system output $\bar{y}(k)$. It is a dynamic relationship, *i.e.* the current output $y(k)$ does not depend on the current input $u(k)$ but on the previous inputs $\{u(k - 1), u(k - 2), \cdots\}$. Moreover, this relationship can be described either by a linear or nonlinear model. Here, we use the state-space model as an example to illustrate the step-by-step development of the MPC algorithm.

Prediction Stage

By recursion of the state-space model in Equations (21.33a)–(21.33b), the prediction of the system state $\{x(k + t), t = 1, 2, \cdots H_p\}$ can be obtained as follows:

$$x(k + 1) = Ax(k) + u(k)$$

$$x(k + 2) = A^2 x(k) + Au(k) + u(k + 1)$$

$$x(k + 3) = A^3 x(k) + A^2 u(k) + Au(k + 1) + u(k + 2)$$

$$\vdots$$

$$x(k + H_u) = A^{H_u} x(k) + \sum_{j=0}^{H_u - 1} \left(A^{H_u - j - 1} u(k + j) \right)$$

$$x(k + H_u + 1) = A^{H_u + 1} x(k) + \sum_{j=1}^{H_u} \left(A^{H_u - j} u(k + j) \right) + (A + I) u(k + H_u - 1)$$

$$\vdots$$

$$x(k + H_p) = A^{H_p} x(k) + \sum_{j=1}^{H_p - 1} \left(A^{H_p - j - 1} u(k + j) \right)$$

$$+ \sum_{j=0}^{H_p - H_u} \left(A^j + I \right) u(k + H_u - 1).$$

Here, the integers H_p and H_u are design parameters representing the prediction and control horizons, respectively. More specifically, H_p defines the number of future steps to be predicted, while H_u defines the discrete-time length of the MPC sequence. By definition, the following relation holds: $H_u \le H_p$.

At each sampling instant k, the recursion is started with $t = 0$ and $x(k + t)$, $t = 1, 2 \cdots H_p$ is computed using the model input vector $[x(k + t - 1)\, x(k + t - 2) \cdots u(k + t)\, u(k + t - 1) \cdots]$, which contains values from the past, thus known at time k. Note that $y(k)$ has to be saved in the database for further use at sampling instant $k + 1$.

In this way, the future state of the system can be predicted H_p steps ahead, as:

$$X(k + 1) = P_A X(k) + P_U U(k), \tag{21.34}$$

with

$$X^T(k + 1) = [x^T(k + 1), \cdots, x^T(k + H_p)] \in \mathbb{R}^{1 \times H_p N}$$
$$U^T(k) = [u^T(k) \cdots u^T(k + H_u - 1))] \in \mathbb{R}^{1 \times H_u N}$$
$$P_A^T = [A\ A^2\ \cdots\ A^{H_u}\ \cdots\ A^{H_p}] \in \mathbb{R}^{1 \times H_p N},$$

and the expression of P_U is given by Equation (21.35):

$$P_U = \begin{bmatrix} I_N & & & & \\ A & I_N & & & \\ \vdots & \vdots & \ddots & & \\ A^{H_u-1} & A^{H_u-2} & \vdots & I_N & \\ A^{H_u} & \vdots & A^2 & A + I_N & \\ \vdots & \vdots & \vdots & \vdots & \\ A^{H_p-1} & \cdots & A^{H_p-H_u-1} & \sum_{j=0}^{H_p-H_u} A^j \end{bmatrix}. \tag{21.35}$$

Control Stage

For practical reasons, one hopes that the future output $y(k)$ will track a desired reference signal, or the internal state $x(k)$ reaches a steady-state. Various different cost functions can be selected. Once the cost function is determined, the optimal control law $U(k)$ can be calculated by turning to the optimization algorithm.

objective Function

Once the future output is predicted, we shall design the input $\{u(k), u(k + 1), \ldots, u(k + H_u - 1)\}$ via minimizing the following cost function:

$$J = \min \sum_{k=0}^{H} \|x(k)\|_Q^2 + \|u(k)\|_R^2, \tag{21.36}$$

where Q and R are compatible real, symmetric, positive definite weighting matrices, and $\|M\|_Q^2 = M^T Q M$. The first term enforces that the state arrives at steady-state, while the second term takes the control effort into account. Notice that the cost function of Equation (21.36) can be easily extended to many alternative cost functions in the optimal control theory. For simplicity, let $H_p = H_u = H$ and we have

$$X(k + 1) = P_A X(k) + P_U U(k),$$

with

$$X^T(k+1) = [x^T(k+1) \cdots x^T(k+H)] \in \mathbb{R}^{1 \times HN},$$
$$U^T(k) = [u^T(k) \cdots u^T(k+H-1)] \in \mathbb{R}^{1 \times HN},$$

where $P_A^T = \left[A^T \cdots \left(A^H \right)^T \right] \in \mathbb{R}^{N \times HN}$ and the expression of P_U given by Equation (21.37):

$$P_U = \begin{bmatrix} I_N & & & \\ A & I_N & & \\ \vdots & \vdots & \ddots & \\ A^{H-1} & A^{H-2} & \cdots & I_N \end{bmatrix} \tag{21.37}$$

Remark If $H \to \infty$ in Equation (21.36), then it is equivalent to a Linear Quadratic Regulator (LQR) Problem. If H is a finite integer, then it is called finite horizon optimal control or, in general, MPC:

$$J(x, (u(0), \ldots, u(H-1))) = x^T(H)Qx(H) + \sum_{k=0}^{H-1} (x^T(k)Qx(k) + u^T(k)Ru(k)).$$

Calculating the Immediate Control

In order to minimize the cost function, we compute the values of $U(k)$ that yield $\partial J(k) = \partial U(k) = 0$ to obtain the optimal MPC action. If this is a convex and quadratic function of U, then the following is obtained:

$$\partial J(k)/\partial U(k) = \partial((P_X X(k) + P_U U(k))^T Q(P_X X(k) + P_U U(k)) + U(k)^T RU(k))/\partial U(k)$$
$$= 2P_U^T Q P_X X(k) + 2(P_U^T Q P_U + R)U(k)$$
$$= 0.$$

Thus $U(k) = -(P_U^T Q P_U + R)^{-1} P_U^T Q P_X X(k)$. The actual control action applied to the real system at current time k is:

$$u(k) = P_{MPC}x(k), \tag{21.38}$$

with

$$P_{MPC} = -\underbrace{[I_N, 0_N, \ldots, 0_N]}_{N \times HN}(P_U^T Q P_U + R)^{-1} P_U^T Q P_X,$$

where I_N is a unit matrix of dimension $N \times N$ and 0_N is the zero matrix of dimension $N \times N$. Note that in order to compute the MPC law in Equation (21.38), the parameters H, Q, R should be available. Once the immediate control $u(k)$ is obtained, we can evaluate the future output under the control $u(k)$ as follows:

$$y(k+1) = x(k+1) = (A + P_{MPC})x(k). \tag{21.39}$$

Feedback Correction

As no model-plant mismatch was present, the measured output were same as the state variables. When the state is real-time measurable and no disturbance exists, the measured $x(k)$ can be regarded as the initial prediction value at the $k + 1$-th step, which amounts to real-time feedback compensation.

When the state is not measurable and the output is measurable, the state observer is required. We compare the measured output $\bar{y}(k+1)$ with the predicted one $y(k+1|k)$ in Equation (21.39), which forms the error given as follows:

$$e(k + 1) = \bar{y}(k + 1) - y(k + 1|k).$$

The observers are stated as:

$$\hat{x}(k + 1|k + 1) = \hat{x}(k + 1|k) + Ce(k + 1)$$

$$\hat{x}(k + 2|k + 1) = A\hat{x}(k + 1|k + 1) + u(k)$$

$$\hat{y}(k + 1|k) = \hat{x}(k + 1|k)$$

where C is the observation matrix. By substituting and eliminating, we can obtain:

$$\hat{x}(k + 2|k + 1) = A(I - AC)\hat{x}(k + 1|k) + u(k) + AC\hat{y}(k + 1). \tag{21.40}$$

Based on this prediction model, we can carry out the prediction stage and then obtain an equation similar to Equation (21.34). The idea of MPC can be summarized by the following steps:

Step 1. Choose the mathematical model and calculate the future output based on the current measured state $x(k)$, as in Equation (21.34)

Step 2. Choose the objective function for a certain objective and compute the optimal finite horizon input sequence $\{u_0^*(k), u_1^*(k), \ldots, u_{N-1}^*(k)\}$, as in Equation (21.36)

Step 3. Implement the first part of the optimal input sequence $k(x) := u_0^*(k)$, as in Equation (21.38)

Step 4. Obtain the initial predictor by feedback correction and return to Step 1, as in Equation (21.40)

21.4.2 Some Remarks for MPC

In the previous subsection, we introduced MPC based on the state-space model. In general, different model structures, optimizing strategies, as well as feedback measures will result in different MPC algorithms.

- **Model selection:** in Section 21.2, the models can be used for MPC including the difference equation, the transfer function model or the impulse response model. Some other approaches to MPC exist, such as generalized prediction control (GPC) [23] and Dynamical Matrix Control (DMC) [24]. These names are only stated to be indicative and represent a few earlier MPC algorithms, and for further reading see the survey in [25]. In particular, when the model for describing the dynamical system is unknown, SI is necessary (see [26] and Section 21.2).
- **MPC with constraints:** in Equation (21.36), the control problem has been formulated assuming that all signals have an unlimited range (unconstrained control). However, this is not realistic in practical applications. In fact, when safety reasons and sensor range and more importantly economic targets are considered, some constraints should be included for the state variables and the control variables. MPC handles constraints in a straightforward manner, leading to optimization

problems with constraints [27]. This is one of the primary reasons for its wide acceptance in industrial applications. Rewrite due to physical limitations are:

- Input constraints: *e.g.,* actuator limits, $\underline{u}(k) \leq u(k) \leq \overline{u}(k)$;
- State constraints: *e.g.,* reservoir capacities $\underline{x}(k) \leq x(k) \leq \overline{x}(k)$.

These can be added as linear constraints in the optimization of cost function J.

- **Nonlinear MPC:** When MPC strategy is intended for nonlinear systems, there are some challenges in the control process. The main difficulty is to obtain the mathematical expression for the prediction model. Another main difficulty lies in solving the nonlinear receding horizon optimization problem. Even so, some adaptive methods have been developed (see [28] and references therein).

21.5 DISCUSSION ON THE IMPACT OF OPTIMIZATION ALGORITHMS ON IDENTIFICATION AND MPC

The techniques presented for SI and MPC lead to parameter estimation and optimization problems that can be solved by methods discussed in this book. However, after devising a numerical method to determine a parameter estimate from data or the optimal control law, the complexity and convergence of the optimization algorithm are considered.

- **Complexity:** in practice, one expects that the algorithm is easily executed. In SI or MPC optimization, the complexity is related to the number of the parameters to be identified as well as the model structure. As described previously, the predicted output, the gradient, or its Hessian for the optimization algorithm need to be calculated.
- **Convergence:** the convergence is related to the properties of the objective function, which determines whether the algorithm converges globally or locally. Thus we should consider the types of of model chosen carefully.
- **Choice of initial value:** the choice of initial value is important for the algorithm execution. When the absorption neighborhood of the algorithm is not the whole plane, an inappropriate initial value may cause divergence.

21.6 REFERENCES

[1] Morari, M. and Lee, J. H. "Model predictive control: Past, present and future." Computers & Chemical Engineering. 1999;23(4):p. 667–682.

[2] Evans, L. Partial Differential Equations. Wadsworth & Brooks/cole Mathematics. 2010;19(1):p. 211–223.

[3] Li, J. and Chen, Y. T. Computational Partial Differential Equations Using MATLAB. CRC Press. 2015.

[4] Li, J. and Chen, Y. T. "A new approach to linear filtering and prediction problems." Journal of basic Engineering. 1960;82(1):p. 35–45.

[5] Kalman, R. E. and Bucy, R. S. "New results in linear filtering and prediction theory." Journal of Basic Engineering. 1961;83(1):p. 109.

[6] Boyd, S. and Vandenberghe, L. Convex Optimization. Cambridge University Press. 2004.

[7] Ljung, L. System Identification: Theory for the User. Prentice Hall, Upper Saddle River, NJ. 1999.

[8] Tikhonov, A. N. and Arsenin, V. Y. Methods for Solving Ill-Posed Problems. Wiley. 1977.

[9] James, W. and Stein, C. "Estimation with quadratic loss. Proceedings of the fourth Berkeley symposium on mathematical statistics and probability." 1961;1:p. 361–379.

[10] Bishop, C. M. Pattern Recognition and Machine Learning. Springer. 2006.

[11] Pillonetto, G., Dinuzzo, F., Chen, T., De Nicolao, G., and Ljung L. "Kernel methods in system identification, machine learning and function estimation: A survey." Automatica. 2014;50(3):p. 657–682.

[12] Chen, T., Ohlsson, H., and Ljung, L. "On the estimation of transfer functions, regularizations and Gaussian processes: Revisited." Automatica. 2012;48(8):p. 1525–1535.

[13] Chen, T. and Ljung, L. "Constructive state space model induced kernels for regularized system identification." IFAC Proceedings Volumes. 2014;47(3):p. 1047–1052.

[14] Chen, T. and Ljung, L. "Sparse multiple kernels for impulse response estimation with majorization minimization algorithms." Conference on Decision and Control (CDC). 2012:p. 1500–1505.

[15] Hastie, T., Tibishirani, R., and Wainwright, M. Statistical Learning with Sparsity: The Lasso and Generalizations. CRC Press. 2015.

[16] Pillonetto, G. and De Nicolao G. "Kernel selection in linear system identification Part I: A Gaussian process perspective." Conference on Decision and Control and European Control Conference (CDC-ECC). 2011:p. 4318–1325.

[17] Bai, E.-W. An optimal two-stage identification algorithm for Hammerstein–Wiener nonlinear systems. Automatica. 1998;34(3):p. 333–338.

[18] Han, Y. and Callafon R. "Closed-loop identification of Hammerstein systems using iterative instrumental variables." 18th IFAC World Congress. 2011:p. 13930–13935.

[19] Goethals, I., Pelckman, K., Suykens, J. A. K., and De Moor, B. "Subspace identification of Hammerstein systems using least squares support vector machines." IEEE Transactions on Automatic Control. 2005;50(10):p. 1509–1519.

[20] Pintelon, R. and Schoukens, J. System Identification: A Frequency Domain Approach. Wiley. 2012.

[21] Wills, A., Schön, T. B., Ljung, L., and Ninness, B. "Identification of hammerstein–wiener models." Automatica. 2013;49(1):p. 70–81.

[22] Pillonetto, G., Quang M. H., and Chiuso, A. "A new kernel-based approach for nonlinearsystem identification." IEEE Transactions on Automatic Control. 2011;56(12):p. 2825–2840.

[23] Clarke, D. W., Mohtadi, C., and Tuffs, P. S. "Generalized predictive control-Part I. The basic algorithm." Automatica. 1987;23(87):p. 137–148.

[24] Cutler, C. R. "Dynamic matrix control – A computer control algorithm." Proceedings of the Acc San Francisco. 1980.

[25] Qin, S. J. and Badgwell, T. A. "A survey of industrial model predictive control technology." Control Engineering Practice. 2003;11(7):p. 733–764.

[26] Maciejowski, J. M. Predictive Control with Constraints. Prentice-Hall. 2002.

[27] Camacho, E. F. and Bordons, C. Model Predictive Control. Springer. 2007:p. 575–615.

[28] Findeisen, R. and Allgöwer, F. "An Introduction to Nonlinear Model Predictive Control." Benelux Meeting on Systems and Control. 2002:p. 1–23.

21.6.1 Further Reading

Introduction to Various Types of Dynamic Models

Giri, F. and Bai, E.-W. (2010). Block-Oriented Nonlinear System Identification. Springer.

Hunter, I. W. and Korenberg, M. J. (1986). "The identification of nonlinear biological systems: Wiener and Hammerstein cascade models." Biological cybernetics. 55(2):pp. 135–144.

Westwick, D. T. and Kearney, R. E. (2001). "Separable least squares identification of nonlinear Hammerstein models: Application to stretch reflex dynamics." Annals of Biomedical Engineering. 29(8):pp. 707–718.

Bai, E.-W., Cai, Z., Dudley-Javorosk, S., and Shields, R. K. (2009). "Identification of a modified Wiener–Hammerstein system and its application in electrically stimulated paralyzed skeletal muscle modeling." Automatica. 45(3):pp. 736–743.

21.7 EXERCISES

1. Consider the following chemical reaction:

$$\xrightarrow{k_1 u^2} A \xrightarrow{k_2 [A]} B \xrightarrow{k_3 [B]}$$

where u is the control variable, and all expressions on top of the arrows represent reaction rates.

(i) Construct a dynamical model for the concentrations of components A and B, denoted by [A] and [B], respectively. Is this description a valid state-space description?

(ii) For the dynamical system, find the set of algebraic equations describing the system at steady-state. Denote the variable for concentration and the control by $\Delta [A]_{ss}$, $\Delta [B]_{ss}$ and u_{ss}, respectively.

(iii) If the control at steady-state, u_{ss}, is perturbed by a small amount, Δu, use the full dynamical model from part (i), together with the introduction of perturbations $\Delta [A]$ and $\Delta [B]$ from steady-state, to formulate a valid state-space representation (Hint: assume small perturbations for the control. Further use the forward finite differences scheme from Chapter 6).

(iv) Given values of $k_1 = 1.0$, $k_2 = 0.5$, and $k_3 = 2.0$, find the value of the control at steady-state u_{ss} to maintain a concentration of $[A]_{ss} = 2.0$. Find the resulting concentration in B at steady-state. Using this information, use the full dynamical model from (i) and the state-space model from (iii), using the steady-state as initial conditions, to simulate both models: vary the control according to $u(t) = 1 + \frac{t}{50} \sin(t)$. For the state-space model vary the length of the time step for the forward finite differences scheme used in (iii). Observe when the full model and the state-space models diverge and explain why this occurs.

2. Based on the models obtained from the previous question, MPC techniques for the reacting system follow the trajectory for the concentration of A. Explain which model can be used for linear MPC and nonlinear MPC.

Let objective function be:

$$J = \min \sum_{k=0}^{H} x(k)^T Q x(k) + u(k)^T R u(k).$$

where H is the horizon length, Q is a weighting matrix for set-point tracking, and R is a weighting for changes in control variables. Explain what happens in the following cases:

(i) $Q = 0$, $R \neq 0$,

(ii) $R = 0$, $Q \neq 0$,

(iii) $H = 1$,

(iv) $H = \infty$.

Index

Index

Kalman filter, 317, 318, 320
kernel learning, 326
KKT conditions, 85, 87, 89, 90, 95, 102, 109, 110

Lagrange
  function (Lagrangian), 92, 96–98, 100
  interpolation, 306, 307
  multiplier, 96, 98–100, 295
Lagrange multiplier, 78, 82–85, 88, 101–103, 109, 110, 144, 163–166, 247
Lagrangian, 82, 83, 85, 86, 88, 101–103, 106–110, 164
Laplace transform, 317
least squares, 9, 52–56, 146, 319, 324
Levenberg-Marquardt, 69
linear programming (LP), 101, 115, 116, 118–121, 123, 125, 128, 130–133, 150, 152, 153, 161–164, 166, 202, 204, 261
linesearch
  exact, 66, 74, 77
  inexact, 74, 77

maximum likelihood method, 319, 325
mixed-integer
  linear programming (MILP), 228–230, 233–235, 237, 239, 241, 243, 244, 252, 261, 264
  nonlinear programming (MINLP), 228–230, 232, 233, 244–246, 248, 250, 252, 254, 256–259, 261, 262, 264, 285
  programming (MIP), 228, 229, 231, 232
model predictive control, 9, 27, 312, 316, 326, 331, 332
molecular design, 26
Monte Carlo, 19, 206, 208–210, 212, 217, 225
multicriteria optimization, 168
multimodal, 4–6, 8, 256, 263–266
multiobjective optimization (MOO), 22, 28, 168, 169
multistage problem, 287, 289, 291

network flow problem, 8, 151–154
Newton's method, 60–62, 67, 68, 70–72, 74, 77, 101, 102, 106, 322, 324
Nonlinear Programming (NLP), 261, 262, 264, 283, 284
nonlinear programming (NLP), 31, 85, 92, 96, 101, 102, 105–107, 109, 110, 143, 153, 161, 224, 226, 296
norm
  l_1, 31, 147, 149, 150, 250
  Euclidean, 31, 56, 146, 147, 149
  infinity, 31, 148–150

optimal control problem, 9, 20, 22, 28, 287, 288, 291, 292, 296, 304, 310
optimality gap, 206, 207, 218

optimization under uncertainty, 195, 197, 205, 218, 225

parameter
  estimation, 8, 9, 27, 322–324, 332
  uncertainty, 23, 195, 197
Pareto
  frontier, 169
  optimal, 169
penalty function
  absolute, 91
  quadratic, 91–94, 97
penalty method, 88, 91, 95, 96, 98, 99
protein folding, 6, 26, 261

rate of convergence, 69
receding horizon control, 327, 332
regression, 146
residual, 31, 52–54, 318
robust optimization, 195–197, 199, 203, 205, 206, 225

Sample LP problems
  Assignment problem, 158
  Diet problem, 134
  Product mix problem, 133
  Transportation problem, 136, 144, 155, 156
  Transshipment problem, 157
scheduling problem, 11, 27, 144, 155, 233, 258
sensitivity
  analysis, 161, 163, 165, 166
  Lagrange multiplier, 84, 85, 145
shooting method, 295
Simplex
  method, 89, 123, 126, 128, 130–132, 152, 238
  tableau, 124, 125, 129, 238–241, 243
state-space model, 316, 317, 319–322, 327, 328
steepest descent, 66, 68, 72, 77, 78, 324
stochastic model, 317, 319
stochastic optimization, 23, 28, 196, 209, 225
successive linear programming (SLP), 88
successive quadratic programming (SQP), 88
system identification, 312, 319, 321, 326, 331, 332

Taylor expansion, 323
Taylor series expansion, 63, 67, 70, 77, 230
travelling salesman problem, 12, 13
tunneling, 280–285
two-point boundary value problem (TPBVP), 294, 295

underestimator, 230, 264–267, 273–276, 278, 279, 283
unimodal, 4

vehicle routing problem, 12, 13